Transport Phenomena
for Biological and
Agricultural Engineers

About the Author

Praveen Kolar, PhD, is a professor of Biological and Agricultural Engineering (BAE) at North Carolina State University. He obtained a bachelor's degree in Civil Engineering (Sri Venkateswara University, India) and a master's degree in Aquacultural Engineering (Indian Institute of Technology, Kharagpur). Subsequently, he worked in the aquaculture industry for seven years as a technical manager where he was responsible for water quality management and processing of seafood and value-added products. In 2002, he moved to the United States and obtained a master's degree in Biological and Agricultural Engineering (Louisiana State University) and a doctoral degree in Biological and Agricultural Engineering (University of Georgia). At North Carolina State University, Professor Kolar's research program is focused on novel agricultural waste management, waste valorization engineering, and bioprocessing engineering. So far, he has directed 6 doctoral dissertations and 10 masters' theses and served on several graduate student committees, which resulted in over 65 refereed publications in various international journals. As a professor of BAE, he currently teaches Transport Phenomena to all the undergraduate students of biological engineering curriculum and Introduction to Food Processing Engineering to bioprocessing engineering students. Professor Kolar is a recipient of the 2021 North Carolina State University Outstanding Teacher Award and is a member of NC State's Academy of Outstanding Teachers. He has professional memberships in the American Society of Agricultural and Biological Engineers and Air & Waste Management Association.

Transport Phenomena for Biological and Agricultural Engineers

A Problem-Based Approach

Praveen Kolar, PhD

New York   Chicago   San Francisco
Athens   London   Madrid
Mexico City   Milan   New Delhi
Singapore   Sydney   Toronto

Library of Congress Cataloging-in-Publication Data

Names: Kolar, Praveen, author.
Title: Transport phenomena for biological and agricultural engineers : a
    problem-based approach / Praveen Kolar.
Description: New York : McGraw Hill, [2023] | Includes index. | Summary:
    "This hands-on guide lays out core principles and practices of heat,
    mass, and momentum transfer in one useful resource. Written by a
    seasoned biological and agricultural engineering professor, Transport
    Phenomena for Biological and Agricultural Engineers: A Problem-Based
    Approach includes examples and problems sets reflecting real-world
    applications. You will explore fluid, mass, and heat transfer; pressure
    measurements; Fick's and Kirchhoff's Laws; and much more. This textbook
    is designed to be the singular resource for biological and agricultural
    engineering students studying transport phenomena"— Provided by
    publisher.
Identifiers: LCCN 2023010085 | ISBN 9781264268221 (hardcover)
Subjects: LCSH: Transport theory. | Agricultural engineering. |
    Bioengineering. | Textbooks.
Classification: LCC TP156.T7 K65 2023 | DDC 660.6—dc23/eng/20230311
LC record available at https://lccn.loc.gov/2023010085

Transport Phenomena for Biological and Agricultural Engineers: A Problem-Based Approach

1 2 3 4 5 6 7 8 9   LCR   28 27 26 25 24 23

ISBN 978-1-264-26822-1
MHID 1-264-26822-X

Sponsoring Editor
Robin Najar

Editorial Supervisor
Janet Walden

Project Manager
Tasneem Kauser,
KnowledgeWorks Global Ltd.

Acquisitions Coordinator
Olivia Higgins

Copy Editor
Manish Kumar

Proofreader
Alekha Jena

Indexer
Edwin Durbin

Production Supervisor
Lynn M. Messina

Composition
KnowledgeWorks Global Ltd.

Illustration
KnowledgeWorks Global Ltd.

Art Director, Cover
Jeff Weeks

To my parents,
Venugopala Rao Kolar and Vidya Kolar

Contents at a Glance

Contents

Preface

Transport phenomena is a required course for students majoring in Agricultural and Biological Engineering. Typically, offered at a junior level, the course covers the principles of heat, mass, and fluid transport and their applications to common agricultural situations. After graduation, agricultural and biological engineers use transport phenomena to analyze and design insulation walls, HVAC systems, pumps and fans, drying processes, and other facilities. It is noted that several excellent textbooks on heat, mass, and momentum transfer including by distinguished authors, Cengel and Ghajar, Incropera, et al., Ozisik, Henderson, et al., and others are already available on the market. These books are outstanding sources of information for mechanical and chemical engineers around the world. However, many agricultural and biological engineering curricula allow only one semester to cover all the concepts of transport phenomena including the problem sessions. Therefore, as an instructor, I had to consult several of these aforementioned books and other resources to compile our discipline-relevant instructional material that could be covered within 15–16 weeks. More importantly, I felt that it was necessary and appropriate for an instructor to provide students with several numerical examples that are directly related to the constantly evolving field of agricultural and biological engineering. Therefore, this book is the result of my efforts to compile agricultural and biological engineering-relevant information from already available textbooks such as Cengel and Ghajar, Incropera, et al., and Henderson, et al., my own research and industry experience, and several freely available online sources. It is my sincere hope that this textbook will serve two purposes: (1) act as a one-stop source for biological and agricultural engineering instructors involved in teaching transport phenomena and (2) provide students with a conceptual understanding of heat, mass, and fluid transport, supported with direct applications in biological and agricultural engineering via numerical examples.

Organization of the Book

The book is divided into 15 chapters such that the entire material could be covered within a semester. The concepts of heat transfer are discussed first followed by fluid and mass transport. Chapter 1 provides a general introduction to heat transfer followed by an in-depth discussion of one-dimensional steady-state and unsteady-state conduction, convection, and radiation through Chapters 2–7. Subsequently, the principles of fluid transport including pumps and fans are discussed in Chapters 8–10. Chapter 11 deals with the foundational principles of mass transfer and its relationship with heat transfer.

Further, the most common industrial applications of heat and mass transfer such as psychrometrics, drying, refrigeration, and adsorption are discussed in Chapters 12–15. Considering the needs and learning styles of twenty-first-century engineers and consolidating the students' physical understanding of the engineering concepts, each idea presented in the book is followed by a relevant agricultural-related numerical example, using the international system of units. Besides, each chapter is equipped with 10 multiple-choice problems for students taking the fundamentals of engineering exam. Similarly, for engineers interested in the principles and practice of engineering, several practice problems are included at the end of each chapter. It is recommended for the students to complete all the example problems provided in the book to acquire a thorough understanding of the underlying engineering concepts.

An appendix consisting of unit conversions, engineering properties of water and air, and a pressure–enthalpy diagram for ammonia, and the solutions to the problems presented in the book are available online at https://www.mhprofessional.com/transportphenomena.

Acknowledgments

Several people have contributed to writing this book. First of all, I would like to thank Dr. Garey Fox, the head of the Department of Biological and Agricultural Engineering at NC State University for strongly encouraging me to write this book. I also would like to thank Dr. Sanjay Shah, professor and extension specialist for constantly engaging me in research discussions on the applications of transport phenomena in animal agriculture, which led to several example problems presented in this book. Thanks to Dr. Ratna Sharma (Professor Emeritus) for gracefully sharing her instructional material with me. I also want to express gratitude to Ms. Robin Najar, Senior Acquisitions Editor, McGraw Hill and Ms. Tasneem Kauser, Project Manager, KnowledgeWorks Global Ltd., for patiently working with me even when the COVID-19 pandemic slowed things down. My students, Nitesh Kasera and Victoria Augoustides, deserve thanks for helping me with the schematics in this book. Finally, special thanks to my beloved wife, Chandrika Patwari, and my son Sujay, for their unwavering support and encouragement while writing this book.

CHAPTER 1

Modes of Heat Transfer

Chapter Objectives

After studying this chapter, students will be able to

- Identify and explain the three modes of heat transfer.
- Solve simple numerical problems related to conduction, convection, and radiation heat transfer.
- Extend the concepts of heat transfer to multimodal heat transfer processes.

1.1 Motivation

We recall from basic physics that heat is a form of energy and it is always transferred from a higher energy level to a lower energy level until an equilibrium is reached. For example, we all have experienced our hot food getting cold over time because the heat is lost from the hot food to the surrounding air until food reaches the air temperature (Fig. 1.1). Similarly, we have invariably observed the warming of a can of ice-cold soda.

As engineers, we are interested in determining how fast the energy is being transferred within a system. In addition, we also would like to know how the temperature varies within the body subjected to cooling or heating. Thermodynamics alone will not be able to predict the rates of heating or cooling and the temperature profiles. That is where the concepts of heat transfer are needed. Heat transfer is a branch of engineering that allows us to determine how fast the heat is transferred between two bodies at different temperatures using thermodynamics. Studying heat transfer allows an engineer to predict temperature within a body after a specific time, design equipment for heating and cooling, and select the correct insulation for homes.

The goal of this chapter is to provide a general introduction to the modes of heat transfer and their basic mathematical expressions. In the subsequent chapters, we will systematically tackle each of the heat transfer processes with ample examples.

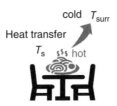

FIGURE **1.1** Hot food gets cold over time due to heat transfer from the food (T_s) to the surroundings (T_{surr}).

1.2 Conduction

Conduction involves transfer of heat from hotter molecules in a medium to the colder molecules via random molecular collisions. To understand the concept of conduction heat transfer, imagine a metal bar being heated at one end by a heat source (Fig. 1.2). As you can imagine, with the passage of time, the other end of the metal bar will also heat up. This transfer of heat from the hot end of the metal rod to the cold end of the metal rod is said to occur via conduction.

The molecules at the hottest portion of the metal rod possess higher energies than the molecules in the adjacent portion of the metal rod. Therefore, these molecules vibrate randomly and collide with the adjacent molecules. As a result, the energy is transferred to the adjacent molecules and this process continues until a thermal equilibrium is attained. Conduction occurs in solids, liquids, and gases as well. However, because the molecules are tightly packed in solids when compared to liquids and gases, the conduction heat transfer is fastest in solids, followed by liquids and gases.

In solids, besides the molecular vibrations, the electron configuration also plays a significant role in conduction heat transfer. Metals, for example, have free electrons in their outer shells. When exposed to heat, these free electrons also vibrate and collide, and thereby transfer the energy to the adjacent atoms. Therefore, metals conduct heat better than their nonmetal counterparts. It is important to note that there is no physical molecular movement in conduction and the entire heat transfer occurs via atomic and electron vibrations and collisions during those vibrations.

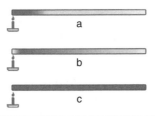

FIGURE **1.2** Heat transfer via conduction. The metal bar closest to the heat source will heat up first (a) and with passage of time (b and c) transfers heat to the colder regions of the bar via conduction.

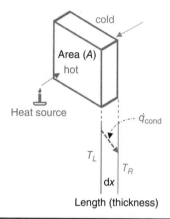

Length (thickness)

FIGURE 1.3   Illustration of conduction heat transfer in a rectangular body.

1.3   Mathematical Description of Conduction—Fourier's Law

Based on experiments, it was observed that the rate of conduction ($\dot{q}_{cond}$) was directly proportional to (1) the surface area normal to the heat transfer, (2) the temperature difference between the two heat transfer surfaces, and (3) inversely proportional to the distance (thickness) between the surfaces. To describe this process mathematically, let us consider a rectangular slab (Fig. 1.3) being heated on the left face.

Based on the above observations, the heat transfer due to conduction may be expressed as

$$\dot{q}_{cond} \propto A \frac{dT}{dx} \tag{1.1}$$

Besides, it was also observed that the heat transfer rate was different for different materials even when the surface areas, thicknesses, and temperature gradients were identical, suggesting that a specific inherent material property also influenced the rate of heat transfer. Therefore, Eq. (1.1) is modified to accommodate the material property, which is called thermal conductivity ($k = W/m°C$) as below:

$$\dot{q}_{cond} = -kA \frac{dT}{dx} \tag{1.2}$$

1.4   The Interpretation of the Negative Sign

We might ask ourselves the rationale behind the negative sign in Eq. (1.2). As can be seen from Fig. 1.4, the slope of the curve, that is, $\frac{dT}{dx}$ is negative because $T_R < T_L$.

The negative sign will counteract the already present negative temperature gradient in the equation and the heat transfer rate will become positive. On the other hand, when $T_R > T_L$, the slope is positive and the heat transfer in the forward (right) direction becomes negative, suggesting that the direction of heat flow is backward (from right to left).

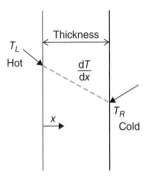

FIGURE 1.4   The negative sign in the Fourier equation will counter the negative temperature gradient $\left(\frac{dT}{dx}\right)$.

1.5   The Concept of Thermal Conductivity

As an intrinsic material property, it is important to understand the physical meaning of thermal conductivity. Mathematically, Eq. (1.2) can be rewritten as:

$$k = -\frac{\dot{q}_{cond}}{A} \times \frac{dx}{dT} \tag{1.3}$$

If we perform an experiment by carefully controlling the physical properties and the temperature gradient such that

$$A = dx = dT = 1 \text{ unit} \tag{1.4}$$

Equation (1.3), when combined with Eq. (1.4) and ignoring the negative sign for now, will now yield

$$k = \dot{q}_{cond} \tag{1.5}$$

Therefore, thermal conductivity of a given material is numerically equivalent to the rate of heat transfer that occurs when such a material possessing a unit area, with unit thickness, is subjected to a unit temperature gradient. This interpretation will help us compare different materials under identical conditions and help us select appropriate materials in engineering design. Table 1.1 contains thermal conductivities of some of the common solids used in biological and agricultural engineering under standard conditions.

It is to be noted that air due to low thermal conductivity $\left(k = 0.026 \, \frac{W}{m \cdot K}\right)$ is an excellent insulating material. By creating a layer of air pockets around an object, one can significantly reduce the heat transfer from the body to the surroundings. Mother nature employed this very idea in a very versatile manner. For example, the fluffy down feathers in certain birds, such as waterfowls, trap air between their fibers. Because the thermal conductivity of air is low, these fibers provide excellent insulation and minimize heat transfer from the bird to the water in winter. Similarly, wool—a hairy fibrous material—that grows on the skin of the lamb also holds air

Solids	
Material	**k (W/m·K)**
Asphalt	0.75
Aluminum	235
Brass	109
Brick (fire)	0.47
Brick (insulation)	0.15
Concrete	0.8
Copper	401
Diamond	2300
Feather (chicken)	0.02
Glass (pyrex)	1.05–1.14
Glass (window)	1
Granite	3.2
Gold	314
Fiberglass	0.048
Ice	2.22
Iron	67
Limestone	1.3
Meat (beef)	0.43–0.48
Meat (chicken)	0.41
Meat (fish)	0.58
Meat (pork)	0.44–0.49
PVC	0.12–0.25
Sand	0.25
Rubber	0.13
Silver	428
Skin (cow)	0.049
Skin (human)	0.37
Skin (Pig)	0.15
Skin (fish)	0.58
Soil	0.33–1.4
Steel	43
Stainless Steel	14
Vegetable oils	0.17
Wood	0.17

TABLE 1.1   Thermal Conductivities of Some of the Commonly Used Materials in Agricultural and Biological Engineering (Data Adapted from Engineering Toolbox with Permission)

Liquids	
Material	**k (W/m·k)**
Acetic acid	0.193
Acetone	0.18
Ammonia	0.5
Blood (human)	0.52
Ethanol	0.171
Glycerin	0.285
Propylene glycol	0.147
Sulfuric acid (conc)	0.5
Vegetable oils	0.17
Water	0.598
Gases	
Material	**k (W/m·k)**
Air	0.026
Carbon dioxide	0.0146
Helium	0.15
Hydrogen	0.18
Methane	0.03
Nitrogen	25.83
Water (steam)	0.0184

TABLE 1.1 Thermal Conductivities of Some of the Commonly Used Materials in Agricultural and Biological Engineering (Data Adapted from Engineering Toolbox with Permission) (*Continued*)

pockets that can minimize heat transfer in winter and summer. The same idea is used in several biological and agricultural engineering applications. For example, commonly used insulation materials are made of several layers of air sandwiched between light weight synthetic fibers (Fig. 1.5).

Such multilayered fibers can be very effective in reducing heat loss from the walls and ceilings of the buildings. Similarly, double-pane windows are another example in which a thin layer of stagnant air is trapped between two layers of glass resulting in an increased resistance to heat transfer.

FIGURE 1.5 Simplified schematic of commonly used insulation materials.

Example 1.1 (Fermentation Engineering—conduction heat transfer through the base of the glass reactor unit):
A laboratory-scale fermenter is being used to convert glucose into ethanol using an engineered yeast. The fermenter, which is made of 1-cm-thick glass, is placed on a flat heating element rated at 750 W. If the measured temperature of the outer glass wall (that was in contact with the heating element) was 150°C, determine the temperature of inner glass wall if the efficiency of the heating is 88% and the surface area of the fermenter base in contact with the heating element is 0.07 m².

Solution This is a simple conduction heat transfer through the base of the fermenter.

Step 1: Schematic
The problem is represented in Fig. 1.6.

Step 2: Given data
Contact surface area for heat transfer, $A = 0.07$ m²; Thickness of the glass wall, $x = 1$ cm; Rated power of the heating element, $P_R = 750$ W; Efficiency of the heating, $\eta = 88\%$; Actual power supplied = actual heat transfer rate = $\dot{q}_{cond} = 0.88 * 750 = 660$ W; and Temperature of the outer glass wall = $T_o = 150°C$

To be determined: Temperature of the inner glass wall (T_i).

Assumptions: Steady state heat transfer occurs in one dimension.

Approach: This is simple case of conduction heat transfer and therefore Fourier's equation can be directly used.

Step 3: Calculations
The Fourier's equation is given by

$$\dot{q}_{cond} = -kA\frac{dT}{dx} \tag{E1.1A}$$

From Table 1.1, the thermal conductivity of glass is $1.05 \dfrac{W}{m \cdot K}$.
  Equation (E1.1A) after integrating is expressed as

$$\dot{q}_{cond} = kA\frac{(\text{Temperature gradient})}{\text{Thickness}} \tag{E1.1B}$$

$$660 = 1.05 \times 0.07 \times \frac{(150 - T_i)}{1/100} \tag{E1.1C}$$

Solving for T_i, we obtain

$$T_i = 60.2°C$$

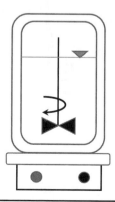

Figure 1.6 Schematic to describe the conduction heat transfer through the base of a fermenter.

1.6 Convection

Unlike conduction, convection heat transfer always involves motion of a fluid. To help understand this phenomenon, let us consider a can of ice-cold soda placed on a table (Fig. 1.7). Now let us use a fan and circulate the air (at room temperature) around the can of soda. As the air flows around the soda can, heat transfer occurs, and the air gets cooled while the can of soda gets warm.

If we investigate further, we realize that the process involves a direct contact between the cold soda can and the air (Fig. 1.8), resulting in conduction heat transfer between the air and the cold surface of the soda can. After the conduction heat transfer is completed, the air molecules are forced away from the can of soda and new molecules take their place and the process will continue until an equilibrium (i.e., the soda will reach the air temperature) is attained. This type of heat transfer in which there is a bulk motion of fluid is called convection heat transfer.

If we further increase the speed of the fan (faster air flow rate), we will find that the cold soda can gets warmer faster, suggesting that bulk flow rate is one of the most

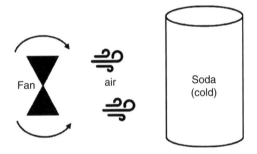

FIGURE **1.7**   A cold can of soda warms up via convection heat transfer.

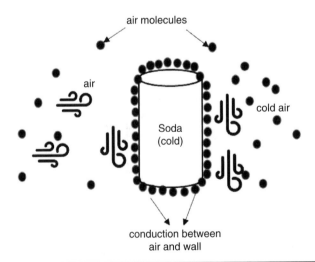

FIGURE **1.8**   Mechanism of convection heat transfer that includes conduction followed by the flow of fluid.

important factors that governs convection heat transfer. Because we are using an external force (fan in this case) to move the air, this type of heat transfer is called forced convection. Some common examples of forced convection in our daily lives include fans in personal computers, air conditioners, hair dryers, convection ovens, and car radiators.

Convection heat transfer also occurs without an external force. To elucidate this idea, let us use the same example as before, but without the fan. In the current scenario, the air molecules that come in contact with the cold soda can become cold as a result of conduction heat transfer. As a result, the molecules get denser and will move down along the walls of the can allowing the next batch of air molecules to contact walls of the cold can. As this process continues, a downward air current is formed, which facilitates convection heat transfer and is usually called free convection or natural convection. Some common examples of free convection in our daily lives include cooling of electronics in mobile phones, radiator heaters, cooling of a hot cup of coffee with the passage of time, and even the formation of sea breeze.

1.7 Mathematical Description of Convection—Newton's Law of Cooling

Convection heat transfer is one of the simplest ideas to experience but one of the most difficult ideas to express mathematically. It was observed that the convection heat transfer is proportional to the contact surface area of the solid–fluid interface and the difference in temperatures. However, complication arises from the fact that the fluid properties (such as density, viscosity, and velocity), the flow regime (such as laminar and turbulent), the geometry of the system (such as sphere, cylinder-hollow or open, plate), the orientation (such as tilted, horizontal, and vertical), and the physical properties (such as thermal conductivity, density, surface roughness) of the solid–liquid interface also significantly affect the convection heat transfer. Therefore, for simplicity, a comprehensive parameter, called heat transfer coefficient $\left(h = \dfrac{W}{m^2 \cdot K}\right)$, is defined to accommodate the properties of the fluid, solid, configuration, and geometry. Now, the convection heat transfer rate is expressed as

$$\dot{q}_{conv} = hA(T_{Surr} - T_S) \qquad (1.6)$$

A similar idea was originally proposed by Isaac Newton in 1701 and therefore the above expression is also known as Newton's law of cooling.

1.8 The Concept of Heat Transfer Coefficient (*h*)

To understand the physical meaning of the heat transfer coefficient, let us revisit the governing equation for convection.

$$\dot{q}_{conv} = hA(T_{Surr} - T_S) \qquad (1.7)$$

Again, if we perform an experiment by carefully controlling the physical properties and the temperature gradient such that

$$\text{Area} = \text{temperature gradient} = 1 \text{ unit} \qquad (1.8)$$

Equation (1.7) will yield

$$h = \dot{q}_{conv} \tag{1.9}$$

Therefore, the heat transfer coefficient can be viewed as the numerical equivalent of the rate of heat transfer rate that would occur between a solid and a fluid when a unit interface area is subjected to a unit temperature gradient. However, unlike thermal conductivity, which is a property of the medium transporting the heat, the heat transfer coefficient is the overall property of the entire system involved in the heat transfer process. As discussed earlier, several factors influence the magnitude of heat transfer coefficient that makes its mathematical treatment from first principles extremely difficult. Therefore, it is usually obtained via empirical approaches which will be discussed in Chap. 5.

Heat transfer coefficients are also influenced by the thermal conductivities of the fluids involved in the heat transfer process. Generally, the thermal conductivities of the liquids are higher than those of the gases. Even in our experience, we notice that water at 15°C feels much cooler than air at 15°C. This is because, for the same temperature gradient and the same exposed surface area, water transfers higher amounts of heat from our body than air due to its higher heat transfer coefficient. We can also extend this conclusion to several engineering problems involving convection and employ liquids for heating or cooling whenever possible. The general range of heat transfer coefficients is summarized in Table 1.2.

Example 1.2 (Bioenergy Systems Engineering—Convective cooling of biochar):   Biochar is a type of charcoal that is produced by heating organic carbon-rich materials such as wood in the presence of nitrogen. After production, the hot biochar is cooled by purging the biochar chamber with an inert gas to minimize combustion and ash formation. For this example problem, imagine a batch of just produced biochar at 300°C that is exposed to an inert environment of helium gas at 30°C. Ignoring the effects of radiation, determine the convective heat transfer rate between the biochar and the helium environment if the exposed area of the biochar and the heat transfer coefficient of the system is 0.25 m² and 500 W/m²·K.

Step 1: Schematic
The schematic is shown in Fig. 1.9.

Step 2: Given data
Area, $A = 0.25$ m²; Heat transfer coefficient, $h = 500\ \dfrac{W}{m^2 \cdot K}$; Surface temperature, $T_s = 300°C$; Temperature of the surrounding medium, $T_{Surr} = 30°C$

Step 3: Calculations
Governing equation: Newton's law of cooling $\dot{q}_{conv} = hA(T_{Surr} - T_s)$ (E1.2A)

$$\dot{q}_{conv} = 500 * 0.25 * (30 - 300) \tag{E1.2B}$$

$$\dot{q}_{conv} = -33,750\ W = -33.75\ kW$$

Convection Process	Liquid (W/m²·K)	Gas (W/m²·K)
Natural	50–1000	2–25
Forced	100–20,000	25–250

TABLE 1.2   General Range of Heat Transfer Coefficients for Fluids

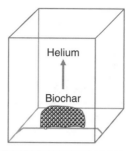

Figure 1.9   Convection cooling of biochar using helium gas in a biochar chamber.

Note: The negative sign indicates that the heat is transferred from the surface (biochar) to the surroundings (helium).

Additional comments: In this problem, the temperature gradient experienced by the system is rather large (270°C). Therefore, radiation heat transfer effects may be significant, and it is strongly encouraged to determine the radiation effects as well when solving such problems (see Example 1.5).

Example 1.3 (Bioenvironmental Engineering–heating of aquaculture tank):   Water temperature is one of the most important water quality parameters that governs the health and growth of aquatic species. In this example, consider a small shrimp larvae holding tank of 100 L in an indoor recirculating aquacultural facility ($h = 750$ W/m²·°C) that is to be heated (Fig. 1.10). The intake water is at 10°C and is to be heated to 22°C in 15 minutes. Four cylindrical immersion heaters (1000 W each) of 2.5 cm diameter and 30 cm in length are available. Is the available heating adequate? Provide suitable recommendations to the producer. Also, determine the temperature the heater will reach during the steady-state heat transfer process for this system assuming that the performance efficiency of the heater is 80%.

Solution

Step 1: Schematic
The problem is depicted in Fig. 1.10.

Step 2: Given data
Volume of the water, $V = 100$ L $= 0.1$ m³; Heat transfer coefficient, $h = 750 \dfrac{\text{W}}{\text{m}^2 \cdot {}^\circ\text{C}}$; Intake water temperature, $T_{\text{Surr}} = 10°C$; Required water temperature $= 22°C$; Temperature gradient, $\Delta T = 12°C$; Available time, $\Delta t = 15$ min $= 900$ s; Heat transfer area $= 2\pi r L \times 4$ heaters $= 2 \times 3.14 \times \dfrac{2.5}{(2 \times 100)} \times 0.3 \times 4$ m² $= 0.0942$ m²

Step 3: Calculations
Assuming the density (ρ) and specific heat (C_p) of water as 1000 kg/m³ and 4180 J/kg·K

Mass of the water, $m = \rho \times V = 1000 \times 0.1 = 100$ kg

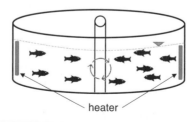

Figure 1.10   Schematic for the heat transfer in an aquacultural tank.

Required heat transfer rate, $\dot{q} = mC_p\Delta T/\Delta t$ (E1.3A)

Required heat transfer rate, $\dot{q} = \dfrac{100\times 4180\times 12}{900} = 5573.3$ W

Analysis and recommendation: The facility has four heaters of 1000 W each with an efficiency of 80%. The total available heating capacity is $4\times 1000\times 0.80 = 3200$ W. Therefore, there is an additional requirement of 2373 W. Hence, it is recommended to procure additional three identical heaters of 1000 W to meet the heating demand.

Calculation of heater temperature

$$\dot{q} = \dot{q}_{conv} = hA(T_s - T_{Surr})$$ (E1.3B)

$$5573 = 750\times 0.0942(T_s - 10)$$ (E1.3C)

$$T_s = 88.8°C$$

1.9   Radiation

Radiation involves transfer of heat between objects via electromagnetic waves. From physics, we recall that all bodies above absolute zero emit heat in the form of infrared waves and these infrared waves are responsible for radiation heat transfer (Fig. 1.11). Radiation heat transfer does not require any medium and is the fastest mode (equivalent to speed of light in vacuum) of heat transfer relative to conduction and convection. Just like conduction heat transfer occurs in solids, liquids, and gases as well, radiation heat transfer also occurs from solids, liquids, and gases although at different rates. Some examples of radiation in our daily lives include cooking food in convection oven, toaster oven, fireplaces to heat our living rooms.

1.10   Mathematical Description of Radiation—The Stefan–Boltzmann Law

The radiation heat transfer from an ideal body (also called blackbody) was found to be a function of (1) the surface area of the body and (2) the fourth power of the absolute temperature and can be expressed as below:

$$\dot{q}_{rad} \propto AT^4$$ (1.10)

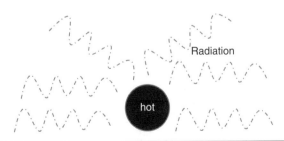

FIGURE 1.11   Heat is transferred from a hot object to the surroundings via electromagnetic (infrared) radiation in all directions.

In the above equation, the proportionality sign may be replaced by a constant called Stefan–Boltzmann constant, σ, whose value is 5.67×10^{-8} W/m²·K⁴.

$$\dot{q}_{\text{rad}} = \sigma A T^4 \tag{1.11}$$

$$\dot{q}_{\text{rad}} = 5.67 \times 10^{-8} A T^4 \tag{1.12}$$

1.11 The Concept of Emissivity (ε)

The above equation applies for blackbodies only. A blackbody is an idealized body that is capable of absorbing 100% of all the energy it receives and radiating 100% of all the energy and therefore used as a reference only. All other bodies, that is, real bodies, however, are not as efficient as a blackbody, and the radiation heat transfer from a real body is always less than a blackbody with the same area and at the same temperature. Hence, for a real body we will use a dimensionless efficiency factor called emissivity ε ($0 \le \varepsilon \le 1$) to describe the extent at which the efficiency from a real body approaches that of a blackbody. Therefore, now the radiation heat transfer is expressed as

$$\dot{q}_{\text{rad}} = 5.67 \times 10^{-8} \varepsilon A T^4 \tag{1.13}$$

The emissivity of commonly used materials in engineering applications is summarized in Table 1.3.

In several engineering problems, we are interested in determining the radiation heat transfer rate between an object enclosed in a large surface such as thermal processing of biomass in a large reactor (Fig. 1.12).

In such cases, the net radiation between the body of a surface area A with an emissivity of ε, maintained at a temperature T_s, and the enclosure maintained at T_{surr} (assuming $T_s > T_{\text{surr}}$) is given as

$$\dot{q}_{\text{rad}} = \varepsilon \sigma A (T_s^4 - T_{\text{surr}}^4) \tag{1.14}$$

$$\dot{q}_{\text{rad}} = 5.67 \times 10^{-8} \varepsilon A (T_s^4 - T_{\text{surr}}^4) \tag{1.15}$$

Material	Emissivity
Construction materials	0.9–0.95
Water	0.96
Human and animal skin	0.95–0.97
Metals (oxidized)	0.25–0.75
Metals (unpolished)	0.1–0.4
Metals (polished)	0.05–0.18
Carbon	0.8–0.95

TABLE **1.3** Emissivity of Common Materials Used in Biological and Agricultural Engineering at Room Temperature (Adapted from Cengel and Ghajar)

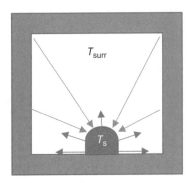

FIGURE 1.12   Net radiation heat transfer between two objects—one enclosed in another.

Example 1.4 (Agricultural Engineering–radiation heat transfer in a hog production unit):   A finishing hog barn holds 350 pigs (emissivity = 0.92) with each pig weighing about 120 kg and an average body surface area of 1.6 m²/animal. The average skin temperature of the pigs was measured to be 40°C. Estimate the radiation heat transfer rate from the pigs when the average wall temperature was found to be 27°C (Fig. 1.13).

Step 1: Schematic
The problem is described in Fig. 1.13.

Step 2: Given data

Surface area of each pig = 1.6 m²

Number of animals = 350

Total surface area = 350 × 1.6 = 560 m²

Emissivity of pig skin $\varepsilon = 0.92$

Surface temperature $T_s = 40°C$

Temperature of the walls $T_{Surr} = 27°C$

Stefan–Boltzmann constant $\sigma = 5.67 \times 10^{-8}\ \dfrac{W}{m^2K^4}$

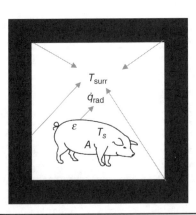

FIGURE 1.13   Radiation heat loss from pigs to the surroundings in a pig barn.

Step 3: Calculations

Governing equation: Stefan–Boltzmann's law $\dot{q}_{rad} = \varepsilon\sigma A(T_S^4 - T_{Surr}^4)$ (E1.4A)

$$\dot{q}_{rad} = 0.92 \times 5.67 \times 10^{-8} \times 560 \times ((273+40)^4 - (273+27)^4)$$ (E1.4B)

$$\dot{q}_{rad} = 43{,}757.14 \text{ W} = 43.76 \text{ kW}$$

Additional comments: During the periods of high ventilation, the hogs also lose significant quantity of heat via connection. Hence, it is strongly encouraged to determine the effects of convection as well when solving such problems. In this example, the convection heat transfer rate (assuming a convection heat transfer coefficient of 50 W/m² · °C of air at 27°C) is calculated as

$$\dot{q}_{conv} = hA(T_S - T_{surr})$$ (E1.4C)

$$\dot{q}_{conv} = 50 \times 560 \times (40 - 27)$$ (E1.4D)

$$\dot{q}_{conv} = 364 \text{ kW}$$

In this situation, the radiation heat loss from the hogs is about 12% relative to convection heat loss.

1.12 Multimodal Heat Transfer

Many situations that we encounter usually involve multimodal heat transfer. For example, if we step out on a hot summer day when the outside temperature is greater than body temperature, our bodies are subjected to radiation and convective heat transfers from the surroundings (Fig. 1.14).

A common hair dryer that we use frequently is another great example of a bimodal heat transfer. The coil within the hair dryer, when heated, will transfer heat via radiation and along with the fan which will facilitate the heat transfer away from the coil to the hair. Similarly, a design of a cooking oven will involve all three heat transfer mechanisms. The total heat transferred to the food during cooking will be the summation of individual heat transfers: conduction from the plate on which the food is placed, heat transferred via convection between the hot air and the surface of the food, and the radiation heat transfer from the hot coil to the surface of the food.

1.13 Heat Transfer Nomenclature

Heat transfer rate is expressed in several ways: The most common way is to express in watts. However, in engineering literature, the heat transfer is sometimes expressed as heat flux rate ($\dot{q}''$), which is the heat transfer rate per unit area. At times, heat rate

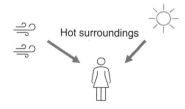

Hot surroundings

Figure 1.14 The heat is transferred from the ambient (hot) via convection and radiation simultaneously resulting in an additive effect.

Variable	Expression	Units
Heat transfer rate	$\dot{q}$	Watts
Heat flux rate	$\dot{q}''$	$\dfrac{\text{Watts}}{\text{m}^2}$
Heat rate per length	$\dot{q}'''$	$\dfrac{\text{Watts}}{\text{m}}$

TABLE 1.4   Various Ways to Express Heat Transfer Rates in Engineering Literature

per length ($\dot{q}'''$) is also used in lieu of heat transfer rate. The easiest way to differentiate is to analyze the units (Table 1.4). Similarly, the conductivity, k and heat transfer coefficient, h can be either expressed in either Kelvin or degrees Celsius such that

$$k = W/m \cdot K = W/m \cdot {}^\circ C$$
$$h = W/m^2 \cdot K = W/m^2 \cdot {}^\circ C.$$

Example 1.5 (Multimodal heat transfer):   Solve Example 1.2 by considering the simultaneous effects of radiation and convection. In Example 1.2, we only considered the effects of convection and determined the convection heat transfer rate as 33.75 kW. Now, we will investigate if radiation effects are also significant. Here, we will assume that the biochar chamber is large and acts as a blackbody. Therefore, the net rate of radiation heat transfer is given by

$$\dot{q}_{rad} = \varepsilon \sigma A(T_s^4 - T_{surr}^4) \tag{E1.5A}$$

Using the same data as in Example 1.2 and the emissivity of biochar as 0.85, we obtain

$$\dot{q}_{rad} = 0.85 \times 5.6710^{-8} A(T_s^4 - T_{surr}^4) \tag{E1.5B}$$

$$\dot{q}_{rad} = 0.85 \times 5.6710^{-8} \times 0.25 \times ((273+300)^4 - (273+30)^4) \tag{E1.5C}$$

$$\dot{q}_{rad} = 1197.3 \ W = 1.19 \ kW$$

Total heat transfer rate $= \dot{q}_{total} = \dot{q}_{conv} + \dot{q}_{rad} = 33.75 + 1.19 \ kW = 34.94 \ kW$

Practice Problems for the FE Exam

1. Which of the following most closely represents conduction heat transfer?
   a. Warming of water in a stagnant pond
   b. Roasting food in an oven
   c. Cooling of a hot object in flowing water
   d. Cooling of electronics in mobile phones

2. Which of the following statements is incorrect?
   a. Conduction heat transfer does not require a medium.
   b. Radiation can occur in vacuum.
   c. Radiation is the fastest mode of heat transfer.
   d. Liquids convect heat faster than gases.

3. The rate of conduction heat transfer across a 15-cm wall $\left(k = 0.71 \dfrac{W}{m \cdot K}\right)$ exposed to a temperature difference of 20°C is closest to:

   a. 95 $\dfrac{W}{m^2}$.

   b. 50 $\dfrac{W}{m^2}$.

   c. 100 $\dfrac{W}{m^2}$.

   d. None of the above.

4. The hourly heat loss from a turkey (area = 0.9 m²) at average temperature of 27°C subjected to a cold stream of air at 15°C with a heat transfer coefficient of 100 W/m² · K is nearly:

   a. 1000 W.

   b. 5000 W.

   c. 3.9 MW.

   d. None of the above.

5. The drop in radiation energy emitted from a body after its temperature is decreased by 25% is approximately:

   a. 70%.

   b. 31%.

   c. 50%.

   d. None of the above.

6. The maximum possible radiation from a 2-cm diameter polished stainless steel (emissivity = 0.07) sphere at 100°C is:

   a. 100 W.

   b. 55.5 W.

   c. 0.38 W.

   d. 2 W.

7. A cake batter whose initial temperature is 27°C placed in an oven whose environment is maintained at 400°C with a heat transfer coefficient of 200 W/m² · °C. Assuming the emissivity of the batter is 0.9, the total heat transferred to the batter is:

   a. 100 W.

   b. 55.5 W.

   c. 0.38 W.

   d. 2 W.

8. The average heat transfer to a pint of ice cream (mass = 454 g and 70% moisture) that experiences a temperature increase of 5°C is:

   a. 6674 J.

   b. 11,000 J.

   c. 5 kW.

   d. 1500 W.

9. The increase in conduction heat transfer rate when the thickness is halved (all other things remaining the same) is:

   a. 100%.

   b. 200%.

   c. 50%.

   d. 45%.

10. The rate of energy needed to heat 5 kg of air (specific heat of air = 1.02 kJ/kg·K) at 5°C by 10°C in 30 seconds is closest to:

   a. 30 W.

   b. 560 J.

   c. 800 J.

   d. 1 kW.

Practice Problems for the PE Exam

1. Estimate heat transfer across a brick wall of 10.5 cm thick whose inner and outer surfaces are measured to be 35°C and 17°C. It is assumed that the average thermal conductivity of the wall was 0.7 W/m·K.

2. Consider a concrete wall ($k = 1.1$ W/m·K) whose outer surface temperature was maintained at 40°C. How thick the wall should be built so that the heat transfer flux and the temperature gradient shall not exceed 140 W/m² and 20°C, respectively?

3. In aquacultural product processing, blast freezers are used to cool fish rapidly. In one of such given situations, estimate the heat transfer rate experienced by a carton of fish (50 cm × 50 cm × 50 cm) originally at 5°C that was subjected to an air flow whose temperature and heat transfer coefficient were determined to be −20°C and 700 W/m²·°C, respectively. Assume that all six sides are available for heat transfer.

4. A solid surface of 0.8 m² is placed a large enclosure with a forced convection ($h = 1000$ W/m²·°C) at 30°C. If the total heat transfer rate (ignoring conduction) is limited to 500 W, determine the maximum surface temperature of the solid.

5. The air (density = 1.2 kg/m³) in a poultry barn (1 atm) of dimensions 50 m × 10 m × 4 m is to be heated from 12 to 23°C in 2 hours by a radiator. Assuming the emissivity of the radiator to be 0.95, determine the temperature to which the radiator is to be heated.

6. A spherical catalyst pellet of 1-cm diameter is heated to a temperature of 200°C. Determine the maximum rate of radiation by this sphere assuming an emissivity of 0.8.

7. For a body of area A at temperature T_1 with an emissivity ε, placed in a large enclosure at a temperature T_2 $(T_1 > T_2)$, shows that

$$\dot{q}_{rad} = \frac{T_1 - T_2}{\dfrac{1}{\varepsilon \sigma A(T_1^2 + T_2^2)(T_1 + T_2)}}$$

8. Describe heat transfer mechanisms in a thermos flask.

9. A box of fresh chicken at 25°C is placed in a large freezer that was maintained at −10°C with a heat transfer coefficient of 100 W/m²·°C. Determine the total heat transfer flux rate, assuming the emissivity of the box is 0.9.

10. Assuming the sun to be a blackbody with a surface temperature of 5500°C, determine the total radiation from the sun if the diameter of the sun is estimated to be 1,392,800 km.

CHAPTER 2

Conduction Heat Transfer

Chapter Objectives

The content presented in this chapter will equip students to

- Describe heat conduction through simple geometries via differential equations.
- Solve the differential equations by applying the boundary and equilibrium conditions.
- Determine the temperature profiles and heat transfer rates for simple geometries.

2.1 Motivation

As we recall from the previous chapter, conduction heat transfer occurs due to the molecular and electronic vibrations that transfer the energy from a high temperature region to a low temperature region within a system. We also described this type of heat transfer via Fourier's law. However, Fourier's law can only provide the heat transfer rate between two chosen locations within a body. In several engineering problems, we are interested in temperature distribution within the body. For example, imagine a 5-cm thick metal plate whose left face is being heated. In addition to the overall heat transfer rate, we would also like to know the temperature at a location 2 cm from the left surface and how the temperature varies with the plate thickness. To solve such problems, we need to expand Fourier's law and develop equations to analyze and predict the temperatures and heat transfer rates within the given system. Therefore, this chapter deals with developing heat conduction equations in all three coordinate systems, namely, rectangular, spherical, and cylindrical, using the first principles.

2.2 The Concept of Thermal Diffusivity (α)

In heat transfer problems, the second most important property of a given material, after thermal conductivity (k), is its thermal diffusivity (α). When a body is subjected to a thermal gradient, the molecules within the body can either store the heat energy or can transfer the heat energy to the adjacent molecules via conduction. The storage of energy within the body is given by its volumetric heat capacity (ρC_p), which is the product of density (ρ) and specific heat (C_p). The ability of the body to transfer of heat via conduction is described by its thermal conductivity (k). Thus, thermal diffusivity is mathematically defined as the ratio of thermal conductivity (k) and volumetric heat capacity (ρC_p).

$$\alpha = \frac{k}{\rho C_p} \qquad (2.1)$$

In a physical sense, thermal diffusivity indicates how fast the heat is propagated relative to how much heat is stored within the material. Generally, materials with high thermal conductivities will have higher thermal diffusivities. For example, heat sinks are made of metal alloys equipped with high thermal conductivities and usually have high thermal diffusivities and therefore are employed in electronic and mechanical equipment to dissipate the heat away to prevent the heat damage to the components. The thermal diffusivities of some commonly used engineering materials are compiled in Table 2.1.

The units of thermal diffusivity are $\frac{m^2}{s}$. Therefore, thermal diffusivity may also be viewed as the speed at which a given material can spread the heat and therefore can serve an indicator of how fast the material will attain a steady state when subjected to a temperature differential. To illustrate the idea, let us consider two different materials of identical dimensions at room temperature: metal and plastic. If we place these two materials in a hot water bath, we will observe that the metal will gain heat and reach an equilibrium more quickly than the plastic. This is because the heat is spread within the metal at a faster rate than plastic. The molecules in plastic, however, are prone to storing more heat relative to transferring and conducting heat to the neighboring molecules.

Material	Thermal Conductivity, k $\left(\dfrac{W}{m^\circ C}\right)$	Density, ρ $\left(\dfrac{kg}{m^3}\right)$	Specific Heat, C_p $\left(\dfrac{kJ}{kg^\circ C}\right)$	Thermal Diffusivity, α $\left(\dfrac{m^2}{s}\right)$
Air	0.026	1.2	1000	2.16667E-05
Aluminum	235	2500	890	0.000105618
Fire brick	0.47	2000	840	2.79762E-07
Concrete	0.8	2400	1000	3.33333E-07
Copper	401	8950	400	0.000112011
Glass	1.14	2200	750	6.90909E-07
Granite	3.2	2700	790	1.50023E-06
Ice (0°C)	2.22	916	2050	1.18223E-06
Iron	67	7900	450	1.88467E-05
Meat (avg)	0.5	1050	3000	1.5873E-07
Steel	43	2700	470	3.38849E-05
Stainless steel	14	8000	500	0.0000035
Wood	0.6	1000	4200	1.42857E-07

TABLE 2.1 Thermal Diffusivities of Commonly Used Engineering Materials (Averages)

Thermal diffusivity is frequently used in conduction heat transfer problems, especially in situations where the heat transfer occurs in an unsteady state.

2.3   Derivation of Three-Dimensional Heat Conduction Equation in Rectangular Coordinate System

Let us consider an isotropic and infinitesimally small cuboid element of dimensions dx, dy, and dz subjected to heat transfer in all three directions (Fig. 2.1). Let the thermal conductivity, density, specific heat be k, ρ, and C_p, respectively. We will also denote temperature as T and time as t. Other variables associated with the cuboid are listed in Table 2.2.

Let us assume that heat transfer rates entering the cubiod in all three directions to be $\dot{q}_x$, $\dot{q}_y$, and $\dot{q}_z$. Using Taylor series, we can express the heat transfer rates leaving the system as

$$\dot{q}_{x+dx} = \dot{q}_x + \frac{\partial \dot{q}_x}{\partial x} dx + \frac{\partial^2 \dot{q}_x}{\partial x^2}\frac{(dx)^2}{2!} + \cdots \tag{2.2}$$

$$\dot{q}_{y+dy} = \dot{q}_y + \frac{\partial \dot{q}_y}{\partial y} dy + \frac{\partial^2 \dot{q}_y}{\partial y^2}\frac{(dy)^2}{2!} + \cdots \tag{2.3}$$

$$\dot{q}_{z+dz} = \dot{q}_z + \frac{\partial \dot{q}_z}{\partial z} dz + \frac{\partial^2 \dot{q}_z}{\partial z^2}\frac{(dz)^2}{2!} + \cdots \tag{2.4}$$

Considering the infinitesimally small element (dx, dy, and $dz \to 0$), the higher order terms in the above equations can be ignored because $\frac{(dx)^2}{2!}$, $\frac{(dy)^2}{2!}$, and $\frac{(dz)^2}{2!}$ will be infinitesimally small as well. Therefore, the conduction heat transfer rates exiting in the x, y, and z directions can be written as

$$\dot{q}_{x+dx} = \dot{q}_x + \frac{\partial \dot{q}_x}{\partial x} dx \tag{2.5}$$

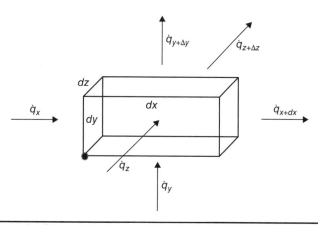

Figure 2.1   Schematic of the cuboid with the dimensions $dx \times dy \times dz$.

Variable	Description	Value/Nomenclature
V	Volume of the element	$dx \cdot dy \cdot dz$
m	Mass	$\rho \cdot dx \cdot dy \cdot dz$
A_x	Area in the x-direction	$dy \cdot dz$
A_y	Area in the y-direction	$dx \cdot dz$
A_z	Area in the z-direction	$dx \cdot dy$
$\dot{q}_x$	Heat transfer rate in the x-direction at the entrance of the cuboid	$-kA_x \dfrac{\partial T}{\partial x} = -k \cdot dy \cdot dz \dfrac{\partial T}{\partial x}$
$\dot{q}_y$	Heat transfer rate in the y-direction at the entrance of the cuboid	$-kA_y \dfrac{\partial T}{\partial y} = -k \cdot dx \cdot dz \dfrac{\partial T}{\partial y}$
$\dot{q}_z$	Heat transfer rate in the z-direction at the entrance of the cuboid	$-kA_z \dfrac{\partial T}{\partial z} = -k \cdot dx \cdot dy \dfrac{\partial T}{\partial z}$
$\dot{q}_{x+dx}$	Heat transfer rate in the x-direction at the exit of the cuboid	$\dot{q}_x + \dfrac{\partial \dot{q}_x}{\partial x} dx$
$\dot{q}_{y+dy}$	Heat transfer rate in the y-direction at the exit of the cuboid	$\dot{q}_y + \dfrac{\partial \dot{q}_y}{\partial y} dy$
$\dot{q}_{z+dz}$	Heat transfer rate in the z-direction at the exit of the cuboid	$\dot{q}_z + \dfrac{\partial \dot{q}_z}{\partial z} dz$

TABLE 2.2 Summary of the Variables Describing the Heat Transfer in a Cuboid

$$\dot{q}_{y+dy} = \dot{q}_y + \frac{\partial \dot{q}_y}{\partial y} dy \tag{2.6}$$

$$\dot{q}_{z+dz} = \dot{q}_z + \frac{\partial \dot{q}_z}{\partial z} dz \tag{2.7}$$

Performing an energy balance around the element, we obtain

Rate of heat conduction entering the element – Rate of heat conduction exiting the element + Rate of heat generation within the element
= Rate of change of heat content of the element   (2.8)

Now we can express each of the component mathematically as under:

$$\text{Rate of heat conduction entering the element} = \dot{q}_x + \dot{q}_y + \dot{q}_z \tag{2.9}$$

$$\text{Rate of heat conduction exiting the element} = \dot{q}_{x+dx} + \dot{q}_{y+dy} + \dot{q}_{z+dz} \tag{2.10}$$

$$\text{Rate of heat generation within the element} = \dot{g} \cdot dx \cdot dy \cdot dz \tag{2.11}$$

where $\dot{g}$ is the rate of energy generation per unit volume.

$$\text{Rate of change of heat content of the element} = \frac{\partial E_{cubiod}}{\partial t}$$

$$= m \cdot C_p \frac{\partial T}{\partial t} = \rho \cdot dx \cdot dy \cdot dz \cdot C_p \cdot \frac{\partial T}{\partial t} \tag{2.12}$$

$$\text{Because mass, } m = \text{density} * \text{volume} = \rho \cdot dx \cdot dy \cdot dz \tag{2.13}$$

Now, after substituting all the above equations in Eq. (2.8), we obtain

$$\dot{q}_x - \dot{q}_{x+dx} + \dot{q}_y - \dot{q}_{y+dy} + \dot{q}_z - \dot{q}_{z+dz} + \dot{g} \cdot dx \cdot dy \cdot dz = \rho \cdot dx \cdot dy \cdot dz \cdot C_p \cdot \frac{\partial T}{\partial t} \tag{2.14}$$

Substituting the expressions for $\dot{q}_{x+dx}$, $\dot{q}_{y+dy}$, and $\dot{q}_{z+dz}$ into Eq. (2.14), we obtain

$$\dot{q}_x - \dot{q}_x - \frac{\partial \dot{q}_x}{\partial x} dx + \dot{q}_y - \dot{q}_y - \frac{\partial \dot{q}_y}{\partial y} dy + \dot{q}_z - \dot{q}_z$$
$$- \frac{\partial \dot{q}_z}{\partial z} dz + \dot{g} \cdot dx \cdot dy \cdot dz = \rho \cdot dx \cdot dy \cdot dz \cdot C_p \cdot \frac{\partial T}{\partial t} \tag{2.15}$$

or

$$- \frac{\partial \dot{q}_x}{\partial x} dx - \frac{\partial \dot{q}_y}{\partial y} dy - \frac{\partial \dot{q}_z}{\partial z} dz + \dot{g} \cdot dx \cdot dy \cdot dz = \rho \cdot dx \cdot dy \cdot dz \cdot C_p \cdot \frac{\partial T}{\partial t} \tag{2.16}$$

Substituting the expressions for $\dot{q}_x$, $\dot{q}_y$, and $\dot{q}_z$ into Eq. (2.16), we obtain

$$- \frac{\partial\left(-k \cdot dy \cdot dz \cdot \frac{\partial T}{\partial x}\right)}{\partial x} dx - \frac{\partial\left(-k \cdot dx \cdot dz \cdot \frac{\partial T}{\partial y}\right)}{\partial y} dy - \frac{\partial\left(-k \cdot dx \cdot dy \cdot \frac{\partial T}{\partial z}\right)}{\partial z} dz$$
$$+ \dot{g} \cdot dx \cdot dy \cdot dz = \rho \cdot dx \cdot dy \cdot dz \cdot C_p \cdot \frac{\partial T}{\partial t} \tag{2.17}$$

or

$$\frac{\partial\left(\frac{\partial T}{\partial x}\right)}{\partial x}(k \cdot dx \cdot dy \cdot dz) + \frac{\partial\left(\frac{\partial T}{\partial y}\right)}{\partial y}(k \cdot dx \cdot dy \cdot dz) + \frac{\partial\left(\frac{\partial T}{\partial z}\right)}{\partial z}(k \cdot dx \cdot dy \cdot dz)$$
$$+ \dot{g} \cdot dx \cdot dy \cdot dz = \rho \cdot dx \cdot dy \cdot dz \cdot C_p \cdot \frac{\partial T}{\partial t} \tag{2.18}$$

After dividing the equation by $k \cdot dx \cdot dy \cdot dz$ and rearranging we arrive at the following equation which is famously called as Fourier-Biot equation for heat conduction.

$$\frac{\partial^2 T}{\partial x^2} + \frac{\partial^2 T}{\partial y^2} + \frac{\partial^2 T}{\partial z^2} + \frac{\dot{g}}{k} = \frac{1}{\alpha} \frac{\partial T}{\partial t} \tag{2.19}$$

Because $\alpha = \dfrac{k}{\rho C_p}$, where α is the thermal diffusivity $\left(\dfrac{m^2}{s}\right)$.

When the system is operating in a steady state, the Fourier-Biot equation can be expressed as

$$\frac{\partial^2 T}{\partial x^2} + \frac{\partial^2 T}{\partial y^2} + \frac{\partial^2 T}{\partial z^2} + \frac{\dot{g}}{k} = 0 \tag{2.20}$$

Several situations that we commonly encounter (such as heat transfer through a residential wall) operate under steady state and do not involve heat generation. For such cases, the Fourier-Biot equation becomes

$$\frac{\partial^2 T}{\partial x^2} + \frac{\partial^2 T}{\partial y^2} + \frac{\partial^2 T}{\partial z^2} = 0 \tag{2.21}$$

Some materials, especially of biological origin, however, usually do not follow steady-state conditions and do not generate heat. For such cases, we can simplify the Fourier-Biot equation as below:

$$\frac{\partial^2 T}{\partial x^2} + \frac{\partial^2 T}{\partial y^2} + \frac{\partial^2 T}{\partial z^2} = \frac{1}{\alpha}\frac{\partial^2 T}{\partial t} \tag{2.22}$$

2.4 Applications of Heat Conduction Equations

It will be easier to recognize the usefulness of the heat equation via solving the example problems that closely mimic practical situations. In the first set, we will solve the heat conduction equation for a few commonly occurring situations in general form. Subsequently, we will use a few additional numerical examples to demonstrate the applicability of heat equation. In all cases, the heat equation will be solved using the appropriate boundary conditions (BCs).

Example 2.1 (General Engineering—All concentrations):   A wall of thickness a is subjected to a temperature gradient of $(T_L - T_R)$ where T_L and T_R are the temperatures of the wall on the left and right faces, respectively (Fig. 2.2). Assuming a one-dimensional steady-state heat transfer process with no energy generation, (i) formulate a differential equation to describe the heat transfer process and (ii) solve the differential equation to determine the temperature profile within the wall.

Step 1: Schematic
The problem is depicted in Fig. 2.2.

Step 2: Given data
Thickness of the wall = a; Temperature of the left face of the wall = T_L; Temperature of the right face of the wall = T_R; and $T_L > T_R$

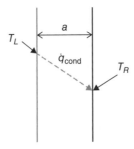

FIGURE 2.2 Schematic describing heat transfer across the wall whose left and right faces are maintained at T_L and T_R, respectively.

To be determined: Temperature distribution across the wall.

Given assumptions

Assumption 1. The heat flow is steady state $\Rightarrow \frac{\partial T}{\partial t} = 0$

Assumption 2. The heat flow is one dimensional $\Rightarrow \frac{\partial^2 T}{\partial y^2} = 0 = \frac{\partial^2 T}{\partial z^2}$

Assumption 3. There is no energy generation $\Rightarrow \dot{g} = 0 \Rightarrow \frac{\dot{g}}{k} = 0$

Approach: In this problem, heat is transferred from the left face to the right face of the wall via conduction. We will first write the general heat equation, apply the assumptions, and subsequently determine the temperature distribution across the wall.

Step 3: Calculations

The heat equation for a rectangular section is written as

$$\frac{\partial^2 T}{\partial x^2} + \frac{\partial^2 T}{\partial y^2} + \frac{\partial^2 T}{\partial z^2} + \frac{\dot{g}}{k} = \frac{1}{\alpha} \frac{\partial T}{\partial t} \tag{E2.1A}$$

Applying assumptions 1–3 into the above heat equation, we obtain

$$\frac{\partial^2 T}{\partial x^2} = 0 \tag{E2.1B}$$

The above partial differential equation (PDE) is only in one dimension and therefore we can directly use an ordinary differential equation (ODE) as below:

$$\frac{d^2 T}{dx^2} = 0 \tag{E2.1C}$$

To solve a second-order ODE, we require two BCs, which are as below:

BC 1: When $x = 0$, $T = T_L$

BC 2: When $x = a$, $T = T_R$

After integrating Eq. (E2.1C) twice successively, we obtain

$$\frac{dT}{dx} = C_1 \tag{E2.1D}$$

$$T = C_1 x + C_2 \tag{E2.1E}$$

By substituting BC 1 and BC 2 in Eq. (E2.1E) will result in

$$T = T_L = C_1(0) + C_2$$

$$C_2 = T_L \tag{E2.1F}$$

$$T = T_R = C_1(a) + C_2$$

$$T_R = C_1(a) + T_L$$

$$C_1 = \frac{(T_R - T_L)}{a} \tag{E2.1G}$$

Substituting the values of C_1 and C_2 in Eq. (E2.1E), we obtain the temperature variation across the wall as below:

$$T = \frac{(T_R - T_L)}{a} x + T_L \tag{E2.1H}$$

Example 2.2 (General Engineering—All concentrations): Now let us consider the same wall as in Example 2.1 (thickness "a" and subjected to a temperature gradient of $(T_L - T_R)$ (Fig. 2.3). Assuming a one-dimensional steady-state heat transfer process with an energy generation rate of $\dot{g}$, (i) formulate a differential equation to describe the heat transfer process and (ii) solve the differential equation to determine the temperature profile within the wall.

Step 1: Schematic
The problem is described in Fig. 2.3.

Step 2: Given data
Thickness of the wall = a; Temperature of the left face of the wall = T_L; Temperature of the right face of the wall = T_R, and $T_L > T_R$

To be determined: Temperature distribution across the wall.

Given assumptions

Assumption 1. The heat flow is steady state $\Rightarrow \dfrac{\partial T}{\partial t} = 0$

Assumption 2. The heat flow is one dimensional $\Rightarrow \dfrac{\partial^2 T}{\partial y^2} = 0 = \dfrac{\partial^2 T}{\partial z^2}$

Assumption 3. There is energy generation term of $\dfrac{\dot{g}}{k}$

Approach: This problem is similar to Example 2.1 except for the fact that the wall is generating energy from inside. We will again first write the general heat equation, apply the assumptions, and subsequently determine the temperature distribution across the wall.

Step 3: Calculations
The heat equation for a rectangular section is written as

$$\frac{\partial^2 T}{\partial x^2} + \frac{\partial^2 T}{\partial y^2} + \frac{\partial^2 T}{\partial z^2} + \frac{\dot{g}}{k} = \frac{1}{\propto}\frac{\partial T}{\partial t} \tag{E2.2A}$$

Applying assumptions 1–3 into the above heat equation, we obtain

$$\frac{\partial^2 T}{\partial x^2} + \frac{\dot{g}}{k} = 0 \tag{E2.2B}$$

As discussed previously, the above PDE can be expressed

$$\frac{d^2 T}{dx^2} + \frac{\dot{g}}{k} = 0 \tag{E2.2C}$$

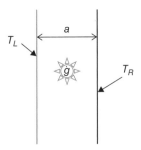

FIGURE 2.3 Heat transfer across the wall generating energy and whose left and right faces are maintained at T_L and T_R, respectively.

which after integrating twice will yield Eqs. (E2.2D) and (E2.2E)

$$\frac{dT}{dx} = -\frac{\dot{g}}{k}x + C_1 \tag{E2.2D}$$

$$T = -\frac{\dot{g}}{k}\frac{x^2}{2} + C_1 x + C_2 \tag{E2.2E}$$

The two BCs needed to solve these equations are

BC 1: When $x = 0$, $T = T_L$

BC 2: When $x = a$, $T = T_R$

Using these BCs, we can rewrite Eq. (E2.2E) as

$$T = T_L = -\frac{\dot{g}}{k}\frac{(0)^2}{2} + C_1(0) + C_2$$

$$C_2 = T_L \tag{E2.2F}$$

and

$$T = T_R = -\frac{\dot{g}}{k}\frac{(a)^2}{2} + C_1(a) + T_L \qquad \text{Because } C_2 = T_L$$

Which after rearranging will result in

$$C_1 = \frac{(T_R - T_L)}{a} + \frac{\dot{g}a}{2k} \tag{E2.2G}$$

Substituting the expressions for C_1 and C_2 in Eq. (E2.2E), we obtain the temperature distribution across the wall

$$T = -\frac{\dot{g}}{k}\frac{x^2}{2} + \left[\frac{(T_R - T_L)}{a} + \frac{\dot{g}a}{2k}\right]x + T_L$$

Which after factoring out common terms can be rearranged as

$$T = T_L + \frac{(T_R - T_L)x}{a} + \frac{\dot{g}x}{2k}(a - x) \tag{E2.2H}$$

Example 2.3 (General Engineering—All concentrations): Imagine a wall of thickness a, whose left face is subjected to a heat flux of $\dot{q}''$ and maintained at a temperature of T_L. If the wall has a conductivity of k and an area of A (Fig. 2.4). Assuming a one-dimensional steady-state heat transfer process with no energy generation, (i) formulate a differential equation to describe the heat transfer process and (ii) solve the differential equation to determine the temperature profile within the wall.

Solution

Step 1: Schematic
The problem is depicted in Fig. 2.4.

Step 2: Given data
Thickness of the wall = a; Heat flux on the left face of the wall = $\dot{q}'' = -k\frac{dT}{dx}$; Temperature of the left face of the wall = T_L; Temperature of the right face of the wall = T_R

To be determined: Temperature distribution across the wall.

Given assumptions

Assumption 1. The heat flow is steady state $\Rightarrow \frac{\partial T}{\partial t} = 0$

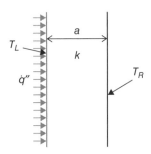

Figure 2.4   Schematic for heat transfer across the wall with a constant heat flux on the left face.

Assumption 2. The heat flow is one dimensional $\Rightarrow \dfrac{\partial^2 T}{\partial y^2} = 0 = \dfrac{\partial^2 T}{\partial z^2}$

Assumption 3. There is no energy generation $\Rightarrow \dot{g} = 0 \Rightarrow \dfrac{\dot{g}}{k} = 0$

Approach: This problem deals with the heat flux that propagates heat through the wall via conduction. As usual, we will first write the general heat equation, apply the assumptions, and subsequently determine the temperature distribution across the wall.

Step 3: Calculations
The heat equation for a rectangular section is written as

$$\frac{\partial^2 T}{\partial x^2} + \frac{\partial^2 T}{\partial y^2} + \frac{\partial^2 T}{\partial z^2} + \frac{\dot{g}}{k} = \frac{1}{\alpha}\frac{\partial T}{\partial t} \tag{E2.3A}$$

Applying assumptions 1–3 into the above heat equation, we obtain

$$\frac{\partial^2 T}{\partial x^2} = 0 = \frac{d^2 T}{dx^2} \tag{E2.3B}$$

To solve a second-order ODE, we require two BCs as below:

BC 1: When $x = 0$, $T = T_L$

BC 2: When $x = a$, $T = T_R$

After integrating Eq. (E2.3B) twice successively, we obtain

$$\frac{dT}{dx} = C_1 \tag{E2.3C}$$

$$T = C_1 x + C_2 \tag{E2.3D}$$

Using the BC described above in Eq. (E2.3D) will result in

$$T = T_L = C_1(0) + C_2 \tag{E2.3E}$$

$$C_2 = T_L \tag{E2.3F}$$

$$T = T_R = C_1(a) + C_2 \tag{E2.3G}$$

$$T_R = C_1(a) + T_L \tag{E2.3H}$$

Now we also know that the flux, $\dot{q}'' = -k\dfrac{dT}{dx}$ \hfill (E2.3I)

Therefore,

$$\frac{dT}{dx} = -\frac{\dot{q}''}{k}$$ (E2.3J)

From Eqs. (E2.3C) and (E2.3J) we obtain

$$C_1 = -\frac{\dot{q}''}{k}$$ (E2.3K)

Substituting the expressions for C_1 and C_2 in Eq. (E2.3D), we obtain the temperature distribution across the wall as below:

$$T = -\frac{\dot{q}''}{k}x + T_L$$ (E2.3L)

Example 2.4 (Bioprocessing Engineering—Heat transfer through a container wall): A rectangular stainless steel container ($k = 16$ W/m·K) of 5 cm thickness is being used to cure tobacco. The outer and inner faces of the wall are measured to be 100°C and 70°C, respectively (Fig. 2.5). Determine the temperature at a section 1 cm from the inner surface (4 cm from the outer wall) of the wall by assuming that the heat transfer is one dimensional, steady state with no heat generation.

Step 1: Schematic
The problem is depicted in Fig. 2.5.

Step 2: Given data
Thickness of the wall = a = 5 cm; Temperature of the outer face of the wall = T_L = 100°C; Thermal conductivity, $k = 16$ W/m·K; Temperature of the inner face of the wall = T_R = 70°C

To be determined: Temperature at a point 1 cm from inner surface.

Given assumptions

Assumption 1. The heat flow is steady state $\Rightarrow \dfrac{\partial T}{\partial t} = 0$

Assumption 2. The heat flow is one dimensional $\Rightarrow \dfrac{\partial^2 T}{\partial y^2} = 0 = \dfrac{\partial^2 T}{\partial z^2}$

Assumption 3. There is no energy generation term of $\dfrac{\dot{g}}{k} = 0$

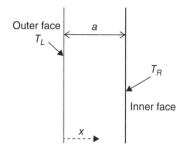

Figure 2.5 Schematic describing heat transfer across the wall whose left and right faces are maintained at T_L = 100°C and T_R = 70°C, respectively.

Step 3: Calculations

With these assumptions, the temperature profile across the wall and the temperature at any point within the wall can be directly estimated via Eq. (E2.1H) derived in Example 2.1 as reproduced below:

$$T = \frac{(T_R - T_L)}{a}x + T_L \tag{E2.4A}$$

Substituting the given values in the above equation, we obtain

$$T_{x=4\ cm} = \frac{(70 - 100)}{0.05} \times 0.04 + 100 \tag{E2.4B}$$

$$T_{x=4\ cm} = \frac{(70 - 100)}{0.05} \times 0.04 + 100 \tag{E2.4C}$$

$$T_{x=4\ cm} = 76°C$$

Example 2.5 (Agricultural Engineering–Heat transfer through a wall):   A 10-cm thick wall ($k = 25$ W/m·°C) is exposed to a thermal flux of 500 W/m². If the outer and the inner faces of the wall are to be maintained at 150°C, determine the rate of energy generation within the wall assuming a one-dimensional steady-state heat transfer (Fig. 2.6).

Step 1: Schematic
The problem is depicted in Fig. 2.6.

Step 2: Given data
Thickness of the wall = 10 cm; Temperature of the left face of the wall = 150°C; Temperature of the right face of the wall = 150°C; Thermal conductivity, $k = 25\dfrac{W}{m \cdot K}$; Heat flux on the left wall, $\dot{q}'' = 500\dfrac{W}{m^2 \cdot °C}$

To be determined: Rate of energy generation.

Given assumptions

Assumption 1. The heat flow is steady state $\Rightarrow \dfrac{\partial T}{\partial t} = 0$

Assumption 2. The heat flow is one dimensional $\Rightarrow \dfrac{\partial^2 T}{\partial y^2} = 0 = \dfrac{\partial^2 T}{\partial z^2}$

Assumption 3. There is energy generation term of $\dfrac{\dot{g}}{k}$

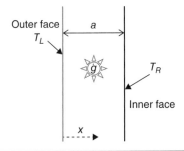

Figure 2.6   Heat transfer across the wall generating energy and whose left and right faces are maintained at 150°C each.

Step 3: Calculations

The heat equation for a rectangular section is written as

$$\frac{\partial^2 T}{\partial x^2} + \frac{\partial^2 T}{\partial y^2} + \frac{\partial^2 T}{\partial z^2} + \frac{\dot{g}}{k} = \frac{1}{\alpha}\frac{\partial T}{\partial t}$$

(E2.5A)

Applying assumptions 1–3 into the above heat equation, we obtain

$$\frac{\partial^2 T}{\partial x^2} + \frac{\dot{g}}{k} = 0$$

(E2.5B)

The above PDE can be expressed as

$$\frac{d^2 T}{dx^2} + \frac{\dot{g}}{k} = 0$$

(E2.5C)

Which after integrating twice will yield Eqs. (E2.5D) and (E2.5E).

$$\frac{dT}{dx} = -\frac{\dot{g}}{k}x + C_1$$

(E2.5D)

$$T = -\frac{\dot{g}}{k}\frac{x^2}{2} + C_1 x + C_2$$

(E2.5E)

The two BCs needed to solve these equations are

BC 1: When $x = 0$, $T = T_L$

BC 2: When $x = a$, $T = T_R$

Using these BCs, we can rewrite Eq. (E2.5E) as

$$T = T_L = -\frac{\dot{g}}{k}\frac{(0)^2}{2} + C_1(0) + C_2$$

$$C_2 = T_L$$

(E2.5F)

and

$$T = T_R = -\frac{\dot{g}}{k}\frac{(a)^2}{2} + C_1(a) + T_L$$

(E2.5G)

Because $C_2 = T_L$

Now we will also know that the flux,

$$\dot{q}'' = -k\frac{dT}{dx}$$

(E2.5H)

$$-\frac{\dot{q}}{k} = \frac{dT}{dx}$$

(E2.5I)

Substituting the expression for $\frac{dT}{dx}$ in Eq. (E2.5D), we obtain

$$-\frac{\dot{q}''}{k} = -\frac{\dot{g}}{k}x + C_1$$

(E2.5J)

and

$$C_1 = \frac{\dot{g}}{k}x - \frac{\dot{q}''}{k}$$

(E2.5K)

Plugging the expressions for C_1 and C_2 into Eqs. (E2.5G) and (E2.5E), following expressions can be obtained for the determination of the temperature on the right face of the wall and the temperature profile across the wall.

$$T_R = -\frac{\dot{g}}{k}\frac{a^2}{2} + \left[\frac{\dot{g}}{k}x - \frac{\dot{q}''}{k}\right]a + T_L = \frac{\dot{g}}{k}\frac{a^2}{2} - \frac{\dot{q}''a}{k} + T_L \qquad \text{(E2.5L)}$$

$$T = -\frac{\dot{g}}{k}\frac{x^2}{2} + \left[\frac{\dot{g}}{k}x - \frac{\dot{q}''}{k}\right]x + T_L = \frac{\dot{g}}{k}\frac{x^2}{2} - \frac{\dot{q}''x}{k} + T_L \qquad \text{(E2.5M)}$$

Substituting the given data into Eq. (E2.5L), we can easily obtain the rate of energy generation as

$$150 = \frac{\dot{g}}{25} \times \frac{0.1^2}{2} - \frac{500 \times 0.1}{25} + 150 \qquad \text{(E2.5N)}$$

$$\dot{g} = 10{,}000 \frac{\text{W}}{\text{m}^3}.$$

All the previous examples were for rectangular coordinates. For spherical and cylindrical-shaped bodies, the same heat balance equation can be applied to derive the conduction equation in relevant coordinate systems as described in the following sections.

2.5 Derivation of Three-Dimensional Heat Conduction Equation in Spherical Coordinate System

Again, we will consider an isotropic and infinitesimally small spherical element whose dimensions are as shown in Fig. 2.7. Let us also assume that the element has a thermal conductivity of k, specific heat of C_p, and density of ρ. All the other variables are listed in Table 2.3.

Let us assume that heat transfer rates entering the spherical element in all three directions to be $\dot{q}_r$, $\dot{q}_\varnothing$, and $\dot{q}_\theta$. Using Taylor series, we can express the heat transfer rates leaving the system as

$$\dot{q}_{r+dr} = \dot{q}_r + \frac{\partial \dot{q}_r}{\partial r}dr + \frac{\partial^2 \dot{q}_r}{\partial r^2}\frac{(dr)^2}{2!} + \cdots \qquad \text{(2.23)}$$

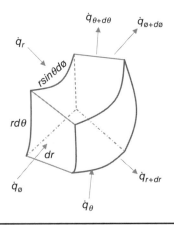

FIGURE 2.7 Schematic of an infinitesimally small element of a sphere.

Variable	Description	Value/Nomenclature
V	Volume of the element	$rd\theta \times dr \times r\sin\theta d\varnothing$
A_r	Area in the r-direction	$rd\theta \cdot r\sin\theta d\varnothing$
$A_\varnothing$	Area in the $\varnothing$-direction	$dr \cdot rd\theta$
A_θ	Area in the θ-direction	$dr \cdot r\sin\theta d\varnothing$
m	Mass	$\rho \cdot rd\theta \cdot dr \cdot r\sin\theta d\varnothing$
$\dot{q}_r$ Fourier's law	Heat transfer rate in the r-direction at the entrance of the element	$-kA_r \dfrac{\partial T}{\partial r} = -k \cdot rd\theta \cdot r\sin\theta d\varnothing \cdot \dfrac{\partial T}{\partial r}$
$\dot{q}_\varnothing$ Fourier's law	Heat transfer rate in the $\varnothing$-direction at the entrance of the element	$-kA_\varnothing \dfrac{\partial T}{r\sin\theta\,\partial\varnothing} = -k \cdot dr \cdot rd\theta \cdot \dfrac{\partial T}{r\sin\theta\,\partial\varnothing}$
$\dot{q}_\theta$ Fourier's law	Heat transfer rate in the θ-direction at the entrance of the element	$-kA_\theta \dfrac{\partial T}{r\,\partial\theta} = -k \cdot dr \cdot r\sin\theta d\varnothing \cdot \dfrac{\partial T}{r\,\partial\theta}$
$\dot{q}_{r+dr}$ Taylor series	Heat transfer rate in the r-direction at the exit of the element	$\dot{q}_r + \dfrac{\partial \dot{q}_r}{\partial r}(dr)$
$\dot{q}_{\varnothing+d\varnothing}$ Taylor series	Heat transfer rate in the $\varnothing$-direction at the exit of the element	$\dot{q}_\varnothing + \dfrac{\partial \dot{q}_\varnothing}{r\sin\theta\,\partial\varnothing}(r\sin\theta d\varnothing)$
$\dot{q}_{\theta+d\theta}$ Taylor series	Heat transfer rate in the θ-direction at the exit of the element	$\dot{q}_\theta + \dfrac{\partial \dot{q}_\theta}{r\,\partial\theta}(rd\theta)$

TABLE 2.3 Summary of the Variables Describing the Heat Transfer in a Spherical Element

$$\dot{q}_{\theta+\Delta\theta} = \dot{q}_\theta + \frac{\partial \dot{q}_\theta}{r\,\partial\theta}rd\theta + \frac{\partial^2 \dot{q}_\theta}{r^2\,\partial\theta^2}\frac{(r^2 d\theta)^2}{2!} + \cdots \tag{2.24}$$

$$\dot{q}_{\varnothing+d\varnothing} = \dot{q}_\varnothing + \frac{\partial \dot{q}_\varnothing}{r\sin\theta\,\partial\varnothing}r\sin\theta d\varnothing + \frac{\partial^2 \dot{q}_r}{r^2\sin^2\theta\,\partial\varnothing^2}\frac{(r^2\sin^2\theta d\varnothing)^2}{2!} + \cdots \tag{2.25}$$

If we assume that the element is infinitesimally small ($r\sin\theta\,d\varnothing$, dr, and $rd\theta \to 0$), the higher-order terms in the above equation can be ignored and the conduction heat transfer rates exiting in the r, θ, and $\varnothing$ directions can be written as

$$\dot{q}_{r+dr} = \dot{q}_r + \frac{\partial \dot{q}_r}{\partial r}dr \tag{2.26}$$

$$\dot{q}_{\theta+\Delta\theta} = \dot{q}_\theta + \frac{\partial \dot{q}_\theta}{r\,\partial\theta}rd\theta \tag{2.27}$$

$$\dot{q}_{\varnothing+d\varnothing} = \dot{q}_\varnothing + \frac{\partial \dot{q}_\varnothing}{r\sin\theta\,\partial\varnothing}r\sin\theta d\varnothing \tag{2.28}$$

Now we can write the energy balance equation as

Rate of heat conduction entering the element – Rate of heat conduction exiting the element + Rate of heat generation within the element = Rate of change of heat content of the element. (2.29)

$$\text{Rate of heat conduction entering the element} = \dot{q}_r + \dot{q}_\varnothing + \dot{q}_\theta \qquad (2.30)$$

$$\text{Rate of heat conduction exiting the element} = \dot{q}_{r+dr} + \dot{q}_{\varnothing+d\varnothing} + \dot{q}_{\theta+d\theta} \qquad (2.31)$$

Rate of heat generation within the element

$$= \dot{g} \times \text{Volume} = \dot{g} \cdot r \cdot d\theta \cdot dr \cdot r \cdot \sin\theta \cdot d\varnothing \qquad (2.32)$$

Where $\dot{g}$ is the rate of energy generation per unit volume.

Rate of change of heat content of the element

$$= m \cdot C_p \cdot \frac{\partial T}{\partial t} = \rho \cdot r \cdot d\theta \cdot dr \cdot r \cdot \sin\theta \cdot d\varnothing \cdot C_p \cdot \frac{\partial T}{\partial t} \qquad (2.33)$$

Substituting the above expressions in equation 2.29, we obtain

$$\dot{q}_r + \dot{q}_\varnothing + \dot{q}_\theta - \dot{q}_{r+dr} - \dot{q}_{\varnothing+d\varnothing} - \dot{q}_{\theta+d\theta} + \dot{g} \cdot r \cdot d\theta \cdot dr \cdot r \cdot \sin\theta \cdot d\varnothing$$

$$= \rho \cdot r \cdot d\theta \cdot dr \cdot r \cdot \sin\theta \cdot d\varnothing \cdot C_p \cdot \frac{\partial T}{\partial t} \qquad (2.34)$$

$$\dot{q}_r - \dot{q}_{r+dr} + \dot{q}_\varnothing - \dot{q}_{\varnothing+d\varnothing} + \dot{q}_\theta - \dot{q}_{\theta+d\theta} + \dot{g} \cdot r \cdot d\theta \cdot dr \cdot r \cdot \sin\theta \cdot d\varnothing$$

$$= \rho \cdot r \cdot d\theta \cdot dr \cdot r \cdot \sin\theta \cdot d\varnothing \cdot C_p \cdot \frac{\partial T}{\partial t} \qquad (2.35)$$

Now we will substitute the expressions for $\dot{q}_{r+dr}$, $\dot{q}_{\varnothing+d\varnothing}$, and $\dot{q}_{\theta+d\theta}$ (Table 2.2) that were evaluated using Taylor's series

$$\dot{q}_r - \left[\dot{q}_r + \frac{\partial \dot{q}_r}{\partial r}(dr) \right] + \dot{q}_\varnothing - \left[\dot{q}_\varnothing + \frac{\partial \dot{q}_\varnothing}{r\sin\theta\,\partial\varnothing}(r\sin\theta d\varnothing) \right] + \dot{q}_\theta - \left[\dot{q}_\theta + \frac{\partial \dot{q}_\theta}{r\,\partial\theta}(rd\theta) \right]$$

$$+ \dot{g} \cdot r \cdot d\theta \cdot dr \cdot r \cdot \sin\theta \cdot d\varnothing = \rho \cdot r \cdot d\theta \cdot dr \cdot r \cdot \sin\theta \cdot d\varnothing \cdot C_p \cdot \frac{\partial T}{\partial t} \qquad (2.36)$$

$$\left[-\frac{\partial \dot{q}_r}{\partial z}(dr) \right] + \left[-\frac{\partial \dot{q}_\varnothing}{r\sin\theta\,\partial\varnothing}(r\sin\theta d\varnothing) \right] + \left[-\frac{\partial \dot{q}_\theta}{r\,\partial\theta}(rd\theta) \right]$$

$$+ \dot{g} \cdot r \cdot d\theta \cdot dr \cdot r \cdot \sin\theta \cdot d\varnothing = \rho \cdot r \cdot d\theta \cdot dr \cdot r \cdot \sin\theta \cdot d\varnothing \cdot C_p \cdot \frac{\partial T}{\partial t} \qquad (2.37)$$

Now, substituting the expressions for $\dot{q}_r$, $\dot{q}_\varnothing$, and $\dot{q}_\theta$ in the above equation

$$-\frac{\partial\left(-krd\theta \cdot r\sin\theta d\varnothing \dfrac{\partial T}{\partial r} \right)}{\partial r}(dr) + -\frac{\partial\left(-kdr \cdot rd\theta \dfrac{\partial T}{r\sin\theta\,\partial\varnothing} \right)}{r\sin\theta\,\partial\varnothing}\left(r\sin\theta d\varnothing \right)$$

$$+ -\frac{\partial\left(-kdr \cdot r\sin\theta d\varnothing \dfrac{\partial T}{r\,\partial\theta} \right)}{r\,\partial\theta}(rd\theta) + \left(\dot{g} \cdot r \cdot d\theta \cdot dr \cdot r \cdot \sin\theta \cdot d\varnothing \right)$$

$$= \left(\rho \cdot r \cdot d\theta \cdot dr \cdot r \cdot \sin\theta \cdot d\varnothing \cdot C_p \cdot \frac{\partial T}{\partial t} \right) \qquad (2.38)$$

$$\frac{k \cdot rd\theta \cdot dr \cdot r\sin\theta \cdot d\varnothing \dfrac{\partial\left(r^2\dfrac{\partial T}{\partial r}\right)}{\partial r}}{r^2} + \frac{k \cdot rd\theta \cdot dr \cdot r\sin\theta \cdot d\varnothing \cdot \dfrac{\partial\left(-kdr \cdot rd\theta \dfrac{\partial T}{\partial\varnothing}\right)}{\partial\varnothing}}{r^2\sin^2\theta}$$

$$+\frac{k \cdot rd\theta \cdot dr \cdot r\sin\theta \cdot d\varnothing \dfrac{\partial\left(\sin\theta\dfrac{\partial T}{\partial\theta}\right)}{\partial\theta}}{r^2\sin\theta}$$

$$+\dot{g}\cdot r\cdot d\theta \cdot dr \cdot r \cdot \sin\theta \cdot d\varnothing = \rho \cdot r \cdot d\theta \cdot dr \cdot r \cdot \sin\theta \cdot d\varnothing \cdot C_p \cdot \frac{\partial T}{\partial t} \qquad (2.39)$$

Because the material is isotropic (k = constant)

Dividing the above equation by the volume of the element ($rd\theta \cdot dr \cdot r\sin\theta \cdot d\varnothing$) will result in

$$k\frac{\partial\left(r^2\dfrac{\partial T}{\partial r}\right)}{\partial r}\Big/r^2 + k\frac{\partial\left(\dfrac{\partial T}{\partial\varnothing}\right)}{\partial\varnothing}\Big/r^2\sin^2\theta + k\frac{\partial\left(\sin\theta\dfrac{\partial T}{\partial\theta}\right)}{\partial\theta}\Big/r^2\sin\theta + \dot{g} = \rho\cdot C_p \cdot \frac{\partial T}{\partial t} \qquad (2.40)$$

or

$$\frac{\partial\left(r^2\dfrac{\partial T}{\partial r}\right)}{\partial r}\Big/r^2 + \frac{\partial\left(\dfrac{\partial T}{\partial\varnothing}\right)}{\partial\varnothing}\Big/r^2\sin^2\theta + \frac{\partial\left(\sin\theta\dfrac{\partial T}{\partial\theta}\right)}{\partial\theta}\Big/r^2\sin\theta + \frac{\dot{g}}{k} = \frac{\rho C_p}{k}\cdot\frac{\partial T}{\partial t} \qquad (2.41)$$

or

$$\frac{1}{r^2}\frac{\partial}{\partial r}\left(r^2\frac{\partial T}{\partial r}\right)+\frac{1}{r^2\sin^2\theta}\frac{\partial}{\partial\varnothing}\left(\frac{\partial T}{\partial\varnothing}\right)+\frac{1}{r^2\sin\theta}\frac{\partial}{\partial\theta}\left(\sin\theta\frac{\partial T}{\partial\theta}\right)+\frac{\dot{g}}{k}=\frac{1}{\alpha}\frac{\partial T}{\partial t} \qquad (2.42)$$

Since

$$\alpha = \frac{k}{\rho C_p}$$

Example 2.6 (General Engineering—All concentrations): Consider an isotropic (constant thermal conductivity) spherical body with inner and outer radii of r_i and r_o held at temperatures T_i and T_o ($T_i > T_o$), respectively (Fig. 2.8). Assuming a steady-state one-dimensional heat flow and no energy generation, determine the temperature profile within the spherical body.

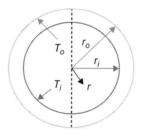

FIGURE 2.8 Heat transfer through the walls of a spherical shell whose inner and outer radii are maintained at T_i and T_o, respectively.

Solution

Approach: This problem involves conduction heat transfer in a sphere. We will again write the general heat equation in spherical coordinates, apply the assumptions, and subsequently determine the temperature distribution across the sphere.

Step 1: Schematic
The problem is depicted in Fig. 2.8.

Step 2: Given data
Inner radius = r_i; Temperature of the inner radius = T_i; Outer radius = r_o; Temperature of the inner radius = T_o, and $T_i > T_o$

To be determined: Temperature profile.

Given assumptions

Assumption 1. The heat flow is steady state $\Rightarrow \dfrac{\partial T}{\partial t} = 0$

Assumption 2. The heat flow is one dimensional $\Rightarrow \dfrac{\partial T}{\partial \varnothing} = 0 = \dfrac{\partial T}{\partial \theta}$

Assumption 3. There is no energy generation $\Rightarrow \dot{g} = 0 \Rightarrow \dfrac{\dot{g}}{k} = 0$

The heat equation for a spherical section is written as

$$\frac{1}{r^2}\frac{\partial}{\partial r}\left(r^2\frac{\partial T}{\partial r}\right)+\frac{1}{r^2\sin^2\theta}\frac{\partial}{\partial\varnothing}\left(\frac{\partial T}{\partial\varnothing}\right)+\frac{1}{r^2\sin\theta}\frac{\partial}{\partial\theta}\left(\sin\theta\frac{\partial T}{\partial\theta}\right)+\frac{\dot{g}}{k}=\frac{1}{\alpha}\frac{\partial T}{\partial t} \qquad \text{(E2.6A)}$$

After applying the assumptions listed above will yield

$$\frac{1}{r^2}\frac{\partial}{\partial r}\left(r^2\frac{\partial T}{\partial r}\right)=0 \qquad \text{(E2.6B)}$$

or

$$\frac{\partial}{\partial r}\left(r^2\frac{\partial T}{\partial r}\right)=0=\frac{d}{dr}\left(r^2\frac{dT}{dr}\right) \qquad \text{(E2.6C)}$$

Now we will solve this PDE and apply the appropriate BCs to obtain the temperature profile in the spherical body.
   After integrating Eq. (E2.6C) twice, we obtain

$$\frac{dT}{dr}=\frac{C_1}{r^2} \qquad \text{(E2.6D)}$$

And

$$T=\frac{-C_1}{r}+C_2 \qquad \text{(E2.6E)}$$

Now we know the BCs
BC 1: When $r = r_i$; $T = T_i$ and BC 2: When $r = r_o$; $T = T_o$
Using the BC 1 and BC 2 in Eq. (E2.6E) will result in

$$T_i=\frac{-C_1}{r_i}+C_2 \qquad \text{(E2.6F)}$$

$$T_o=\frac{-C_1}{r_o}+C_2 \qquad \text{(E2.6G)}$$

Eliminating C_2 from the above two equations, we obtain

$$T_i - T_o = \frac{C_1}{r_o} - \frac{C_1}{r_i} = C_1 \frac{(r_i - r_o)}{r_i r_o} \tag{2.6H}$$

$$C_1 = \frac{(T_i - T_o) r_i r_o}{(r_i - r_o)} \tag{E2.6I}$$

Similarly, C_2 can be determined as below:

$$T_o = \frac{-C_1}{r_o} + C_2 \tag{E2.6J}$$

$$T_o = \frac{-(T_i - T_o) r_i r_o}{(r_i - r_o) r_o} + C_2 \tag{E2.6K}$$

From which

$$C_2 = T_o + \frac{(T_i - T_o) r_i}{(r_i - r_o)} \tag{E2.6L}$$

Substituting the values of C_1 and C_2 in Eq. (E2.8E), we obtain the temperature profile in the spherical body in the radial direction

$$T = T_o - \frac{(T_i - T_o) r_i r_o}{(r_i - r_o) r} + \frac{(T_i - T_o) r_i}{(r_i - r_o)} \tag{E2.6M}$$

Example 2.7 (Bioenergy Systems Engineering—Heat transfer in a catalyst pellet): An exothermic oxidative chemical reaction is occurring between methane and oxygen to produce methanol within spherical catalyst pellets. The catalyst pellet consists of platinum supported on activated carbon with a thermal conductivity of 0.5 $\frac{W}{m \cdot ^\circ C}$ and 10 mm in diameter. Experimental measurements have indicated that at steady-state conditions, the surface temperature was 200°C with an energy generation of 25 $\frac{W}{m^3}$. For these experimental conditions, estimate the temperature of the pellet at its center (Fig. 2.9).

Step 1: Schematic
The problem is illustrated in Fig. 2.9.

Step 2: Given data
Outer radius $= r_s = 5$ mm; Temperature of the outer radius $= T_s = 200°C$

$$k = 0.5 \ \frac{W}{m \cdot ^\circ C}; \dot{g} = 25 \ \frac{W}{m^3}$$

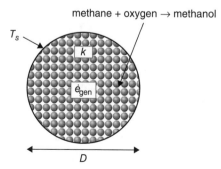

methane + oxygen → methanol

T_s

k

$\dot{e}_{gen}$

D

FIGURE 2.9 Schematic of methane oxidation within a spherical pellet.

To be determined: Temperature at the center of the pellet.

Assumptions:

Assumption 1. The heat flow is steady state $\Rightarrow \dfrac{\partial T}{\partial t} = 0$

Assumption 2. The heat flow is one dimensional $\Rightarrow \dfrac{\partial T}{\partial \varnothing} = 0 = \dfrac{\partial T}{\partial \theta}$

Assumption 3. There is energy generation of $\dot{g} = 25 \dfrac{W}{m^3}$

Approach: This problem involves conduction heat transfer due to an exothermic reaction in a spherical pellet. We will again write the general heat equation in spherical coordinates, apply the assumptions, and solve the PDE. Subsequently we will use the given data and determine the temperature at the center of the sphere.

Step 3: Calculations

The heat equation for a sphere is written as

$$\frac{1}{r^2}\frac{\partial}{\partial r}\left(r^2\frac{\partial T}{\partial r}\right)+\frac{1}{r^2\sin^2\theta}\frac{\partial}{\partial\varnothing}\left(\frac{\partial T}{\partial\varnothing}\right)+\frac{1}{r^2\sin\theta}\frac{\partial}{\partial\theta}\left(\sin\theta\frac{\partial T}{\partial\theta}\right)+\frac{\dot{g}}{k}=\frac{1}{\alpha}\frac{\partial T}{\partial t} \tag{E2.7A}$$

Applying the assumptions 1–3 will result in

$$\frac{1}{r^2}\frac{\partial}{\partial r}\left(r^2\frac{\partial T}{\partial r}\right)+\frac{\dot{g}}{k}=0 \tag{E2.7B}$$

or

$$\frac{1}{r^2}\frac{d}{dr}\left(r^2\frac{dT}{dr}\right)+\frac{\dot{g}}{k}=0 \tag{E2.7C}$$

or

$$\frac{d}{dr}\left(r^2\frac{dT}{dr}\right)=-\frac{r^2\dot{g}}{k} \tag{E2.7D}$$

Integrating Eq. (E2.7D) once will yield

$$r^2\frac{dT}{dr}=-\frac{r^3\dot{g}}{3k}+C_1 \tag{E2.7E}$$

$$\frac{dT}{dr}=-\frac{r\dot{g}}{3k}+\frac{C_1}{r^2} \tag{E2.7F}$$

Integrating Eq. (E2.7E) will result in

$$T=-\frac{r^2\dot{g}}{6k}-\frac{C_1}{r}+C_2 \tag{E2.7G}$$

Again, we will use the appropriate BCs to evaluate the values of the constants C_1 and C_2.

BC #1: When $r=0$, $\dfrac{dT}{dr}=0$, because the maximum temperature always occurs at the center of the sphere.

BC #1: When $r=r_s$, $T=T_s$, because at the outer radius, the temperature reaches the surface temperature.

$$T=-\frac{r^2\dot{g}}{6k}-\frac{C_1}{r}+C_2 \tag{E2.7H}$$

From Eq. (E2.7E) we obtain

$$0 \times 0 = -\frac{0^3 \dot{g}}{3k} + C_1$$

$$C_1 = 0 \tag{E2.7I}$$

$$T_s = -\frac{r_s^2 \dot{g}}{6k} - \frac{0}{r} + C_2 \tag{E2.7J}$$

$$C_2 = T_s + \frac{r_s^2 \dot{g}}{6k} \tag{E2.7K}$$

After plugging the values of C_1 and C_2 in Eq. (E2.7J), we obtain

$$T = -\frac{r^2 \dot{g}}{6k} - \frac{0}{r} + T_s + \frac{r_s^2 \dot{g}}{6k} \tag{E2.7L}$$

$$T = T_s + \frac{r_s^2 \dot{g}}{6k} - \frac{r^2 \dot{g}}{6k} \tag{E2.7M}$$

or

$$T = T_s + \frac{(r_s^2 - r^2) \dot{g}}{6k} \tag{E2.7N}$$

or

$$T = T_s + \frac{(r_s + r)(r_s - r) \dot{g}}{6k} \tag{E2.7O}$$

Now using Eq. (E2.7O) we can easily determine the temperature at the center of the pellet ($T_{r=0}$)

$$T_{(r=0)} = 200 + \frac{0.005 \times 0.005 \times 25}{6 \times 0.5} \tag{E2.7P}$$

$$T_{(r=0)} = 200.00 °C$$

The results suggest that the pellet had negligible temperature gradient.

2.6 Derivation of Three-Dimensional Heat Conduction Equation in Cylindrical Coordinate System

To analyze the heat conduction in an isotropic cylindrical-shaped body, let us consider a small cylindrical element of dimensions dr, $rd\theta$, and dy subjected to heat transfer in radial (r), azimuthal (θ), and vertical (y) directions as shown in Fig. 2.10 and Table 2.4.

Let us assume that heat transfer rates entering the cylindrical element in all three directions to be $\dot{q}_y$, $\dot{q}_\theta$, and $\dot{q}_r$. Using Taylor series, we can express the heat transfer rates leaving the system as

$$\dot{q}_{y+dy} = \dot{q}_y + \frac{\partial \dot{q}_y}{\partial y} dy + \frac{\partial^2 \dot{q}_y}{\partial y^2} \frac{(dy)^2}{2!} + \cdots \tag{2.43}$$

$$\dot{q}_{\theta+\Delta\theta} = \dot{q}_\theta + \frac{\partial \dot{q}_\theta}{r \cdot \partial\theta} r \cdot d\theta + \frac{\partial^2 \dot{q}_\theta}{r^2 \cdot \partial\theta^2} \frac{(r \cdot d\theta)^2}{2!} + \cdots \tag{2.44}$$

$$\dot{q}_{r+dr} = \dot{q}_r + \frac{\partial \dot{q}_r}{\partial r} dr + \frac{\partial^2 \dot{q}_r}{\partial r^2} \frac{(dr)^2}{2!} + \cdots \tag{2.45}$$

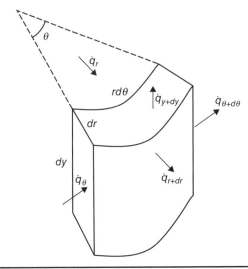

FIGURE 2.10   Schematic of an infinitesimally small element of a cylinder.

Variable	Description	Value/Nomenclature
V	Volume of the element	$r \cdot d\theta \cdot dr \cdot dy$
A_r	Area in the r-direction	$r \cdot d\theta \cdot dy$
A_y	Area in the y-direction	$dr \cdot rd\theta$
A_θ	Area in the θ-direction	$dr \cdot dy$
m	Mass	$\rho \cdot r \cdot d\theta \cdot dr \cdot dy$
$\dot{q}_r$ Fourier's law	Heat transfer rate in the r-direction at the entrance of the element	$-kA_r \dfrac{\partial T}{\partial r} = -k \cdot r \cdot d\theta \cdot dy \cdot \dfrac{\partial T}{\partial r}$
$\dot{q}_y$ Fourier's law	Heat transfer rate in the y-direction at the entrance of the element	$-kA_y \dfrac{\partial T}{\partial y} = -k \cdot dr \cdot r \cdot d\theta \cdot \dfrac{\partial T}{\partial y}$
$\dot{q}_\theta$ Fourier's law	Heat transfer rate in the θ-direction at the entrance of the element	$-kA_\theta \dfrac{\partial T}{r \partial \theta} = -k \cdot dr \cdot dy \cdot \dfrac{\partial T}{r \partial \theta}$
$\dot{q}_{r+dr}$ Taylor series	Heat transfer rate in the r-direction at the exit of the element	$\dot{q}_r + \dfrac{\partial \dot{q}_r}{\partial r}(dr)$
$\dot{q}_{y+dy}$ Taylor series	Heat transfer rate in the y-direction at the exit of the element	$\dot{q}_y + \dfrac{\partial \dot{q}_y}{\partial y}(dy)$
$\dot{q}_{\theta+d\theta}$ Taylor series	Heat transfer rate in the θ-direction at the exit of the element	$\dot{q}_\theta + \dfrac{\partial \dot{q}_\theta}{r \partial \theta}(rd\theta)$

TABLE 2.4   Summary of the Variables Describing the Heat Transfer in a Cylindrical Element

If we assume that element is infinitesimally small (dz, dr, and $rd\theta \to 0$), the higher-order terms in the above equation can be ignored and the conduction heat transfer rates exiting in the y, θ, and r directions can be written as

$$\dot{q}_{y+dy} = \dot{q}_y + \frac{\partial \dot{q}_y}{\partial y} \cdot dy \tag{2.46}$$

$$\dot{q}_{\theta+rd\theta} = \dot{q}_\theta + \frac{\partial \dot{q}_\theta}{r\partial\theta} \cdot r \cdot d\theta \tag{2.47}$$

$$\dot{q}_{r+dr} = \dot{q}_r + \frac{\partial \dot{q}_r}{\partial r} \cdot dr \tag{2.48}$$

Now, the energy balance around the element will yield

Rate of heat conduction entering the element – Rate of heat conduction exiting the element + Rate of heat generation within the element
= Rate of change of heat content of the element (2.49)

Rate of heat conduction entering the element $= \dot{q}_r + \dot{q}_\theta + \dot{q}_y$ (2.50)

Rate of heat conduction exiting the element $= \dot{q}_{r+dr} + \dot{q}_{\theta+rd\theta} + \dot{q}_{y+dy}$ (2.51)

Rate of heat generation within the element $= \dot{g} \cdot r \cdot d\theta \cdot dr \cdot dy$ (2.52)

where $\dot{g}$ is the rate of energy generation per unit volume.

Rate of change of heat content of the element

$$= m \cdot C_p \cdot \frac{\partial T}{\partial t} = \rho \cdot r \cdot d\theta \cdot dr \cdot dy \cdot C_p \cdot \frac{\partial T}{\partial t} \tag{2.53}$$

After substituting the above expression in the energy balance equation, we obtain

$$\dot{q}_r + \dot{q}_\theta + \dot{q}_y - \dot{q}_{r+dr} - \dot{q}_{\theta+rd\theta} - \dot{q}_{y+dy} + \dot{g} \cdot r \cdot d\theta \cdot dr \cdot dy = \rho \cdot r \cdot d\theta \cdot dr \cdot dy \cdot C_p \cdot \frac{\partial T}{\partial t} \tag{2.54}$$

Substituting Eqs. (2.46), (2.47), and (2.48) into Eq. (2.54) will result in

$$-\frac{\partial \dot{q}_y}{\partial y} \cdot dy - \frac{\partial \dot{q}_\theta}{r\partial\theta} \cdot r \cdot d\theta - \frac{\partial \dot{q}_r}{\partial r} \cdot dr + \dot{g} \cdot r \cdot d\theta \cdot dr \cdot dy = \rho \cdot r \cdot d\theta \cdot dr \cdot dy \cdot C_p \cdot \frac{\partial T}{\partial t} \tag{2.55}$$

At this stage, we apply Fourier's law for conduction in each direction

$$\dot{q}_y = -k \cdot dr \cdot r \cdot d\theta \cdot \frac{\partial T}{\partial y}; \dot{q}_r = -k \cdot r \cdot d\theta \cdot dy \cdot \frac{\partial T}{\partial r}; \dot{q}_\theta = -k \cdot dr \cdot dy \cdot \frac{\partial T}{r\partial\theta} \tag{2.56}$$

Substituting the above equations in Eq. (2.55), we obtain

$$-\left(\frac{\partial\left(-k \cdot dr \cdot r \cdot d\theta \cdot \dfrac{\partial T}{\partial y} \right)}{\partial y} \cdot dy \right) - \left(\frac{\partial\left(-k \cdot dr \cdot dy \cdot \dfrac{\partial T}{r\partial\theta} \right)}{r\partial\theta} \times rd\theta \right)$$

$$-\left(\frac{\partial\left(-k \cdot r \cdot d\theta \cdot dy \cdot \dfrac{\partial T}{\partial r} \right)}{\partial r} \cdot dr \right) + \dot{g} \cdot dr \cdot r \cdot d\theta \cdot dy = \rho \cdot dr \cdot r \cdot d\theta \cdot dy \cdot C_p \cdot \frac{\partial T}{\partial t} \tag{2.57}$$

$$k \cdot dr \cdot r \cdot d\theta \cdot dy \cdot \left(\frac{\partial^2 T}{\partial y^2}\right) + k \cdot dr \cdot r \cdot d\theta \cdot dy \cdot \left(\frac{\partial^2 T}{r^2 \, \partial \theta^2}\right)$$

$$+ k \cdot dr \cdot d\theta \cdot dy \frac{\partial \left(r \cdot \frac{\partial T}{\partial r}\right)}{\partial r} + \dot{g} \cdot r \cdot d\theta \cdot dr \cdot dy = \rho \cdot r \cdot d\theta \cdot dr \cdot dy \cdot C_p \cdot \frac{\partial T}{\partial t} \tag{2.58}$$

Dividing both the sides of the equation by $k \cdot r \cdot d\theta \cdot dr \cdot dy$ will yield

$$\frac{\partial^2 T}{\partial y^2} + \frac{\partial^2 T}{r^2 \, \partial \theta^2} + \frac{1}{r} \frac{\partial \left(r \cdot \frac{\partial T}{\partial r}\right)}{\partial r} + \frac{\dot{g}}{k} = \frac{\rho C_p}{k} \cdot \frac{\partial T}{\partial t} \tag{2.59}$$

Or can be expressed in the familiar form

$$\frac{\partial^2 T}{\partial y^2} + \frac{\partial^2 T}{r^2 \, \partial \theta^2} + \frac{1}{r} \frac{\partial}{\partial r}\left(r \cdot \frac{\partial T}{\partial r}\right) + \frac{\dot{g}}{k} = \frac{1}{\alpha} \frac{\partial T}{\partial t} \tag{2.60}$$

Since

$$\alpha = \frac{k}{\rho C_p}$$

Example 2.8 (General Engineering—All concentrations): Now, let us extend our discussion to a long isotropic cylindrical body of length L and inner and outer radii of r_i and r_o held at temperatures T_i and T_o ($T_i > T_o$), respectively (Fig. 2.11). Determine the temperature profile across the radial direction using the assumptions (1) steady-state heat transfer, (2) one-dimensional heat flow, and (3) no energy generation within the cylinder.

Solution

Approach: This problem involves conduction heat transfer in a cylinder. To solve this problem, let us first write the general heat equation in cylindrical coordinates and apply the given assumptions to obtain the temperature distribution within the cylinder.

Step 1: Schematic
The problem is illustrated in Fig. 2.11.

Step 2: Given data
Length = L; Inner radius = r_i; Temperature of the inner radius = T_i; Outer radius = r_o; Temperature of the inner radius = T_o, and $T_i > T_o$

To be determined: Temperature distribution within a cylinder in radial direction.

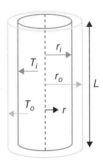

FIGURE 2.11 Schematic for the determination of heat transfer in a cylinder.

Given assumptions

Assumption 1. The heat flow is steady state $\Rightarrow \dfrac{\partial T}{\partial t} = 0$

Assumption 2. The heat flow is one dimensional $\Rightarrow \dfrac{\partial T}{\partial \theta} = 0 = \dfrac{\partial T}{\partial y}$

Assumption 3. There is no energy generation $\Rightarrow \dot{g} = 0 \Rightarrow \dfrac{\dot{g}}{k} = 0$

The heat equation in cylindrical coordinate system for a cylinder is written as

$$\frac{\partial^2 T}{\partial y^2} + \frac{\partial^2 T}{r^2 \partial \theta^2} + \frac{1}{r}\frac{\partial}{\partial r}\left(r \cdot \frac{\partial T}{\partial r}\right) + \frac{\dot{g}}{k} = \frac{1}{\alpha}\frac{\partial T}{\partial t} \tag{E2.8A}$$

After applying the assumptions listed above will yield

$$\frac{1}{r}\frac{\partial}{\partial r}\left(r\frac{\partial T}{\partial r}\right) = 0 \tag{E2.8B}$$

or

$$\frac{\partial}{\partial r}\left(r\frac{\partial T}{\partial r}\right) = 0 \tag{E2.8C}$$

Integration of Eq. (E2.8C) twice successively will yield

$$r\frac{\partial T}{\partial r} = C_1 \tag{E2.8D}$$

$$\frac{\partial T}{\partial r} = \frac{C_1}{r} \tag{E2.8E}$$

$$T = C_1 \ln(r) + C_2 \tag{E2.8F}$$

Using the BC, when $r = r_i$, $T = T_i$, and $r = r_o$, $T = T_o$ and substituting in Eq. (E2.8F), we obtain

$$T_i = C_1 \ln(r_i) + C_2 \text{ and} \tag{E2.8G}$$

$$T_o = C_1 \ln(r_o) + C_2 \tag{E2.8H}$$

Eliminating C_2 from the above two equations,

$$T_i - T_o = C_1 \ln\left(\frac{r_i}{r_o}\right) \tag{E2.8I}$$

$$C_1 = \frac{T_i - T_o}{\ln\left(\dfrac{r_i}{r_o}\right)} \tag{E2.8J}$$

By substituting the value of C_1 in Eq. (E2.8H), we obtain

$$T_o = C_1 \ln r_o + C_2 \tag{E2.8K}$$

$$T_o = \frac{T_i - T_o}{\ln\left(\dfrac{r_i}{r_o}\right)} \ln(r_o) + C_2 \tag{E2.8L}$$

because $C_1 = \dfrac{T_i - T_o}{\ln\left(\dfrac{r_i}{r_o}\right)}$

$$C_2 = T_o - \ln(r_o)\frac{T_i - T_o}{\ln\left(\dfrac{r_i}{r_o}\right)} \tag{E2.8M}$$

Substituting C_1 and C_2 into Eq. (E2.8F), we obtain the expression for temperature distribution across the cylindrical wall.

$$T = C_1 \ln(r) + C_2 \tag{E2.8N}$$

$$T = \frac{T_i - T_o}{\ln\left(\frac{r_i}{r_o}\right)} \ln(r) + T_o - \ln(r_o)\frac{T_i - T_o}{\ln\left(\frac{r_i}{r_o}\right)} \tag{E2.8O}$$

$$T = T_o + \frac{T_i - T_o}{\ln\left(\frac{r_i}{r_o}\right)} \ln\left(\frac{r}{r_o}\right) \tag{E2.8P}$$

Example 2.9 (Bioprocessing Engineering—Hydrothermal carbonization reactor):   A stainless steel cylindrical container $\left(k = 16\ \dfrac{W}{m\,°C}\right)$ is being used to process a slurry of wood powder and water with a metal oxide catalyst to convert wood powder into hydrochar, a multifunctional product. The cylinder of 30 cm in length has an inner diameter of 15 cm and 1 cm thick. During a steady-state heating process the outer and inner temperatures were measured to be 400°C and 300°C, respectively. Plot the temperature profile across the wall of the cylinder and determine the rate of heat transfer through the wall (Fig. 2.12).

Step 1: Schematic
The problem is depicted in Fig. 2.12.

Step 2: Given data
Length $= L = 30$ cm; Thermal conductivity, $k = 16\ \dfrac{W}{m\,°C}$; Inner radius $= r_i = 7.5$ cm; Temperature of the inner radius $= 300°C$; Outer radius $= r_o = 8.5$ cm; Temperature of the outer radius $= 400°C$

To be determined: Temperature distribution and heat transfer rate.

Given assumptions

Assumption 1. The heat flow is steady state $\Rightarrow \dfrac{\partial T}{\partial t} = 0$

Assumption 2. The heat flow is one dimensional $\Rightarrow \dfrac{\partial T}{\partial y} = 0 = \dfrac{\partial T}{\partial \theta}$

Assumption 3. There is no energy generation $\Rightarrow \dot{g} = 0 \Rightarrow \dfrac{\dot{g}}{k} = 0$

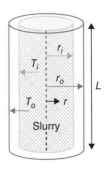

Figure 2.12   Schematic of the slurry reactor with catalyst for hydrothermal carbonization.

Step 3: Calculations

3 (i): Determination of temperature profile

The temperature profile within a cylinder is given by Eq. (E2.8P) that was derived in Example 2.8.

$$T = T_o + \frac{(T_i - T_o)}{\ln\left(\frac{r_i}{r_o}\right)} \ln\left(\frac{r}{r_o}\right) \tag{E2.9A}$$

We will determine and plot the variation of temperature with the radial distance $r = 7.5, 7.75, 8, 8.25,$ and 8.5 from the center. A sample calculation is described for $r = 7.75$ cm.

$$T(r = 7.75 \text{ cm}) = 400 + \frac{(300 - 400)}{\ln\left(\frac{7.5}{8.5}\right)} \ln\left(\frac{7.75}{8.5}\right) \tag{E2.9B}$$

$$T(r = 7.75 \text{ cm}) = \textbf{326.19°C}$$

Similarly, temperatures at various locations within the wall are calculated and plotted in Fig. 2.13.

3 (ii): Calculation of heat transfer rate

The heat transfer rate can be calculated via Fourier's law

$$\dot{q}_{\text{cond}} = -kA\frac{dT}{dr} \tag{E2.9C}$$

The heat transfer area for the cylinder $= 2\pi r L \tag{E2.9D}$

$\frac{dT}{dr}$ can be easily determined by differentiating Eq. (E2.9A).

$$\frac{dT}{dr} = 0 + \frac{(T_i - T_o)}{\ln\left(\frac{r_i}{r_o}\right)}\frac{1}{r} - \frac{(T_i - T_o)}{\ln\left(\frac{r_i}{r_o}\right)}(0) \tag{E2.9E}$$

$$\frac{dT}{dr} = \frac{(T_i - T_o)}{\ln\left(\frac{r_i}{r_o}\right)}\frac{1}{r} \tag{E2.9F}$$

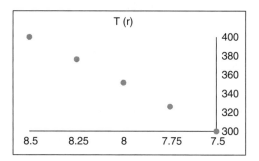

Figure 2.13 Temperature profile across the hydrothermal carbonization reactor cylinder wall.

Substituting the values of A, $\dfrac{dT}{dr}$, and k, and Eq. (E2.9F) in Eq. (E2.9C), we obtain

$$\dot{q}_{\text{cond}} = -k(2\pi r L)\left(\frac{(T_i - T_o)}{\ln\left(\dfrac{r_i}{r_o}\right)}\frac{1}{r}\right) \tag{E2.9G}$$

$$\dot{q}_{\text{cond}} = -k(2\pi L)\left(\frac{(T_i - T_o)}{\ln\left(\dfrac{r_i}{r_o}\right)}\right) \tag{E2.9H}$$

$$\dot{q}_{\text{cond}} = -16 \times (2 \times 3.14 \times 0.3)\left(\frac{(300 - 400)}{\ln\left(\dfrac{0.075}{0.0875}\right)}\right) \tag{E2.9I}$$

$$\dot{q}_{\text{cond}} = -24.08 \text{ kW}$$

The negative sign indicates that the heat transfer is from the outer wall to the inner wall.

The solutions to the above general examples are summarized as below.

Configuration	Conditions	Conditions	Solution
Plane wall	Wall of thickness "a" subjected to a temperature differential.	Left face temp = T_L Right face temp = T_R	$T = \dfrac{(T_R - T_L)}{a}x + T_L$
Plane wall	Wall of thickness "a" subjected to a temperature differential with internal energy generation.	Left face temp = T_L Right face temp = T_R Energy generation = $\dot{g}$	$T = T_L + \dfrac{(T_R - T_L)x}{a} + \dfrac{\dot{g}x}{2k}(a - x)$
Plane wall	Wall of thickness "a" whose left face is subjected to a heat flux.	Left face temp = T_L Flux on the left face = $\dot{q}''$	$T = -\dfrac{\dot{q}''}{k}x + T_L$
Plane wall	Wall of thickness "a" whose left face is subjected to a heat flux with energy generation within the wall.	Left face temp = T_L Flux on the left face = $\dot{q}''$ Energy generation = $\dot{g}$	$T = \dfrac{\dot{g}x^2}{2k} - \dfrac{\dot{q}''}{k}x + T_L$
Sphere	Spherical body with a hot inner wall	Inner radius = r_i; Outer radius = r_o; Inner radius temp = T_i; Outer radius temp = T_o	$T = T_o - \dfrac{(T_i - T_o)r_i r_o}{(r_i - r_o)r} + \dfrac{(T_i - T_o)r_i}{(r_i - r_o)}$
Cylinder	Cylindrical body with a hot inner wall	Inner radius = r_i; Outer radius = r_o; Inner radius temp = T_i; Outer radius temp = T_o	$T = T_o + \dfrac{(T_i - T_o)}{\ln\left(\dfrac{r_i}{r_o}\right)}\ln\left(\dfrac{r}{r_o}\right)$

TABLE 2.5   Temperature Profiles for Standard Geometric Configurations

Example 2.10 (Aquacultural Engineering—Heater selection for a recirculating tank): Water in a recirculating aquacultural system is being heated by a 20-cm long and 2-cm diameter cylindrical heater with a thermal conductivity of 16 W/m·°C. The heater generates an energy of 1×10^7 W/cm³ (10 W/m³) under steady-state conditions. Determine the surface temperature of the heater if the temperature at the center is given as 100°C (Fig. 2.14).

Approach: The energy generated by the coil at the center of the heater is dissipated to the outer surface via conduction. We will solve the heat equation for a cylindrical section in general variables and eventually solve for the surface temperature of the heater for the given conditions.

Step 1: Schematic
The problem is illustrated in Fig. 2.14.

Step 2: Given data
Length of the heater, $L = 20$ cm; Outer radius $r_s = 1$ cm; Temperature at the center $= T_o = 100$°C

$$k = 16 \ \frac{W}{m°C} \ ; \ \dot{g} = 10 \ \frac{W}{m^3}$$

To be determined: Temperature at the heater surface.

Assumption 1. The heat flow is steady state $\Rightarrow \dfrac{\partial T}{\partial t} = 0$

Assumption 2. The heat flow is one dimensional $\Rightarrow \dfrac{\partial T}{\partial y} = 0 = \dfrac{\partial T}{\partial \theta}$

Assumption 3. There is energy generation of $\dot{g} = 10 \ \dfrac{W}{m^3}$

Step 3: Calculations
The heat equation in cylindrical coordinate system for a cylinder is written as

$$\frac{\partial^2 T}{\partial y^2} + \frac{\partial^2 T}{r^2 \partial \theta^2} + \frac{1}{r} \frac{\partial}{\partial r}\left(r \cdot \frac{\partial T}{\partial r} \right) + \frac{\dot{g}}{k} = \frac{1}{\alpha} \tag{E2.10A}$$

Because the heat transfer is in radial direction, we can assume a one-dimensional heat transfer situation. In addition, it was also given that the steady-state condition was valid. Therefore, Eq. (E2.10A) can be simplified into

$$\frac{1}{r} \frac{d}{dr}\left(r \frac{dT}{dr} \right) + \frac{\dot{g}}{k} = 0 \tag{E2.10B}$$

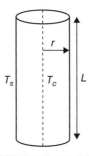

Figure 2.14 Schematic of the cylindrical heater.

and may be rearranged as

$$\frac{d}{dr}\left(r\frac{dT}{dr}\right) = -\frac{\dot{g}r}{k}$$

(E2.10C)

Integrating Eq. (E2.10C) once

$$r\frac{dT}{dr} = -\frac{\dot{g}r^2}{2k} + C_1$$

(E2.10D)

$$\frac{dT}{dr} = -\frac{\dot{g}r}{2k} + \frac{C_1}{r}$$

(E2.10E)

Integrating Eq. (E2.10E) again

$$T = -\frac{\dot{g}r^2}{4k} - \frac{C_1}{r^2} + C_2$$

(E2.10F)

The BCs in this problem are:

BC #1: When $r = 0$, $\dfrac{dT}{dr} = 0$, because the maximum temperature always occurs at the center of the cylinder.

BC #2: When $r = r_s$, $T = T_s$, because the temperature reaches the surface temperature at the outer radius of the cylinder.

Applying these BCs in Eq. (E2.10D) will result in

$$0 \times 0 = 0 = -\frac{\dot{g} \times 0}{2k} + C_1$$

(E2.10G)

$$C_1 = 0$$

(E2.10H)

and

$$T_s = -\frac{\dot{g}r_s^2}{4k} + C_2$$

(E2.10I)

$$C_2 = T_s + \frac{\dot{g}r_s^2}{4k}$$

(E2.10J)

Substituting the values of C_1 and C_2 in Eq. (E2.10F) will yield

$$T = -\frac{\dot{g}r^2}{4k} - \frac{0}{r^2} + T_s + \frac{\dot{g}r_s^2}{4k}$$

(E2.10K)

or

$$T = T_s + \frac{(r_s^2 - r^2)\dot{g}}{4k}$$

(E2.10L)

or

$$T = T_s + \frac{(r_s + r)(r_s - r)\dot{g}}{4k}$$

(E2.10M)

After substituting the given data into the above equation, we see that the temperature of the surface is very close to its center temperature.

In many situations, in addition to the BCs, we can also employ the equilibrium conditions to solve the conduction problems. The following examples will illustrate these ideas.

Example 2.11 (Agricultural Engineering—Heat transfer across a metal plate): Consider a part of an agricultural implement that is of rectangular shape made of 2.5-cm thick iron ($k = 80$ W/m·K) with an area of 1.6 m². The underside of the plate reaches a constant temperature of 110°C. However, the upper face of the plate is exposed to an ambient airflow at 27°C with a heat transfer coefficient of 100 W/m²·°C. Determine the rate of heat loss from the plate (Fig. 2.15).

Step 1: Schematic
The problem is illustrated in Fig. 2.15.

Step 2: Given data:
Thickness of the plate, $a = 2.5$ cm; Thermal conductivity of iron, $k = 80 \dfrac{W}{m \cdot K}$; Heat transfer area, $A = 1.6$ m²; Temperature of the plate at the lower face $T_L = 110$°C; Air flow temperature at the upper plate, $T_\infty = 27$°C; Heat transfer coefficient, $h = 100 \dfrac{W}{m^2 \cdot {}^\circ C}$

To be determined: Heat transfer rate from the surface.

Assumptions: We will make the following assumptions:

Assumption 1. The heat flow is steady state $\Rightarrow \dfrac{\partial T}{\partial t} = 0$

Assumption 2. The heat flow is one dimensional $\Rightarrow \dfrac{\partial^2 T}{\partial y^2} = 0 = \dfrac{\partial^2 T}{\partial z^2}$

Assumption 3. There is no energy generation term $\dfrac{\dot{g}}{k} = 0$

Step 3: Calculations
3 (i). The heat equation for a rectangular section is written as

$$\frac{\partial^2 T}{\partial x^2} + \frac{\partial^2 T}{\partial y^2} + \frac{\partial^2 T}{\partial z^2} + \frac{\dot{g}}{k} = \frac{1}{\alpha} \frac{\partial T}{\partial t} \tag{E2.11A}$$

Applying assumptions 1–3 into the above heat equation, we obtain

$$\frac{\partial^2 T}{\partial x^2} = \frac{d^2 T}{dx^2} = 0 \tag{E2.11B}$$

3 (ii). After integrating Eq. (E2.11B) twice successively, we obtain

$$\frac{dT}{dx} = C_1 \tag{E2.11C}$$

$$T = C_1 x + C_2 \tag{E2.11D}$$

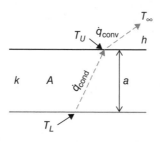

Figure 2.15 Schematic of the hot plate subjected to convection cooling.

Now, we will apply the BCs

$$\text{BC 1: When } x = 0, T = T_L$$

$$\text{BC 2: When } x = a, T = T_U$$

By substituting BC 1 and BC 2 in Eq. (E2.11D), the following equations are obtained

$$T_L = 110°C = C_1 \times 0 + C_2 \tag{E2.11E}$$

$$C_2 = 110°C \tag{E2.11F}$$

$$T_U = C_1(0.025) + 110 \tag{E2.11G}$$

Because $a = 0.025$ m and $C_2 = 110°C$

3 (iii). We can now use the equilibrium condition that the rate of heat conduction across the plate is same as the rate of heat convection from the plate surface to the ambient air. Mathematically,

$$-kA\frac{dT}{dx} = hA(T_U - T_\infty) \tag{E2.11H}$$

We can now cancel the area from both sides and substitute the values of $\frac{dT}{dx}$ and T_U from Eqs. (E2.11C) and (E2.11G) into the above equilibrium Eq. (E2.11H).

$$-kC_1 = h(C_1(0.025) + 110 - T_\infty) \tag{E2.11I}$$

After substituting the values of k, h, and T_∞ into Eq. (E2.11I) and solving for C_1, we obtain

$$C_1 = -100.6 \frac{°C}{m} \tag{E2.11J}$$

By substituting the values of C_1 and C_2 into Eq. (E2.11G) the temperature of the outer face of the plate is determined as

$$T_U = -100.6(0.025) + 110 \tag{E2.11K}$$

$$T_U = 107.48°C \tag{E2.11L}$$

3 (iv). The heat transfer rate can now be easily determined by applying either of the conduction or convection equations.

$$\dot{q}_{conv} = hA(T_U - T_\infty) = 100 \times 1.6 \times (107.48 - 27) \tag{E2.11M}$$

$$\dot{q}_{conv} = 12,877.6 \text{ W}$$

Note: The same answer can also be obtained via Fourier's law.

Example 2.12 (Environmental Engineering—Heat transfer through a metal pipe):   A 10-m long, 4-mm thick copper pipe of 3 cm inner diameter ($k = 400$ W/m²·°C) is carrying hot water at 80°C with a heat transfer coefficient of 200 W/m²·°C inside the pipe. The pipe is exposed to ambient (air) conditions of $T = 27°C$ with an external heat transfer coefficient of 25 W/m²·°C. Determine the heat transfer rate from the pipe into the surroundings.

Step 1: Schematic
The problem is illustrated in Fig. 2.16.

Step 2: Given data
Length = $L = 10$ m; Thermal conductivity = $k = 400 \dfrac{W}{m \cdot k}$; Inner radius = $r_i = 0.015$ m; Temperature of water inside the pipe = $T_{\infty,i} = 80°C$; Outer radius = $r_o = 0.019$ m; Temperature of the ambient air around the pipe = $T_{\infty,o} = 27°C$; Heat transfer coefficient inside the pipe = $h_i = 200 \dfrac{W}{m^2 \cdot °C}$; Heat transfer coefficient outside the pipe = $h_o = 25 \dfrac{W}{m^2 \cdot °C}$

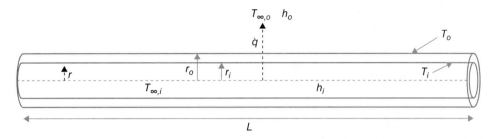

Figure 2.16   Schematic of the copper pipe carrying hot water and losing heat to the surroundings.

To be determined: Heat transfer rate from the pipe.

Given assumptions

Assumption 1. The heat flow is steady state $\Rightarrow \dfrac{\partial T}{\partial t} = 0$

Assumption 2. The heat flow is one dimensional $\Rightarrow \dfrac{\partial T}{\partial \theta} = 0 = \dfrac{\partial T}{\partial y}$

Assumption 3. There is no energy generation $\Rightarrow \dot{g} = 0 \Rightarrow \dfrac{\dot{g}}{k} = 0$

Step 3: Calculations

The heat equation in cylindrical coordinate system is written as

$$\frac{1}{r} \frac{\partial}{\partial r} \left(r \frac{\partial T}{\partial r} \right) + \frac{1}{r^2} \frac{\partial^2 T}{\partial \theta^2} + \frac{\partial^2 T}{\partial y^2} + \frac{\dot{g}}{k} = \frac{1}{\alpha} \frac{\partial T}{\partial t} \tag{E2.12A}$$

After applying the assumptions listed above will yield

$$\frac{1}{r} \frac{d}{dr} \left(r \frac{dT}{dr} \right) = 0 \tag{E2.12B}$$

or

$$\frac{d}{dr} \left(r \frac{dT}{dr} \right) = 0 \tag{E2.12C}$$

Integration of Eq. (E2.12C) twice will yield Eqs. (E2.12E) and (E2.12F)

$$r \frac{dT}{dr} = C_1 \tag{E2.12D}$$

$$\frac{dT}{dr} = \frac{C_1}{r} \tag{E2.12E}$$

$$T = C_1 \ln (r) + C_2 \tag{E2.12F}$$

Using the BCs, when $r = r_i$, $T = T_i$ and $r = r_o$, $T = T_o$ and substituting in Eq. (E2.13F), we obtain

$$T_i = C_1 \ln(r_i) + C_2 \tag{E2.12G}$$

and

$$T_o = C_1 \ln(r_o) + C_2 \tag{E2.12H}$$

Now we will apply the equilibrium conditions namely,

The rate of convection inside the pipe = the rate of conduction at the inner surface of the pipe

and

The rate of convection outside the pipe = the rate of conduction at the outer surface of the pipe. Mathematically,

$$h_i A(T_{\infty,i} - T_i) = -kA\frac{dT}{dr}@r = r_i \text{ and } h_o A(T_o - T_{\infty,o}) = -kA\frac{dT}{dr}@r = r_o \tag{E2.12I}$$

We know that $\dfrac{dT}{dr} = \dfrac{C_1}{r}$, $T_i = C_1 \ln(r_i) + C_2$ and $T_o = C_1 \ln(r_o) + C_2$

Substituting the above equations in Eq. (E2.12I) and cancelling out area on both sides

$$h_i(T_{\infty,i} - C_1 \ln(r_i) - C_2) = -k\frac{C_1}{r_i} \tag{E2.12J}$$

$$h_o(C_1 \ln(r_o) + C_2 - T_{\infty,o}) = -k\frac{C_1}{r_o} \tag{E2.12K}$$

We will algebraically manipulate the above two equations, to evaluate the constants C_1 and C_2

$$(T_{\infty,i} - C_1 \ln(r_i) - C_2) = -k\frac{C_1}{h_i r_i} \tag{E2.12L}$$

$$(C_1 \ln(r_o) + C_2 - T_{\infty,o}) = -k\frac{C_1}{h_o r_o} \tag{E2.12M}$$

$$-k\frac{C_1}{h_i r_i} - k\frac{C_1}{h_o r_o} = T_{\infty,i} - C_1 \ln(r_i) - C_2 + C_1 \ln(r_o) + C_2 - T_{\infty,o} \tag{E2.12N}$$

$$-C_1\left(\frac{k}{h_i r_i} + \frac{k}{h_o r_o}\right) = T_{\infty,i} + C_1 \ln\left(\frac{r_o}{r_i}\right) - T_{\infty,o} \tag{E2.12O}$$

$$-C_1\left(\frac{k}{h_i r_i} + \frac{k}{h_o r_o}\right) - C_1 \ln\left(\frac{r_o}{r_i}\right) = T_{\infty,i} - T_{\infty,o} \tag{E2.12P}$$

$$C_1\left[\frac{k}{h_i r_i} + \frac{k}{h_o r_o} + \ln\left(\frac{r_o}{r_i}\right)\right] = T_{\infty,o} - T_{\infty,i} \tag{E2.12Q}$$

$$C_1 = \frac{T_{\infty,o} - T_{\infty,i}}{\left[\dfrac{k}{h_i r_i} + \dfrac{k}{h_o r_o} + \ln\left(\dfrac{r_o}{r_i}\right)\right]} \tag{E2.12R}$$

Rearranging Eq. (E2.12M) and substituting the expression of C_1, we obtain

$$C_2 = -k\frac{C_1}{h_o r_o} - C_1 \ln(r_o) + T_{\infty,o} \tag{E2.12S}$$

$$C_2 = T_{\infty,o} - C_1\left[\frac{k}{h_o r_o} + \ln(r_o)\right] \tag{E2.12T}$$

$$C_2 = T_{\infty,o} - \frac{T_{\infty,o} - T_{\infty,i}}{\left[\dfrac{k}{h_i r_i} + \dfrac{k}{h_o r_o} + \ln\left(\dfrac{r_o}{r_i}\right)\right]}\left[\frac{k}{h_o r_o} + \ln(r_o)\right] \tag{E2.12U}$$

Now we will substitute the given data into the expressions for C_1 and C_2 to obtain

$$C_1 = \frac{T_{\infty,o} - T_{\infty,i}}{\left[\dfrac{k}{h_i r_i} + \dfrac{k}{h_o r_o} + \ln\left(\dfrac{r_o}{r_i}\right)\right]} = \frac{27 - 80}{\left[\dfrac{400}{200 \times 0.015} + \dfrac{400}{25 \times 0.019} + \ln\left(\dfrac{0.019}{0.015}\right)\right]} \qquad \text{(E2.12V)}$$

$$C_1 = -0.054°C \qquad \text{(E2.12W)}$$

Similarly,

$$C_2 = T_{\infty,o} - \frac{T_{\infty,o} - T_{\infty,i}}{\left[\dfrac{k}{h_i r_i} + \dfrac{k}{h_o r_o} + \ln\left(\dfrac{r_o}{r_i}\right)\right]}\left[\dfrac{k}{h_o r_o} + \ln(r_o)\right]$$

$$= 27 - \frac{27 - 80}{\left[\dfrac{400}{200 \times 0.015} + \dfrac{400}{25 \times 0.019} + \ln\left(\dfrac{0.019}{0.015}\right)\right]}\left[\dfrac{400}{25 \times 0.019} + \ln(0.019)\right] \qquad \text{(E2.12X)}$$

$$C_2 = 72.53°C \qquad \text{(E2.12Y)}$$

The temperature profile can now be expressed using an equation as given below:

$$T = C_1 \ln(r) + C_2 = -0.054 \ln(r) + 72.53°C \qquad \text{(E2.12Z)}$$

We can determine the temperature of the inner and outer surfaces of the pipe using Eqs. (E2.12G) and (E2.12H) by substituting the appropriate values for r.

$$T_i = -0.054 \ln(r_i) + 72.53 = -0.054 \times \ln(0.015) + 72.53 \approx 72.76°C$$

$$T_o = -0.054 \ln(r_o) + 72.53 = -0.054 \times \ln(0.019) + 72.53 \approx 72.74°C$$

Finally, the heat transfer can be calculated in several ways (Fig. 2.17):

(i) Convection inside the pipe

$$\dot{q}_{conv,i} = h_i A_i (T_{\infty,i} - T_i) = 200 \times 2 \times 3.14 \times 0.015 \times 10 \times (80 - 72.76) = 1364.55 \text{ W}$$

(ii) Convection outside the pipe

$$\dot{q}_{conv,o} = h_o A_o (T_o - T_{\infty,o}) = 25 \times 2 \times 3.14 \times 0.019 \times 10 \times (72.74 - 27) = 1364.55 \text{ W}$$

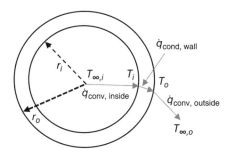

FIGURE 2.17 Modes of heat transfer through the pipe carrying hot water. The heat is transported via convection from the center of the pipe to the inner wall of the pipe followed by conduction through the walls and through convention from the outer wall to the surroundings.

(iii) Conduction at the inner wall of the pipe ($r = r_i = 0.015$ m)

$$\dot{q}_{cond} = -kA_i \frac{dT}{dr} = -k \times 2\pi r_i L \times \frac{C_1}{r_i} = -400 \times 2 \times 3.14 \times 10 \times (-0.054) = 1364.55 \text{ W}$$

(iv) Conduction at the outer wall of the pipe ($r = r_i = 0.019$ m)

$$\dot{q}_{cond} = -kA_o \frac{dT}{dr} = -k \times 2\pi r_o L \times \frac{C_1}{r_o} = -400 \times 2 \times 3.14 \times 10 \times (-0.054) = 1364.55 \text{ W}$$

(v) Conduction through the walls of the pipe (from Eq. (E2.9H) in example Problem 2.9)

$$\dot{q}_{cond} = -2\pi Lk \left(\frac{T_i - T_o}{\ln\left(\frac{r_i}{r_o}\right)} \right) = -400 \times 2 \times 3.14 \times 10 \times \frac{(72.757 - 72.744)}{\ln\left(\frac{0.015}{0.019}\right)} \approx 1364.55 \text{ W}$$

Example 2.13 (Bioprocessing Engineering—Heat transfer through a storage tank): A circular tank made of food-grade stainless steel ($k = 16$ W/m·K) of 2 m in diameter and 1-cm thick is holding hot water that is used for spray cleaning the equipment at the end of the shift. The inner surface temperature of the tank was measured to be 80°C. The outer surface of the tank is uninsulated and exposed to a room temperature of 27°C with a heat transfer coefficient of 50 W/m².°C including radiation. Determine the temperature profile across the wall of the tank and calculate the heat transfer rate from the tank to the room.

Step 1: Schematic
The problem is illustrated in Fig. 2.18.

Step 2: Given data
Inner radius = r_i = 1.0 m; Temperature at the inner radius surface = T_i = 80°C; Outer radius = r_o = 1.01 m; Ambient temperature = T_∞ = 27°C; Thermal conductivity = k = 16 W/m·K; Heat transfer coefficient = h = 50 W/m².°C

To be determined: Temperature profile and heat transfer rate from the sphere.

Step 3: Calculations
The heat equation for a spherical section for a steady-state one-dimensional system is written as

$$\frac{d}{dr}\left(r^2 \frac{dT}{dr} \right) = 0 \tag{E2.13A}$$

Integrating twice successfully we obtain

$$\frac{dT}{dr} = \frac{C_1}{r^2} \tag{E2.13B}$$

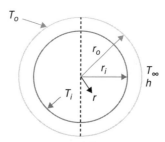

FIGURE 2.18 Schematic of the spherical shell containing hot water.

and

$$T = \frac{-C_1}{r} + C_2 \tag{E2.13C}$$

Applying the BCs:

At $r = r_i$, $T = T_i$ and $r = r_o$, $T = T_o$

$$\frac{dT}{dr} = \frac{C_1}{r^2} \tag{E2.13D}$$

$$T_i = \frac{-C_1}{r_i} + C_2 \tag{E2.13E}$$

$$T_o = \frac{-C_1}{r_o} + C_2 \tag{E2.13F}$$

Using the equilibrium conditions:

At $r = r_o$,

$$-k\frac{dT}{dr} = h(T_o - T_\infty) \tag{E2.13G}$$

From Eqs. (E2.13D) and (E2.13F)

$$-k\frac{C_1}{r_o^2} = h\left(\frac{-C_1}{r_o} + C_2 - T_\infty\right) \tag{E2.13H}$$

$$\frac{-kC_1}{hr_o^2} = \left(\frac{-C_1}{r_o} + C_2 - T_\infty\right) \tag{E2.13I}$$

From Eqs. (E2.13E) and (E2.13I) we can determine the expressions for C_1 and C_2

$$\frac{-kC_1}{hr_o^2} - T_i = \frac{-C_1}{r_o} + \frac{C_1}{r_i} - T_\infty \tag{E2.13J}$$

$$C_1\left[\frac{-k}{hr_o^2} - \frac{1}{r_i} + \frac{1}{r_o}\right] = T_i - T_\infty \tag{E2.13K}$$

$$C_1 = \frac{T_i - T_\infty}{\left[\dfrac{1}{r_o} - \dfrac{1}{r_i} - \dfrac{k}{hr_o^2}\right]} \tag{E2.13L}$$

Similarly,

$$T_i = \frac{-C_1}{r_i} + C_2 \tag{E2.13M}$$

$$C_2 = T_i + \frac{C_1}{r_i} \tag{E2.13N}$$

$$C_2 = T_i + \frac{T_i - T_\infty}{\left[\dfrac{1}{r_o} - \dfrac{1}{r_i} - \dfrac{k}{hr_o^2}\right]}\frac{1}{r_i} \tag{E2.13O}$$

The temperature distribution can now be expressed as

$$T = \frac{-C_1}{r} + C_2 \tag{E2.13P}$$

$$T = \dfrac{-\dfrac{T_i - T_\infty}{\left[\dfrac{1}{r_o} - \dfrac{1}{r_i} - \dfrac{k}{hr_o^2}\right]}}{r} + T_i + \dfrac{T_i - T_\infty}{\left[\dfrac{1}{r_o} - \dfrac{1}{r_i} - \dfrac{k}{hr_o^2}\right]}\dfrac{1}{r_i} \tag{E2.13Q}$$

$$T = T_i + \dfrac{T_i - T_\infty}{\left[\dfrac{1}{r_o} - \dfrac{1}{r_i} - \dfrac{k}{hr_o^2}\right]}\left(\dfrac{1}{r_i} - \dfrac{1}{r}\right) \tag{E2.13R}$$

Now substituting the given data into Eq. (E2.13R)

$$T = 80 + \dfrac{80 - 27}{\left[\dfrac{1}{1.01} - \dfrac{1}{1.0} - \dfrac{16}{50 \times (1.01)^2}\right]}\left(\dfrac{1}{1.0} - \dfrac{1}{r}\right) \tag{E2.13S}$$

$$T = 80 - 163.78\left(1 - \dfrac{1}{r}\right) \tag{E2.13T}$$

From Eq. (E2.13T), we can determine the outer wall temperature, T_o as 78.37°C.

Additionally, the heat transfer rate from the surface to the ambient air can be calculated as

$$\dot{q} = hA(T_o - T_\infty) = 50 \times 4 \times 3.14 \times 1.01^2 \times (78.37 - 27) = 32{,}914.15 \text{ W}$$

Alternatively, the heat transfer rate can also be determined via Fourier's law:

$$\dot{q} = -kA\dfrac{dT}{dr} = -k \times 4\pi(r_0)^2 \dfrac{\dfrac{T_i - T_\infty}{\left[\dfrac{1}{r_o} - \dfrac{1}{r_i} - \dfrac{k}{hr_o^2}\right]}}{r_o^2}$$

$$= -16 \times 4 \times 3.14 \times 1.01^2 \dfrac{80 - 27}{\left[\dfrac{1}{1.01} - \dfrac{1}{1} - \dfrac{16}{50 \times 1.01^2}\right]} \times \dfrac{1}{1.01^2} = 32{,}914.15 \text{ W}$$

Practice Problems for the FE Exam

1. Thermal diffusivity of a material describes:

   a. the ability of the material to store heat.

   b. the ability of the material to transfer heat.

   c. the ratio of heat propagation to the heat storage.

   d. the ratio of heat storage to the heat propagation.

2. The thermal diffusivity of freshly prepared biodiesel whose thermal conductivity, density, and specific heat are given as 0.17 W/m·°C, 850 kg/m³, and 2550 J/kg·°C is about

   a. 6.5×10^{-8} m²/s.

   b. 8×10^{-8} m²/s.

   c. 5×10^{-8} m²/s.

   d. 9.7×10^{-8} m²/s.

3. A 15-cm thick concrete ($k = 1.05$ W/m·°C) wall is subjected to a temperature gradient of 25°C. The heat transfer flux across the wall is

   a. 175 W/m².

   b. 200 W/m².

   c. 100 W/m².

   d. 150 W/m².

4. The same concrete wall described in Problem 2 is exposed to 35°C on the left face and 10°C on the right face. The temperature at a location within the wall 5 cm from the left face is closest to

   a. 17°C.

   b. 12°C.

   c. 27°C.

   d. None of the above.

5. The temperatures on either side of a 30-cm thick wall ($k = 1.6$ W/m·°C) were measured to be 30°C (right face) and 12°C (left face). If a heater placed within the wall generates an energy of 10 kW/m³, the temperature at a location within the wall 5 cm from the right face is nearly

   a. 57°C.

   b. 96°C.

   c. 48°C.

   d. None of the above.

6. A 0.5-cm thick and 10.5-m long stainless ($k = 15$ W/m·°C) pipe of internal diameter of 10 cm is carrying bio-oil from a pyrolysis unit. The temperatures of the inner and outer walls of the pipe were measured to be 175°C and 79°C, respectively. The rate of heat transfer through the wall is nearly

   a. 996 kW.

   b. 49.8 kW.

   c. 4980 W.

   d. 498 kW.

7. An exothermic chemical reaction is being performed in a glass (0.7 W/m·°C) spherical shell with an inner diameter of 25 cm. The inner and outer walls were found to be 85°C and 42°C, respectively. If the heat transfer through the reactor wall is limited to 1670 W, the thickness of the shell is closest to

   a. 1.25 cm.

   b. 2 cm.

   c. 1 cm.

   d. 1.5 cm.

8. The equation $\dfrac{\partial^2 T}{\partial x^2} + \dfrac{\dot{g}}{k} = \dfrac{1}{\alpha}\dfrac{\partial T}{\partial t}$ represents

   a. Steady-state one-dimensional heat transfer with energy generation in a slab.

   b. Steady-state one-dimensional heat transfer with energy generation in a cylinder.

   c. Unsteady-state one-dimensional heat transfer with energy generation in a cylinder.

   d. Unsteady-state one-dimensional heat transfer with energy generation in a slab.

9. The equation $\dfrac{\partial}{\partial r}\left(r^2 \dfrac{\partial T}{\partial r}\right) = 0$ corresponds to

   a. Steady-state one-dimensional heat transfer with energy generation in a slab.

   b. Steady-state one-dimensional heat transfer with energy generation in a sphere.

   c. Steady-state one-dimensional heat transfer with no energy generation in a sphere.

   d. Unsteady-state one-dimensional heat transfer with energy generation in a slab.

10. The equation $\dfrac{1}{r}\dfrac{d}{dr}\left(r\dfrac{dT}{dr}\right) + \dfrac{\dot{g}}{k} = 0$ describes

   a. Steady-state one-dimensional heat transfer with energy generation in a slab.

   b. Steady-state one-dimensional heat transfer with energy generation in a cylinder.

   c. Steady-state one-dimensional heat transfer with energy generation in a sphere.

   d. Steady-state one-dimensional heat transfer with energy generation in a slab.

Practice Problems for the PE Exam

1. Describe the importance of thermal diffusivity in conduction heat transfer. Considering it has a unit of m^2/s, what does it signify? Which other transport variables have the same units? Are they related?

2. Calculate the thermal diffusivity of air, water, and copper. Discuss your answers.

3. A 10-cm thick wall ($k = 1.2$ W/m·°C) in a swine barn is exposed to a temperature differential. The temperature of the outer face of the wall was measured to be 37°C while the inner face of the wall was maintained at 20°C. Assuming a one-dimensional steady-state heat transfer process with no energy generation, (i) formulate a differential equation to describe the heat transfer process and (ii) solve the differential equation to determine the temperature profile within the wall. Also determine the temperature within the wall at 5 cm from the left surface.

4. The left face of the same wall ($k = 1.2$ W/m·°C) described in Problem 1, with a unit area, is now exposed to a heat flux of 250 W/m² while still being maintained at 37°C. Assuming a one-dimensional steady-state heat transfer process with no energy generation, (i) formulate a differential equation to describe the heat transfer process, (ii) solve the differential equation to determine the temperature profile within the wall, and (iii) determine the temperature of right face of the wall.

5. The right face of a plain wall is exposed to air while the left face is insulated using fiberglass. Assuming a one-dimensional steady-state heat transfer process with no energy generation, formulate a differential equation to describe the heat transfer process and solve the differential equation to determine the temperature profile within the wall.

6. Consider a spherical glass (0.76 W/m·°C) shell of inner and outer diameters of 30 cm and 32 cm, respectively. If the inner and outer surfaces are measured to be 40°C and 75°C, respectively, determine the temperature profile within the wall assuming steady-state one-dimensional heat flow and no energy generation.

7. A 1-cm thick, 60-cm long cylindrical reactor made of stainless steel ($k = 15$ W/m·°C) of 30-cm diameter is being used for thermochemically processing swine manure slurry. The reactor was loaded with manure slurry (95% water) and heated in a furnace. During the reaction, the outer and inner walls of the reactor were found to be 550°C and 435°C, respectively. Assuming a one-dimensional steady-state heat transfer process with no energy generation, (i) formulate a differential equation to describe the heat transfer process, (ii) solve the differential equation to determine the temperature profile within the wall, and (iii) plot the temperature profile across the wall of the cylinder and determine the rate of heat transfer through the wall.

8. The right face of a 1-cm thick aluminum ($k = 240$ W/m·°C) plate of 2 m² is heated to a temperature of 200°C while the left face is subjected to an air flow at 25°C with a heat transfer coefficient of 55 W/m²·°C. Assuming steady-state one-dimensional heat transfer, determine the rate of heat transfer through the plate.

9. In a food processing facility that was maintained at a constant temperature of 23°C, hot water at 95°C flows in a stainless steel ($k = 15$ W/m·°C) pipe of 5-cm diameter and 0.5-cm thickness. If the heat transfer coefficients within the pipe and outside the pipe are measured to be 300 W/m²·°C and 10 W/m²·°C, calculate the rate of heat loss from the pipe per unit length.

10. Biodiesel at 65°C is stored in a 1-cm thick stainless steel ($k = 16$ W/m·K) circular tank of 3 m diameter. The inner surface temperature of the tank was measured to be 80°C. The outer surface of the tank exposed to ambient air temperature of 19°C with a heat transfer coefficient of 15 W/m²·°C. Neglecting the effects of radiation, determine the heat transfer rate from the tank to the room.

CHAPTER 3

Steady-State Conduction Heat Transfer

Chapter Objectives

After finishing this chapter, students will be able to

- Determine the rate of one-dimensional conduction heat transfer in simple geometries.

- Visualize the similarities between electrical and thermal resistances.

- Analyze the heat transfer in composite sections and determine the temperatures at any given location.

- Understand the idea of insulation and its practical implications.

3.1 Motivation

One of the tasks engineers frequently undertake is to determine the heat transfer rates through different types of structures such as walls separating two different temperatures, pipes carrying hot or cold fluids, and spheres serving as holding tanks. In Chap. 2, we derived heat conduction equations for different geometries in three dimensions. However, for sufficiently large bodies (e.g., a large plain wall), heat transfer can be approximated as one dimensional (because variation of the temperature in length and width direction is negligible). Besides, in several situations, the walls are composed of multiple layers of materials with different thermal properties. Therefore, we require one-dimensional mathematical solutions to analyze and solve heat conduction problems pertaining to walls, cylinders, sphere, and composite structures.

3.2 One-Dimensional Steady-State Conduction in Simple Geometries

Let us visualize a large isotropic rectangular section of area A and thickness L with a constant thermal conductivity k exposed to a thermal gradient, similar to a wall on a hot summer day. As shown in Fig. 3.1, we will assume that the entire surface of the left face

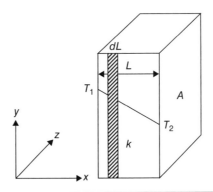

Figure 3.1   Schematic for one-dimensional heat transfer in a rectangular section.

is subjected to a hot temperature T_1 while the right face of the wall is cold at a temperature T_2. As we can see, the temperature gradients in the y and z directions are small, relative to the x-direction and therefore may be neglected. Therefore, the heat flows between the left and right surfaces of the wall in the x-direction.

Now, using Fourier's law, we can describe the heat flow rate $\dot{q}$ for a small section of thickness dL as

$$\dot{q} = -kA \frac{dT}{dL} \tag{3.1}$$

After rearranging Eq. (3.1) and integrating between T_1 and T_2 and 0 and L, we obtain

$$\dot{q} \int_0^L dL = -kA \int_{T_1}^{T_2} dT \tag{3.2}$$

Evaluating the integral and simplifying, we obtain the heat transfer rate across a rectangular section as

$$\dot{q} = kA \frac{(T_1 - T_2)}{L} \tag{3.3}$$

It is often easy to remember the equation when written in words as below:

$$\text{Heat transfer rate across the wall} = \frac{\text{Thermal conductivity} \times \text{Area} \times \text{Temperature gradient}}{\text{Wall thickness}} \tag{3.4}$$

We can also approximate the heat transfer associated with a spherical body as one-dimensional heat transfer. Now, let us consider a spherical-shaped body (e.g., compressed gas storage vessels) with thermal conductivity k that contains a hot fluid and exposed to a colder environment (Fig. 3.2). We will assign the inside radius and inside wall temperature as r_1 and T_1 and outside radius and outside wall temperature as r_2 and T_2, respectively.

The conduction heat transfer from inner wall to the outer wall can be expressed using Fourier's law as

$$\dot{q} = -kA \frac{dT}{dr} \tag{3.5}$$

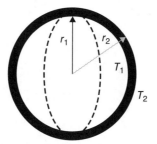

FIGURE 3.2 Schematic for one-dimensional heat transfer in a spherical section ($T_1 > T_2$).

Recall that when $r = r_1$, $T = T_1$ and when $r = r_2$, $T = T_2$ and the area of the sphere, $A = 4\pi r^2$

$$\dot{q} \int_{r_1}^{r_2} \frac{dr}{4\pi r^2} = -k \int_{T_1}^{T_2} dT \tag{3.6}$$

After integrating, evaluating the limits, and simplification, we obtain

$$\dot{q} = \frac{4\pi r_1 r_2 k (T_1 - T_2)}{r_2 - r_1} \tag{3.7}$$

Heat transfer rate in a sphere

$$= \frac{4\pi \times \text{inner radius} \times \text{outer radius} \times \text{thermal conductivity} \times \text{temperature gradient}}{\text{Wall thickness}} \tag{3.8}$$

The same concept of one-dimensional heat transfer applies to a long cylinder (Fig. 3.3) of length L, thermal conductivity k, whose inner and outer radii are r_1 and r_2 at temperatures T_1 and $T_2 (T_1 > T_2)$.

Using Fourier's law, the conduction heat transfer from the inner wall to the outer wall can be written as

$$\dot{q} = -kA \frac{dT}{dr} \tag{3.9}$$

Expressing the heat transfer area of the cylinder as $2\pi r L$ and ignoring the area on the top and bottom of the cylinder, the rate of heat transfer can be determined by integrating

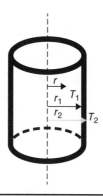

FIGURE 3.3 Schematic for one-dimensional heat transfer in a cylindrical section.

the equation between r_1 and r_2 and T_1 and T_2; thus we obtain

$$\dot{q}\int_{r_1}^{r_2}\frac{dr}{2\pi rL}=-k\int_{T_1}^{T_2}dT \tag{3.10}$$

Evaluating the integral and simplifying, the rate of heat transfer can be expressed as

$$\dot{q}=\frac{2\pi kL(T_1-T_2)}{\ln\left[\dfrac{r_2}{r_1}\right]} \tag{3.11}$$

$$\text{Heat transfer rate in a cylinder}=\frac{2\pi\times\text{thermal conductivity}\times\text{length}\times\text{temperature gradient}}{\ln\left(\dfrac{\text{outer radius}}{\text{inner radius}}\right)}$$

$$\tag{3.12}$$

3.3   Similarity with Flow of Electricity

Conceptually, heat transfer can be related to the flow of current. Recalling that using Ohm's law, the current, I (rate of electron flow) between two points 1 and 2 with a voltage drop of $V_1 - V_2$ between them, can be expressed as

$$I=\frac{V_1-V_2}{R}=\frac{\Delta V}{R} \tag{3.13}$$

In other words, in a conductor, the voltage gradient is the driving force for the flow of electrons while the resistance of the conductor is the opposing force for the electron flow. It follows that the smaller the resistance, the greater the rate of flow of electrons and vice versa. Similarly, heat transfer through a material can be viewed as a competition between the driving force (temperature gradient) and an opposing force (thermal resistance) to the heat transfer. For example, for a rectangular wall, the rate of conduction heat transfer can be expressed in terms of resistance as

$$\dot{q}=\frac{(T_1-T_2)}{\left(\dfrac{L}{kA}\right)}=\frac{\Delta T}{R_{\text{cond,rect}}}=\frac{\text{Temperature Gradient}}{\text{Conduction Resistance}} \tag{3.14}$$

where, conduction resistance of a rectangular section, $R_{\text{cond,rect}}$ is given by

$$R_{\text{cond,rect}}=\frac{L}{kA} \tag{3.15}$$

Similarly, for cylindrical and spherical configurations, heat transfer rates can be expressed as ratios of thermal gradients and thermal resistances, respectively, as below:

$$\dot{q}=\frac{4\pi r_1 r_2 k(T_1-T_2)}{r_2-r_1}=\frac{(T_1-T_2)}{\left(\dfrac{(r_2-r_1)}{4\pi r_1 r_2 k}\right)}=\frac{\Delta T}{R_{\text{cond,Sph}}}=\frac{\text{Temperature Gradient}}{\text{Conduction Resistance}} \tag{3.16}$$

where, the conduction resistance of a sphere, $R_{\text{cond,Sph}}$ is given by

$$R_{\text{cond,Sph}}=\frac{(r_2-r_1)}{4\pi r_1 r_2 k} \tag{3.17}$$

For a cylinder,

$$\dot{q}=\frac{2\pi kL(T_1-T_2)}{\ln\left[\dfrac{r_2}{r_1}\right]}=\frac{(T_1-T_2)}{\left(\dfrac{\ln\left[\dfrac{r_2}{r_1}\right]}{2\pi kL}\right)}=\frac{\Delta T}{R_{\text{cond,Cyl}}}=\frac{\text{Temperature Gradient}}{\text{Conduction Resistance}} \tag{3.18}$$

where, the conduction resistance of a cylinder, $R_{cond,Cyl}$ is given by

$$R_{cond,Cyl} = \frac{\ln\left[\frac{r_2}{r_1}\right]}{2\pi kL} \tag{3.19}$$

Table 3.1 summarizes one-dimensional heat transfer equations in standard form and the thermal resistance form.

It is to be noted that the concept of thermal resistance can also be applied to situations where heat transfer occurs via convection and radiation. Mathematically,

In the case of a convection heat transfer,

$$\dot{q}_{conv} = hA(T_1 - T_2) = \frac{(T_1 - T_2)}{\left(\frac{1}{hA}\right)} = \frac{\Delta T}{R_{conv}} \tag{3.20}$$

where the convection resistance R_{conv} is given by

$$R_{conv} = \frac{1}{hA} \tag{3.21}$$

Similarly, for a radiation heat transfer,

$$\dot{q}_{rad} = \varepsilon\sigma A(T_1^4 - T_2^4) = \varepsilon\sigma A(T_1^2 + T_2^2)(T_1 + T_2)(T_1 - T_2) = \frac{T_1 - T_2}{\dfrac{1}{\varepsilon\sigma A(T_1^2 + T_2^2)(T_1 + T_2)}} \tag{3.22}$$

where the radiation resistance R_{rad} is given by

$$R_{rad} = \frac{1}{\varepsilon\sigma A(T_1^2 + T_2^2)(T_1 + T_2)} \tag{3.23}$$

Configuration	Heat Transfer Rate, $\dot{q}$ (W) (Standard form)	Heat Transfer Rate, $\dot{q}$ (W) (Thermal resistance form) Temperature Gradient / Conduction Resistance	Thermal Resistance, R (°C m⁻¹)
Rectangular	$kA\dfrac{(T_1 - T_2)}{L}$	$\dfrac{T_1 - T_2}{\dfrac{L}{kA}}$	$\dfrac{L}{kA}$
Spherical	$\dfrac{4\pi r_1 r_2 k(T_1 - T_2)}{r_2 - r_1}$	$\dfrac{T_1 - T_2}{\dfrac{(r_2 - r_1)}{4\pi r_1 r_2 k}}$	$\dfrac{(r_2 - r_1)}{4\pi r_1 r_2 k}$
Cylindrical	$\dfrac{2\pi Lk(T_1 - T_2)}{\ln\left[\dfrac{r_2}{r_1}\right]}$	$\dfrac{T_1 - T_2}{\dfrac{\ln\left[\dfrac{r_2}{r_1}\right]}{2\pi Lk}}$	$\dfrac{\ln\left[\dfrac{r_2}{r_1}\right]}{2\pi Lk}$

TABLE 3.1 Heat Transfer Equations for Common Geometrical Configurations

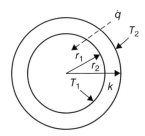

FIGURE **3.4**   Cross section of the pipe carrying a cold fluid as described in Example 3.1.

Example 3.1 (Bioprocessing Engineering—Heat transfer through a cylinder):   A pipe $(k = 0.484 \frac{W}{m \cdot °C})$ of 5 cm internal diameter is transporting a cold fluid. If the measured temperature on the outer wall and inner wall is 25°C and 7°C, respectively, determine the minimum wall thickness if the heat transfer rate should not exceed 300 W per unit length of the pipe (Fig. 3.4).

Step 1: Schematic
The problem is decribed in Fig. 3.4.

Step 2: Given data
Thermal conductivity of the pipe $k = 0.484 \frac{W}{m \cdot °C}$; Internal diameter = 5 cm (i.e., $r_1 = 0.025$ m); Temperature on the inner wall, $T_1 = 7°C$; Temperature on the outer wall, $T_2 = 25°C$; Heat transfer rate per meter, $\dot{q} = 300 \ W/L$

To be determined: Thickness of the pipe.

Assumptions: One-dimensional heat transfer occurs (in radial direction) and thermal conductivity is constant throughout.

Approach: The heat transfer rate that occurs via conduction through the walls of the cylinder is limited to 300 W per meter length of the pipe. We will assume a length of the pipe as 1 m and set the heat transfer equation to 300 W and determine the thickness $(r_2 - r_1)$.

Step 3: Calculations
The rate of conduction heat transfer in a cylinder is given by

$$\dot{q} = \frac{2\pi kL(T_1 - T_2)}{\ln\left[\frac{r_2}{r_1}\right]} \qquad \text{(E3.1A)}$$

Substitute the given data,

$$300 = \frac{2 \times 3.14 \times 0.484 \times 1 \times (25 - 7)}{\ln\left[\frac{r_2}{0.025}\right]} \qquad \text{(E3.1B)}$$

Solving for r_2, we obtain

$$r_2 = 0.03 \text{ m}$$

Therefore, the thickness of the pipe should be at least $0.03 - 0.025 = 0.005$ m.

3.4   Heat Transfer in Composite Sections in Series

Engineers oftentimes encounter situations where different materials with different thermal conductivities are combined to form composite sections. One example that we can readily imagine is a wall that consists of sections of concrete, brick, and plastering cement (Fig. 3.5). Similarly, another classic example is plywood in which several sheets of wood veneers and crossbands are glued via resin to form a wooden composite (Fig. 3.6).

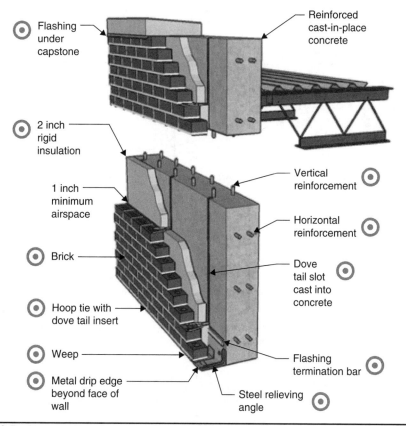

⊙ Flashing under capstone

⊙ Reinforced cast-in-place concrete

⊙ 2 inch rigid insulation

1 inch minimum airspace

⊙ Vertical reinforcement

⊙ Brick

⊙ Horizontal reinforcement

⊙ Hoop tie with dove tail insert

⊙ Dove tail slot cast into concrete

⊙ Weep

⊙ Flashing termination bar

⊙ Metal drip edge beyond face of wall

Steel relieving angle ⊙

FIGURE 3.5 Cross section of a composite wall made of concrete, insulation, and brick. (Image courtesy of Ernest Maier. Reprinted with permission.)

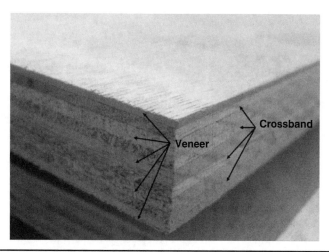

FIGURE 3.6 Cross section of plywood used in construction. (Image courtesy of Home Depot. Reprinted with permission.)

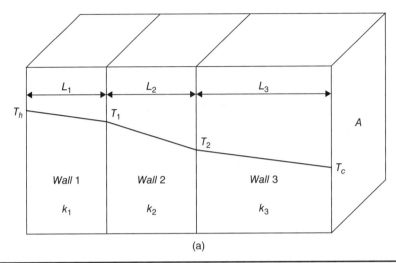

(a)

Figure 3.7A Schematic of a three-section composite arranged in series subjected to a temperature gradient.

Let us consider a composite section of area A that was prepared by combining three different rectangular sections of thicknesses L_1, L_2, and L_3 (Fig. 3.7A) and thermal conductivities k_1, k_2, and k_3. The left surface (T_h) is assumed to be hotter than the right surface (T_c) while the interface temperatures are denoted as T_1 and T_2, respectively.

Assuming a steady-state heat transfer from the hot surface on the left (T_h) to the cold surface on the right (T_c), we can write the 1-D heat transfer equations through each wall as below:

$$\text{Heat transfer through wall 1, } \dot{q}_1 = k_1 A \frac{(T_h - T_1)}{L_1} \tag{3.24}$$

$$\text{Heat transfer through wall 2, } \dot{q}_2 = k_2 A \frac{(T_1 - T_2)}{L_2} \tag{3.25}$$

$$\text{Heat transfer through wall 3, } \dot{q}_3 = k_3 A \frac{(T_2 - T_c)}{L_3} \tag{3.26}$$

because the heat transfer rate is same in each wall. Therefore, the above equations can be combined as

$$\dot{q}_1 = \dot{q}_2 = \dot{q}_3 = \dot{q} = k_1 A \frac{(T_h - T_1)}{L_1} = k_2 A \frac{(T_1 - T_2)}{L_2} = k_3 A \frac{(T_2 - T_c)}{L_3} \tag{3.27}$$

Rewriting Eq. (3.27) in terms of conduction resistances, we obtain

$$\dot{q} = \frac{(T_h - T_1)}{\left(\dfrac{L_1}{k_1 A}\right)} = \frac{(T_1 - T_2)}{\left(\dfrac{L_2}{k_2 A}\right)} = \frac{(T_2 - T_c)}{\left(\dfrac{L_3}{k_3 A}\right)} \tag{3.28}$$

Using the rules of algebra, the equation can be simplified as

$$\dot{q} = \frac{(T_h - T_1) + (T_1 - T_2) + (T_2 - T_c)}{\left(\dfrac{L_1}{k_1 A}\right) + \left(\dfrac{L_2}{k_2 A}\right) + \left(\dfrac{L_3}{k_3 A}\right)} \tag{3.29}$$

$$\dot{q} = \frac{(T_h - T_c)}{\left(\dfrac{L_1}{k_1 A}\right) + \left(\dfrac{L_2}{k_2 A}\right) + \left(\dfrac{L_3}{k_3 A}\right)} \tag{3.30}$$

From Eq. (3.30), we can see that each term in the denominator corresponds to the thermal resistance to heat transfer and sum of the resistances is the total thermal resistance to the heat transfer. Thus, the rate of heat transfer can be written in the general form as

$$\dot{q} = \frac{(T_h - T_c)}{R_1 + R_2 + R_3} = \frac{T_h - T_c}{R_{\text{Total,cond}}} \tag{3.31}$$

where,

$$R_{\text{Total,cond}} = \left(\frac{L_1}{k_1 A}\right) + \left(\frac{L_2}{k_2 A}\right) + \left(\frac{L_3}{k_3 A}\right) \tag{3.32}$$

And the thermal resistance network for the whole system can be represented by Fig. 3.7B.

Additionally, any of the interface temperatures, T_1, or T_2 can also be easily determined by combining the calculated heat transfer rate with Eqs. (3.24)–(3.26).

Similarly, for the same composite wall described previously subjected to convection on both sides (Fig. 3.8), may be viewed as five resistances separated by temperatures $T_{\infty,1}$ and $T_{\infty,2}$.

Therefore, the rate of heat transfer can now be written as

$$\dot{q} = \frac{T_{\infty,1} - T_{\infty,2}}{R_{\text{Total}}} \tag{3.33}$$

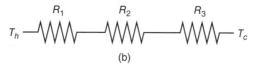

(b)

FIGURE 3.7B Thermal resistance network of a three-section composite subjected to a temperature gradient.

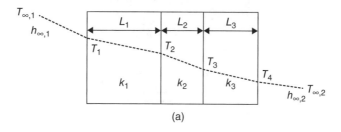

(a)

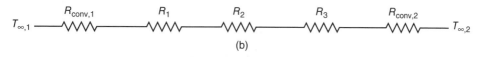

(b)

FIGURE 3.8 Schematic of a three-section composite arranged in series and subjected to a temperature gradient due to convection on both sides (a) along with the thermal resistance network circuit (b).

Or in terms of individual resistances,

$$\dot{q} = \frac{(T_{\infty,1} - T_{\infty,2})}{\frac{1}{h_{\infty,1}A} + \left(\frac{L_1}{k_1A}\right) + \left(\frac{L_2}{k_2A}\right) + \left(\frac{L_3}{k_3A}\right) + \frac{1}{h_{\infty,2}A}} \tag{3.34}$$

Example 3.2 (Agricultural Engineering—Heat transfer in a composite wall with convection): A two-layer composite wall of 6 m², made of brick (11-cm thick with a thermal conductivity $1 \frac{W}{m \cdot °C}$) and wood (5-cm thick with a thermal conductivity $0.15 \frac{W}{m \cdot °C}$) of a sweet potato storage facility is exposed to an outside air temperature of −15°C with an outside heat transfer coefficient of $25 \frac{W}{m^2 \cdot °C}$. Determine the heat loss over an 8-hour period if the inner wall is subjected to an air temperature of 12°C with a heat transfer coefficient of $50 \frac{W}{m^2 \cdot °C}$.

Step 1: Schematic
The problem is depicted in Fig. 3.9.

Step 2: Given data
We will denote brick section as wall 1 and wooden section as wall 2. Area of the wall, $A = 6$ m²; Thickness of the brick section $L_1 = 0.11$ m; Thermal conductivity of the brick section, $k_1 = 1 \frac{W}{m \cdot °C}$;

Thickness of the wood section, $L_2 = 0.05$ m; Thermal conductivity of the wood section, $k_2 = 0.15 \frac{W}{m \cdot °C}$;

Temperature of the outside air $= T_{\infty,2} = -15°C$; Heat transfer coefficient (outside), $h_{\infty,1} = 25 \frac{W}{m^2 \cdot °C}$;

Temperature of the inside air $= T_{\infty,1} = 12°C$; Heat transfer coefficient (inside), $h_{\infty,1} = 50 \frac{W}{m^2 \cdot °C}$; Heat transfer time, $\Delta t = 8$ h

To be determined: Amount of heat lost across the wall over an 8-hour period.

Assumptions: One-dimensional heat transfer occurs (in the x-direction) and thermal conductivity for each wall remains constant throughout.

Analysis: As can be seen from the data, the inside temperature is hotter than the outside environment and the heat is transferred from inside to outside (Fig. 3.9). The two temperatures are separated by 4 resistances in series as shown in Fig. 3.10.

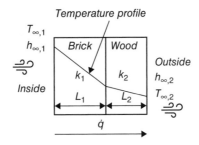

FIGURE 3.9 Schematic of the composite wall of the sweet potato storage facility.

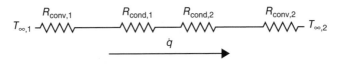

FIGURE 3.10 The thermal resistance network for the wall of the sweet potato storage facility.

Step 3: Calculations

The rate of heat transfer can be calculated from the general equation

$$\dot{q} = \frac{\Delta T}{R_{\text{Total}}}$$ (E3.2A)

or

$$\dot{q} = \frac{(T_{\infty,1} - T_{\infty,2})}{\dfrac{1}{h_{\infty,1}A} + \left(\dfrac{L_1}{k_1 A}\right) + \left(\dfrac{L_2}{k_2 A}\right) + \dfrac{1}{h_{\infty,2}A}}$$ (E3.2B)

Substituting the given data into the above equation,

$$\dot{q} = \frac{(12 - (-15))}{\dfrac{1}{50 \times 6} + \left(\dfrac{0.11}{1 \times 6}\right) + \left(\dfrac{0.05}{0.15 \times 6}\right) + \dfrac{1}{25 \times 6}}$$ (E3.2C)

$$\dot{q} = 321.85 \text{ W}$$

The heat lost in an 8-hour period, $q_{t=8h} = 321.85 \, \dfrac{\text{J}}{\text{s}} \times 8 \times 60 \times 60 \text{ s} = 9.17 \text{ MJ}$.

3.5   Heat Transfer in Composite Sections in Parallel

Sometimes, we encounter problems in which the flow of heat is parallel to the composite wall. Such parallel configurations can be easily analyzed using the electrical resistance analogy. Of course, we will assume that the heat transfer occurs only in the x-direction and the plane in the normal direction to heat flow always stays isothermal. We are familiar that when resistances are arranged in parallel, we can write the total resistance as

$$\frac{1}{R_{\text{Total}}} = \frac{1}{R_1} + \frac{1}{R_2} + \frac{1}{R_3} + \cdots + \frac{1}{R_n}$$ (3.35)

Now for a simple two-section rectangular wall arranged in parallel (Fig. 3.11), the thermal network can be written as two resistances in parallel.

The heat transfer rate, $\dot{q}$ is given by

$$\dot{q} = \frac{\Delta T}{R_{\text{Total}}}$$ (3.36)

where

$$\frac{1}{R_{\text{Total}}} = \frac{1}{R_1} + \frac{1}{R_2}$$ (3.37)

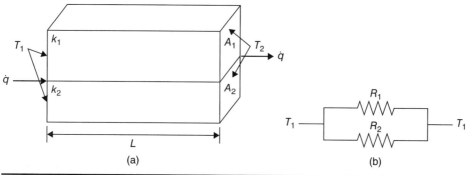

FIGURE 3.11 Schematic of a two-section composite arranged in parallel (a) with the corresponding thermal resistance network (b).

Example 3.3 (Bioprocessing Engineering—Heat transfer in a biocomposite): A biocomposite wall (4 m long × 3 m width × 0.3 m thickness) was built by fusing three carbon-based materials in parallel (Fig. 3.12A). The thermal conductivities and areas of cross section for three sections are 0.03, 0.2, and $2\frac{W}{m \cdot °C}$ and 0.15, 0.375, and 0.375 m², respectively. If the left and right walls are subjected to temperatures of 50°C and 22°C, determine the heat transfer rate.

Step 1: Schematic
The schematic is shown in Fig. 3.12A.

Step 2: Given data
Thermal conductivity of section 1, $k_1 = 0.03\frac{W}{m \cdot °C}$; Area of section 1, $A_1 = 0.15$ m²; Thermal conductivity of section 2, $k_2 = 0.2\frac{W}{m \cdot °C}$; Area of section 1, $A_2 = 0.375$ m²; Thermal conductivity of section 3, $k_3 = 2\frac{W}{m \cdot °C}$; Area of section 1, $A_3 = 0.375$ m²; Temperature of the left face of the biocomposite wall, $T_1 = 50°C$; Temperature of the right face of the biocomposite wall, $T_2 = 22°C$

To be determined: Rate of heat transfer across the wall in the parallel direction.

Assumptions: Heat transfer occurs in the x-direction only and thermal conductivity for each wall remains constant throughout.

Analysis: The given sections are arranged in parallel. The total heat transfer rate is the sum of heat transfer rates in each section. The system may be viewed as two temperatures are separated by three conduction resistances in parallel as shown in Fig. 3.12B.

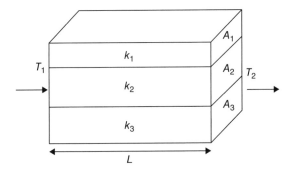

FIGURE 3.12A Schematic of the biocomposite wall arranged in parallel and subjected to a thermal gradient.

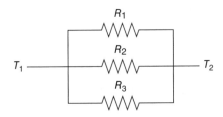

FIGURE 3.12B Thermal resistance network for the biocomposite wall.

Step 3: Calculations

The rate of heat transfer is written in the general form as

$$\dot{q} = \frac{\Delta T}{R_{\text{Total}}} \tag{E3.3A}$$

For a parallel configuration, the total resistance R_{Total} is expressed as

$$\frac{1}{R_{\text{Total}}} = \frac{1}{R_1} + \frac{1}{R_2} + \frac{1}{R_3} \tag{E3.3B}$$

or

$$\dot{q} = \frac{\Delta T}{\dfrac{1}{\left(\dfrac{1}{R_1} + \dfrac{1}{R_2} + \dfrac{1}{R_3}\right)}} \tag{E3.3C}$$

or

$$\dot{q} = \frac{T_1 - T_2}{\left[\dfrac{1}{\dfrac{1}{\left(\dfrac{L_1}{k_1 A_1}\right)} + \dfrac{1}{\left(\dfrac{L_2}{k_2 A_2}\right)} + \dfrac{1}{\left(\dfrac{L_3}{k_3 A_3}\right)}}\right]} \tag{E3.3D}$$

Substituting the given data into the above equation,

Recalling that $L_1 = L_2 = L_3 = 4$ m and substituting the rest of the data into the above equation,

$$\dot{q} = \frac{50 - 22}{\left[\dfrac{1}{\dfrac{1}{\left(\dfrac{4}{0.03 \times 0.15}\right)} + \dfrac{1}{\left(\dfrac{4}{0.2 \times 0.375}\right)} + \dfrac{1}{\left(\dfrac{4}{2 \times 0.375}\right)}}\right]} \tag{E3.3E}$$

$$\dot{q} = 5.8 \text{ W}$$

The rate of heat transfer through the biocomposite wall is 5.8 W.

3.6 Heat Transfer in Composite Sections in Series and Parallel

Oftentimes, composite sections are fabricated by joining several walls in series and parallel (Fig. 3.13A). Once again, the electrical resistance analogy can be extended to solve these problems.

For the composite section shown in Fig. 3.13A, sections 1, 2, and 6 are in series. Sections 3, 4, and 5, while being parallel to each other, are still in series with sections 1, 2, and 6. Now, the thermal resistance network for this combination can be written as shown in Fig. 3.13B.

The heat transfer rate through the composite wall can be written as

$$\dot{q} = \frac{T_1 - T_2}{R_1 + R_2 + \left[\dfrac{1}{R_3} + \dfrac{1}{R_4} + \dfrac{1}{R_5}\right]^{-1} + R_6} \tag{3.38}$$

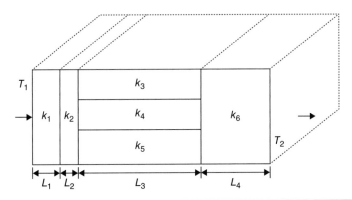

FIGURE **3.13A**   Schematic of a composite section arranged in series and parallel.

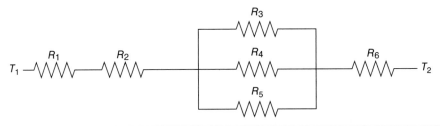

FIGURE **3.13B**   Thermal resistance network for the given composite section arranged in series and parallel.

Assuming equal areas, the heat transfer rate can now be written as

$$\dot{q} = \frac{T_1 - T_2}{\dfrac{L_1}{k_1 A} + \dfrac{L_2}{k_2 A} + \left[\dfrac{1}{\left(\dfrac{L_3}{k_3 A}\right)} + \dfrac{1}{\left(\dfrac{L_4}{k_4 A}\right)} + \dfrac{1}{\left(\dfrac{L_5}{k_5 A}\right)}\right]^{-1} + \dfrac{L_6}{k_6 A}} \tag{3.39}$$

Example 3.4 (Bioprocessing Engineering—Heat transfer in series and parallel network):   A composite wall consisting of rectangular sections is subjected to convection heat transfer from both sides (Fig. 3.14). The temperature of the surrounding air near the left wall was found to be 30°C with a heat transfer

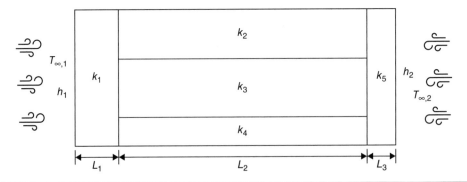

FIGURE **3.14**   Schematic of the series–parallel composite wall subjected to a temperature gradient.

Section	Length (m)	Area (m²)	Thermal Conductivity (k) W/m·°C
1	0.1	6	0.2
2	0.7	2	0.5
3	0.7	3	0.03
4	0.7	1	0.1
5	0.08	6	0.1

TABLE E3.4 Physical and Thermal Properties of the Sections of the Wall

coefficient of 110 $\dfrac{W}{m^2 \cdot °C}$. The physical and thermal properties are summarized in Table E3.4. Determine the heat transfer rate if the temperature of the surrounding air near the right wall is 12°C with a heat transfer coefficient of 35 $\dfrac{W}{m^2 \cdot °C}$.

Step 1: Schematic
The schematic of the composite wall is shown in Fig. 3.14.

Step 2: Given data
Temperature at air near the left wall, $T_{\infty,1} = 30°C$; Temperature at air near the left wall, $T_{\infty,2} = 12°C$; Heat transfer coefficient of the air at the left wall, $h_{\infty,1} = 110 \dfrac{W}{m^2 \cdot °C}$; Heat transfer coefficient of the air at the right wall, $h_{\infty,2} = 35 \dfrac{W}{m^2 \cdot °C}$; The lengths, areas, and the thermal conductivities are provided in Table E3.4

To be determined: Heat transfer across the wall.

Assumptions: One-dimensional heat transfer occurs (in the x-direction) and thermal conductivity is constant throughout.

Analysis: The heat is transferred via convection from the air surrounding the left wall to the surface of the left face. Thereafter, conduction heat transfer prevails through the composite wall made of series and parallel networks. Finally, the heat transfers out to the air around the right wall via convection.

Step 3: Calculations
We can write the resistance network as shown in Fig. 3.15.

The heat transfer rate can be determined via

$$\dot{q} = \frac{\Delta T}{R_{\text{Total}}} \tag{E3.4A}$$

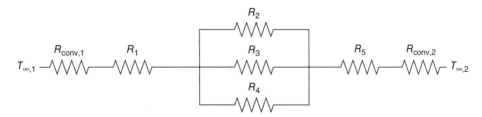

FIGURE 3.15 Thermal resistance network of the given composite wall consisting of a total of six resistances in which sections 2, 3, and 4 are in parallel.

where,

$$R_{\text{Total}} = R_{\text{conv},1} + R_{\text{cond},1} + \left[\frac{1}{R_{\text{cond},2}} + \frac{1}{R_{\text{cond},3}} + \frac{1}{R_{\text{cond},4}} \right]^{-1} + R_{\text{cond},5} + R_{\text{conv},2} \qquad \text{(E3.4B)}$$

Expressing in terms of resistances

$$R_{\text{Total}} = \frac{1}{h_{\infty,1} A_1} + \frac{L_1}{k_1 A_1} + \left[\frac{1}{\frac{L_2}{k_2 A_2}} + \frac{1}{\frac{L_3}{k_3 A_3}} + \frac{1}{\frac{L_4}{k_4 A_4}} \right]^{-1} + \frac{L_5}{k_5 A_5} + \frac{1}{h_{\infty,2} A_5} \qquad \text{(E3.4C)}$$

Substituting the given data into the above equation, we determine the total resistance as

$$R_{\text{Total}} = \frac{1}{(110 \times 6)} + \frac{0.1}{(0.2 \times 6)} + \left[\frac{1}{\left(\frac{0.7}{0.5 \times 2}\right)} + \frac{1}{\left(\frac{0.7}{0.03 \times 3}\right)} + \frac{1}{\left(\frac{0.7}{0.1 \times 1}\right)} \right]^{-1} + \frac{0.08}{(0.1 \times 6)} + \frac{1}{(35 \times 6)} \qquad \text{(E3.4D)}$$

$$R_{\text{Total}} = 0.81 \ \frac{°C}{W}$$

$$\dot{q} = \frac{(30 - 12)}{0.81} = 22.18 \ W$$

3.7   Heat Transfer in Composite Spherical and Cylindrical Bodies

The thermal resistance concepts that we just discussed can also be applied to composite spheres and cylinders as well. Let us first consider a simple case of a sphere whose inner and outer radii are r_1 and r_2, respectively and contains a hot fluid at temperature $T_{\infty,i}$ with a heat transfer coefficient $h_{\infty,i}$. If we assume a colder outside temperature of $T_{\infty,o}$ with a heat transfer coefficient, $h_{\infty,o}$, heat will transfer out of the sphere to the ambient (outside) surroundings (Fig. 3.16) via three steps: (1) the convection heat transfer from

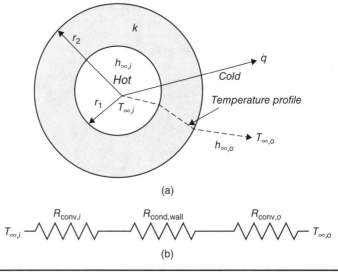

(a)

(b)

Figure 3.16 Schematic of a spherical shell containing a hot fluid and exposed to a cold environment (a) expressed as a thermal resistance network (b).

the center of the sphere to the inner wall of the sphere, (2) the conduction heat transfer through the walls of the sphere, and (3) the convection heat transfer from the outer wall to the ambient air surroundings.

These steps can be viewed as resistances to the heat transfer. These resistances can be mathematically expressed as

$$\text{Resistance to convection, inside, } R_{\text{conv},i} = \frac{1}{h_{\infty,i} A_1} = \frac{1}{h_{\infty,i}(4\pi r_1^2)} \tag{3.40}$$

$$\text{Resistance to conduction, from the wall, } R_{\text{cond,wall}} = \frac{(r_2 - r_1)}{4\pi k r_1 r_2} \tag{3.41}$$

$$\text{Resistance to convection, outside, } R_{\text{conv},o} = \frac{1}{h_{\infty,o} A_2} = \frac{1}{h_{\infty,o}(4\pi r_2^2)} \tag{3.42}$$

Now, the rate of heat transfer can be expressed in the general form as

$$\dot{q} = \frac{\Delta T}{R_{\text{Total}}} \tag{3.43}$$

Rewriting the above equation in terms of temperatures and resistances

$$\dot{q} = \frac{T_{\infty,i} - T_{\infty,o}}{R_{\text{conv},i} + R_{\text{cond,wall}} + R_{\text{conv},o}} \tag{3.44}$$

or
which after substitution of all the resistance will become

$$\dot{q} = \frac{T_{\infty,i} - T_{\infty,o}}{\dfrac{1}{h_{\infty,i}(4\pi r_1^2)} + \dfrac{(r_2 - r_1)}{4\pi k r_1 r_2} + \dfrac{1}{h_{\infty,o}(4\pi r_2^2)}} \tag{3.45}$$

If the sphere is composed of two solid layers as shown in Fig. 3.17, the equation can be modified to include the additional resistance term as below:

$$\dot{q} = \frac{T_{\infty,i} - T_{\infty,o}}{R_{\text{conv},i} + R_{\text{cond,wall,1}} + R_{\text{cond,wall,2}} + R_{\text{conv},o}} \tag{3.46}$$

$$\dot{q} = \frac{T_{\infty,i} - T_{\infty,o}}{\dfrac{1}{h_{\infty,i}(4\pi r_1^2)} + \dfrac{(r_2 - r_1)}{4\pi k_1 r_1 r_2} + \dfrac{(r_3 - r_2)}{4\pi k_2 r_2 r_3} + \dfrac{1}{h_{\infty,o}(4\pi r_3^2)}} \tag{3.47}$$

Example 3.5 (Bioprocess Engineering—Heat transfer in a spherical water tank): A 2.5-cm thick stainless steel ($k = 15 \frac{W}{m \cdot °C}$) spherical tank of 1 m internal diameter is being used to perform an endothermic chemical reaction. The reactants are well mixed with a heat transfer coefficient of 300 $\frac{W}{m^2 \cdot °C}$. To provide the required heat, the outer wall of the reactor is wrapped with an electric heating blanket rated for 7.5 kW that heats up the outer surface of the reactor to 90°C. Assuming an electric efficiency of 80%, determine the suitability of the electric blanket to be able to heat the reactants to a temperature of 72°C.

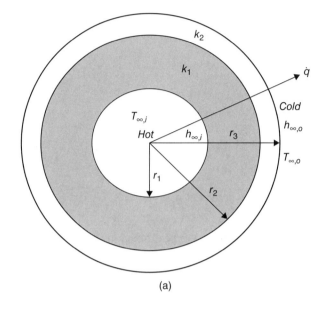

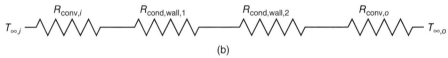

FIGURE 3.17   Schematic of a spherical shell composite containing a hot fluid and exposed to a cold environment (a) expressed as a thermal resistance network (b).

Step 1: Schematic
The schematic is shown in Fig. 3.18.

Step 2: Given data
Internal radius of the tank, $r_1 = 0.5$ m; Thickness of the wall = 2.5 cm; External radius of the tank, $r_2 = 0.525$ m; Thermal conductivity of stainless steel, $k = 15 \, \dfrac{W}{m \cdot {}^\circ C}$; Heat transfer coefficient, $h_{\infty,1} = 300 \, \dfrac{W}{m^2 \cdot {}^\circ C}$; Temperature on the outer surface of the tank, $T_2 = 90^\circ C$; Heat transfer rating, $\dot{q}_{rated} = 7.5$ kW; Efficiency of the heater, $\eta = 0.8$

To be determined: Temperature of the reactants at the center of the tank.

Assumptions: One-dimensional heat transfer occurs (in the radial direction) and thermal conductivity is constant throughout.

Analysis: The tank is heated from the outside. Therefore, the heat is transferred via conduction through the walls and reaches the inner wall. Subsequently, the heat is transferred via convection to the center of the reactor. Thus, we can consider this as a two-resistance system as shown in Fig. 3.18.

Step 3: Calculations
We can write the heat transfer rate for a spherical section as

$$\dot{q} = \frac{T_2 - T_{\infty,i}}{R_{cond,wall} + R_{conv,1}} \qquad \text{(E3.5A)}$$

or

$$\dot{q} = \frac{T_2 - T_{\infty,i}}{\dfrac{(r_2 - r_1)}{4\pi k r_1 r_2} + \dfrac{1}{h_{\infty,i}(4\pi r_1^2)}} \qquad \text{(E3.5B)}$$

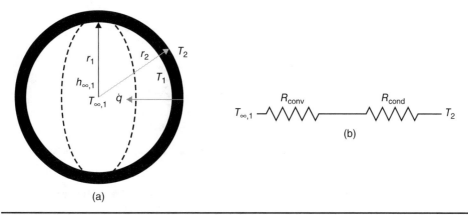

(a)

(b)

FIGURE 3.18 Schematic of a spherical reactor subjected to conduction on the outer surface and convection within the reactor (a) and its thermal resistance network (b).

Substituting the given data, and solving for $T_{\infty,i}$

$$7500 \times 0.8 = \cfrac{(90 - T_{\infty,i})}{\cfrac{(0.525 - 0.5)}{4 \times 3.14 \times 15 \times 0.525 \times 0.5} + \cfrac{1}{300 \times (4 \times 3.14 \times 0.5^2)}} \tag{E3.5C}$$

$$T_{\infty,i} = 80.6°C$$

Therefore, the electrical heating blanket is adequate to raise the temperature of the reactants well beyond 72°C.

The same approach can be employed for cylindrical sections as well. For example, for a simple cylinder (Fig. 3.19) with inner and outer radii of r_1 and r_2, respectively, and carrying a hot fluid at

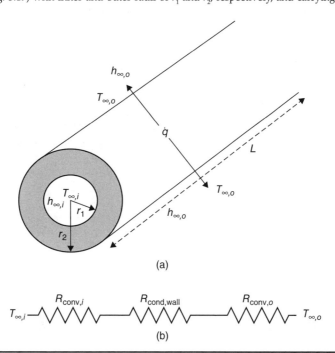

(a)

(b)

FIGURE 3.19 Schematic of a cylinder containing a hot fluid and exposed to a cold environment (a) expressed as a thermal resistance network (b).

temperature $T_{\infty,i}$ with a heat transfer coefficient $h_{\infty,i}$, while losing heat to a colder environment $T_{\infty,o}$ with a heat transfer coefficient $h_{\infty,o}$, the heat transfer rate can be expressed as

$$\dot{q} = \frac{T_{\infty,i} - T_{\infty,0}}{R_{conv,1} + R_{cond,wall} + R_{conv,2}} \qquad (3.48)$$

$$\dot{q} = \frac{T_{\infty,i} - T_{\infty,0}}{\dfrac{1}{h_{\infty,i}(2\pi r_1 L)} + \dfrac{\ln\left(\dfrac{r_2}{r_1}\right)}{2\pi k L} + \dfrac{1}{h_{\infty,0}(2\pi r_2 L)}} \qquad (3.49)$$

Example 3.6 (Aquacultural Engineering—Heat transfer through the pipe):   In a shrimp hatchery, hot water at 75°C is being transported in a 5-cm inner diameter, 30-m long stainless steel ($k = 15\,\dfrac{W}{m \cdot °C}$) pipe of 0.5 cm wall thickness (Fig. 3.20). The pipe is exposed to a constant temperature of 25°C. If the inner and outer heat transfer coefficients are 100 and 50 $\dfrac{W}{m^2 \cdot °C}$, respectively, determine the heat transfer rate and the cost incurred by the producer due to this heat loss in a day.

Step 1: Schematic
The schematic is shown in Fig. 3.20.

Step 2: Given data
Length of the pipe, $L = 30$ m; Inner radius, $r_1 = 2.5$ cm; Outer radius, $r_2 = 3$ cm; Temperature inside the pipe, $T_{\infty,i} = 75°C$; Heat transfer coefficient inside the pipe, $h_{\infty,i} = 100\dfrac{W}{m^2 \cdot °C}$; Temperature outside the pipe, $T_{\infty,0} = 25°C$; Heat transfer coefficient outside the pipe, $h_{\infty,0} = 50\dfrac{W}{m^2 \cdot °C}$; Thermal conductivity, $k = 15\dfrac{W}{m \cdot °C}$

To be determined: Rate of heat transfer from the pipe and the cost.

Assumptions: The heat transfer is steady state and occurs in one dimension (radial). The thermal conductivity remains constant. The cost per kWh is 10 cents.

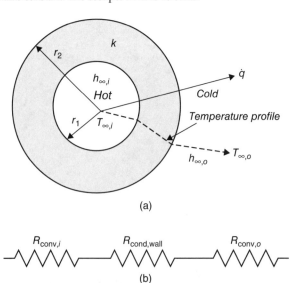

(a)

(b)

FIGURE 3.20   Cross section of the pipe carrying hot water in a shrimp hatchery exposed to cold environment (a) and its thermal resistance network (b).

Analysis: The cylinder is carrying hot water. The heat is transferred from the hot water to the wall of the cylinder via convection followed by conduction through the wall and finally via convection from the outer wall to the surroundings.

Step 3: Calculations

The rate of heat transfer is given by

$$\dot{q} = \frac{T_{\infty,i} - T_{\infty,o}}{R_{conv,1} + R_{cond,wall} + R_{conv,2}} \tag{E3.6A}$$

Or expressing the resistances in their mathematical forms

$$\dot{q} = \frac{T_{\infty,i} - T_{\infty,o}}{\dfrac{1}{h_{\infty,i}(2\pi r_1 L)} + \dfrac{\ln\left(\dfrac{r_2}{r_1}\right)}{2\pi k L} + \dfrac{1}{h_{\infty,o}(2\pi r_2 L)}} \tag{E3.6B}$$

$$\dot{q} = \frac{75 - 25}{\dfrac{1}{100 \times (2 \times 3.14 \times 0.025 \times 30)} + \dfrac{\ln\left(\dfrac{0.03}{0.025}\right)}{2 \times 3.14 \times 15 \times 30} + \dfrac{1}{50 \times (2 \times 3.14 \times 0.03 \times 30)}} \tag{E3.6C}$$

$$\dot{q} = 8731.75 \text{ W} = 8.731 \text{ kW}$$

Cost per day $= 8.731 \text{ kW} \times 24 \text{ h} \times \dfrac{\$0.1}{\text{kWh}} = \$20.95$

The loss of heat through the pipe is costing the producer about \$21 per day. The heat loss can be mitigated by insulating the pipe.

3.8 Controlling Heat Transfer via Insulation

We add insulation to our ceilings and walls to decrease the rate of heat transfer into and out of the building. Insulation materials are made of low conductivity materials including cellulose, fiberglass, cork, polyurethane, aerogels, and other materials that are specially engineered to possess large portion of internal void space. A typical insulation is manufactured by systematically assembling layers of low conductivity materials separated by air (Fig. 3.21).

As a result, an insulation provides resistance to the rate of heat transfer, $\dot{q}$ which can be expressed as

$$\dot{q} = \frac{kA(T_h - T_c)}{L} \tag{3.50}$$

which can also be rewritten as

$$\frac{L}{k} = \frac{(T_h - T_c)}{\left(\dfrac{\dot{q}}{A}\right)} \tag{3.51}$$

where A is the heat transfer area (m²), k is the thermal conductivity of the material $\left(\dfrac{\text{W}}{\text{m} \cdot \text{K}}\right)$, L is the thickness of the material (m), T_h and T_c are the temperatures of the hot and cold surfaces (°C) of the insulation surfaces, respectively.

In Eq. (3.51), the term $\dfrac{L}{k}$ describes the overall resistance to the heat transfer offered by the insulation and is called by its R-value.

$$R\text{-value} = \frac{L}{k} \tag{3.52}$$

Figure 3.21 One of the popular insulations used by the construction industry. (Image courtesy of Johns Manville. Reprinted with permission.)

Therefore, mathematically, the R-value of an insulation of thickness, L is thermal conductivity, k is defined as the temperature gradient resulting from a heat transfer flux of 1 unit $\left(\dfrac{W}{m^2}\right)$. R-values are routinely used in construction industry and their values are usually provided by the insulation manufacturers. Although R-value has a unit, $\dfrac{m^2 \cdot {}^\circ C}{W}$, it is common to denote R-values as unitless.

Example 3.7 (Aquacultural Engineering—Controlling heat transfer through the pipe): Imagine that you were offered an internship in the same shrimp hatchery described in Example 3.6. One of your first tasks is to decrease the rate of heat loss. The producer desires to limit the heat loss to 1 kW and offers to purchase an insulation with a thermal conductivity of 0.03 W/m·K. Using the same data, determine the thickness of the insulation required to limit the heat loss to 1 kW (Fig. 3.22).

Step 1: Schematic
The problem is depicted in Fig. 3.22.

Step 2: Given data
Maximum allowable rate of heat transfer, $\dot{q} = 1$ kW; Length of the pipe, $L = 30$ m; Inner radius, $r_1 = 2.5$ cm; Outer radius, $r_2 = 3$ cm; Temperature inside the pipe, $T_{\infty,i} = 75^\circ C$; Heat transfer coefficient inside the pipe, $h_{\infty,i} = 100 \dfrac{W}{m^2 \cdot {}^\circ C}$; Temperature outside the pipe, $T_{\infty,o} = 25^\circ C$; Heat transfer coefficient outside the pipe, $h_{\infty,o} = 50 \dfrac{W}{m^2 \cdot {}^\circ C}$; Thermal conductivity of the cylinder wall, $k_{wall} = 15 \dfrac{W}{m \cdot {}^\circ C}$; Thermal conductivity of the insulation, $k_{ins} = 0.03 \dfrac{W}{m \cdot {}^\circ C}$

To be determined: Thickness of the insulation needed to limit the heat transfer rate to 1 kW.

Analysis: With the added insulation, the system now has four resistances (namely, convection resistance inside the pipe, conduction resistance of the wall, conduction resistance of the insulation, and the convection resistance outside the pipe). We will set the total heat transfer to 1 kW and solve for the radius of the insulation.

Assumptions: The heat transfer is steady state and occurs in one dimension (radial). The thermal conductivity remains constant.

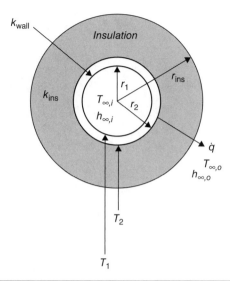

Figure 3.22 The section of the stainless steel pipe carrying hot water whose insulation thickness is to be determined.

Step 3: Calculations

The heat transfer from the inside (center of the pipe) to the outside environment is given by the general equation.

$$\dot{q} = \frac{\Delta T}{R_{Total}} \tag{E3.7A}$$

Or in terms of specific resistances,

$$\dot{q} = \frac{T_{\infty,i} - T_{\infty,o}}{\dfrac{1}{h_{\infty,i}(2\pi r_1 L)} + \dfrac{\ln\left(\frac{r_2}{r_1}\right)}{2\pi k_{wall} L} + \dfrac{\ln\left(\frac{r_{ins}}{r_2}\right)}{2\pi k_{ins} L} + \dfrac{1}{h_{\infty,o}(2\pi r_{ins} L)}} \tag{E3.7B}$$

Substituting the relevant values,

$$1000 = \frac{75 - 25}{\dfrac{1}{100 \times (2 \times 3.14 \times 0.025 \times 30)} + \dfrac{\ln\left(\frac{0.03}{0.025}\right)}{2 \times 3.14 * 15 \times 30} + \dfrac{\ln\left(\frac{r_{ins}}{0.03}\right)}{2 \times 3.14 \times 0.03 \times 30} + \dfrac{1}{100 \times (2 \times 3.14 \times r_{ins} \times 30)}}$$

$$\tag{E3.7C}$$

Solving for r_{ins} using Excel Solver, we obtain

$$r_{ins} = 0.0387 \text{ m} = 3.87 \text{ cm}$$

Therefore, the requited thickness of the insulation is 3.87 cm − 3.0 cm = 0.87 cm.

Example 3.8 (Bioprocessing Engineering—Effect of insulation on heat transfer): In food processing operations, hot water with chemicals is regularly used to clean the contact surfaces. A 2-m radius spherical tank made of 1-cm thick stainless steel $\left(k = 15 \ \dfrac{W}{m \cdot °C} \right)$ is used to store hot water at 90°C (Fig. 3.23). If the tank is exposed to a constant temperature of 25°C, determine the rate of heat transfer and the cost to

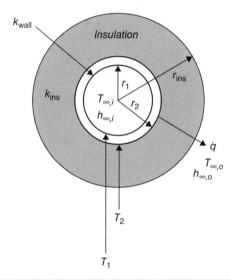

FIGURE 3.23 Schematic of the sphere with insulation for Example 3.8.

the company per day assuming the price of heating as 10 cents/kWh. It was also given that the inner and outer heat transfer coefficients are $15 \dfrac{W}{m^2 \cdot °C}$ each.

Further, during an energy audit, an employee proposed that the costs could be lowered by insulating the tank with insulation. If an insulation material of 2.5 cm thick ($k_{ins} = 0.05 \dfrac{W}{m \cdot °C}$) that costs $\dfrac{\$50}{m^2}$ is available, what will be your assessment regarding the employee's recommendation?

Step 1: Schematic
The problem is depicted in Fig. 3.23.

Step 2: Given data
Inner radius, $r_1 = 2$ m; Outer radius, $r_2 = 2.01$ cm; Thermal conductivity of the spherical wall, $k_{wall} = 15 \dfrac{W}{m \cdot °C}$; Temperature inside the sphere, $T_{\infty,i} = 90°C$; Temperature outside the sphere, $T_{\infty,0} = 25°C$; Heat transfer coefficient inside the sphere, $h_{\infty,i} = 15 \dfrac{W}{m^2 \cdot °C}$; Heat transfer coefficient outside the sphere, $h_{\infty,0} = 15 \dfrac{W}{m^2 \cdot °C}$; Heating cost $= \dfrac{\$0.1}{kWh}$; Thickness of the insulation $= 0.025$ m; Thermal conductivity of the insulation, $k_{ins} = 0.05 \dfrac{W}{m \cdot °C}$; and Cost of the insulation $= \dfrac{\$50}{m^2}$

To be determined: (a) Rate of heat transfer and the cost of the heat loss and (b) analyze the effect of insulation.

Analysis: The sphere is storing hot water wherein the heat is transferred from the hot water to the wall of the wall via convection followed by conduction through the wall and finally via convection from the outer wall to the surroundings.

Assumptions: The heat transfer is steady state and occurs in one dimension (radial). The thermal conductivity remains constant.

Step 3: Calculations
(a) The heat transfer from the inside (center of the sphere) to the outside environment is given by the general equation.

$$\dot{q} = \frac{\Delta T}{R_{Total}} = \frac{T_{\infty,i} - T_{\infty,0}}{R_{conv,i} + R_{cond,wall} + R_{conv,0}} \tag{E3.8A}$$

Expressing in terms of the resistances

$$\dot{q} = \frac{T_{\infty,i} - T_{\infty,o}}{\dfrac{1}{h_{\infty,i}(4\pi r_1^2)} + \dfrac{(r_2 - r_1)}{4\pi k r_1 r_2} + \dfrac{1}{h_{\infty,o}(4\pi r_2^2)}} \tag{E3.8B}$$

After substituting the given data,

$$\dot{q} = \frac{90 - 25}{\dfrac{1}{15 \times (4 \times 3.14 \times 2^2)} + \dfrac{(2.01 - 2)}{4 \times 3.14 \times 15 \times 2.01 \times 2} + \dfrac{1}{15 \times (4 \times 3.14 \times 2.01^2)}} \tag{E3.8C}$$

$$\dot{q} = 24,491 \text{ W} = 24.49 \text{ kW}$$

The cost of this heat loss $= 24.49 \text{ kW} \times 24 \text{ h} \times \dfrac{\$0.1}{\text{kWh}} = \$58.78/\text{day}$

(b) The rate of heat transfer with the insulation wrapped around the sphere

$$\dot{q}_{\text{with ins}} = \frac{\Delta T}{R_{\text{Total}}} = \frac{T_{\infty,i} - T_{\infty,o}}{R_{\text{conv},i} + R_{\text{cond,wall}} + R_{\text{cond,ins}} + R_{\text{conv},o}} \tag{E3.8D}$$

Expressing in terms of the resistances

$$\dot{q}_{\text{with ins}} = \frac{T_{\infty,i} - T_{\infty,o}}{\dfrac{1}{h_{\infty,i}(4\pi r_1^2)} + \dfrac{(r_2 - r_1)}{4\pi k_{\text{wall}} r_2 r_1} + \dfrac{(r_3 - r_2)}{4\pi k_{\text{ins}} r_3 r_2} + \dfrac{1}{h_{\infty,o}(4\pi r_3^2)}} \tag{E3.8E}$$

Substituting the given data

$$\dot{q}_{\text{with ins}} = \frac{90 - 25}{\dfrac{1}{15 \times (4 \times 3.14 \times 2^2)} + \dfrac{(2.01 - 2)}{4 \times 3.14 \times 15 \times 2.01 \times 2} + \dfrac{(2.035 - 2.01)}{4 \times 3.14 \times 0.05 \times 2.035 \times 2.01} + \dfrac{1}{15 \times (4 \times 3.14 \times 2.035^2)}} \tag{E3.8F}$$

$$\dot{q}_{\text{with ins}} = 5,261.33 \text{ W} = 5.26 \text{ kW}$$

The cost of the heat loss with insulation $= 5.26 \text{ kW} \times 24 \text{ h} \times \dfrac{\$0.1}{\text{kWh}} = \$12.63/\text{day}$

Savings/day $= \$58.78 - \$12.63 = \$46.15/\text{day}$

Cost of the insulation $= 4 \times 3.14 \times 2.035^2 \text{ m}^2 \times \dfrac{\$50}{\text{m}^2} = \$2600.69$

Days needed for complete cost recovery $= \dfrac{2600.69}{46.15} = 56.35$ days

The analysis suggests that the employee is right, and the cost of the insulation can be recovered within 57 days and therefore this proposal has merit.

3.9 Critical Radius of Insulation

As discussed previously, the purpose of the insulation is to increase the thermal resistance and decrease the rate of heat transfer. In rectangular sections, such as walls and ceilings, the heat transfer rate will be decreased with an increase in the thickness of the insulation (L_{ins}) as shown in Fig. 3.24 and Eq. (3.53).

$$\dot{q} = \frac{T_2 - T_{\infty}}{R_{\text{cond,ins}} + R_{\text{conv}}} = \frac{\Delta T}{\dfrac{L_{\text{ins}}}{kA} + \dfrac{1}{h_{\infty}A}} \tag{3.53}$$

This is, however, not the case for radial systems. For example, consider a cylinder of length L, with inner and outer radii of r_1 and r_2 carrying a hot fluid at temperature

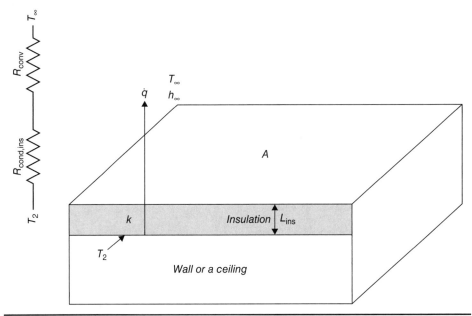

FIGURE 3.24   Illustration of heat transfer through a rectangular wall with insulation on the top face.

$T_{\infty,i}$ (with a heat transfer coefficient $h_{\infty,i}$). If this cylinder is exposed to a colder environment, say at $T_{\infty,o}$ (with a heat transfer coefficient $h_{\infty,o}$), we expect that the heat loss occurs out of the cylinder. Now, in order to decrease the heat transfer rate, let us wrap the cylinder with an insulation material with a conductivity, k_{ins} and radius r_{ins} (Fig. 3.25). If we assume the temperatures of the inner and outer walls of the cylinder as T_1 and T_2, respectively, we can express the rate of heat transfer from the outer wall to the surrounding as

$$\dot{q} = \frac{\Delta T}{R_{Total}} \tag{3.54}$$

The total resistance is a combination of the conduction resistance of the insulation and the convective resistance of the insulation. Therefore,

$$\dot{q} = \frac{(T_2 - T_{\infty,o})}{R_{cond,ins} + R_{conv,ins}} \tag{3.55}$$

$$\dot{q} = \frac{(T_2 - T_{\infty,o})}{\dfrac{\ln\left(\dfrac{r_{ins}}{r_2}\right)}{2\pi k_{ins} L} + \dfrac{1}{h_{\infty,o} 2\pi r_{ins} L}} \tag{3.56}$$

From Eq. (3.56), we note that an increase in the thickness of the insulation will increase

the conduction resistance $\left(\dfrac{\ln\left(\dfrac{r_{ins}}{r_2}\right)}{2\pi k_{ins} L}\right)$ and decrease the conduction heat transfer (favorable).

However, a thicker insulation is also associated with an increased surface area and will result

in decreased convection insulation resistance, $\left(\dfrac{1}{h_{\infty,o} 2\pi r_{ins} L}\right)$ and increased convection

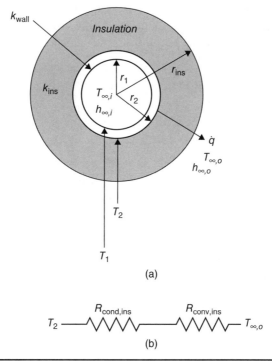

FIGURE 3.25   Cross section of a cylindrical pipe covered with insulation.

heat transfer rate (unfavorable). Therefore, the net effect on heat transfer rate will depend on the thickness of the insulation. If we plot the rate of heat transfer rate as a function of the insulation thickness (Fig. 3.26), we see that the rate of heat transfer increases initially with an increase in the insulation thickness, which appears to be counterproductive.

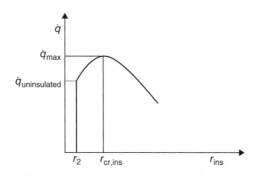

FIGURE 3.26   Effect of insulation thickness on heat transfer in cylindrical and spherical sections. The insulation will increase the transfer rate until the thickness approaches its critical radius.

However, with a further increase in the insulation thickness, the heat transfer rate reaches a maximum value ($\dot{q}_{max}$) and begins to decrease. We can determine the thickness of the insulation at which the heat transfer rate reaches its maximum. This condition is mathematically equivalent to

$$\frac{d\dot{q}}{dr_{ins}} = 0 \tag{3.57}$$

or

$$\frac{d\left[\dfrac{(T_2 - T_{\infty,o})}{\dfrac{\ln\left(\dfrac{r_{ins}}{r_2}\right)}{2\pi k_{ins} L} + \dfrac{1}{h_{\infty,o} 2\pi r_{ins} L}}\right]}{dr_{ins}} = 0 \tag{3.58}$$

Using the quotient rule,

$$\frac{\left[\dfrac{\ln\left(\dfrac{r_{ins}}{r_2}\right)}{2\pi k_{ins} L} + \dfrac{1}{h_{\infty,o} 2\pi r_{ins} L}\right]\dfrac{d(T_2 - T_{\infty,o})}{dr_i} - (T_2 - T_{\infty,o})\left[\dfrac{d}{dr_{ins}}\left(\dfrac{\ln\left(\dfrac{r_{ins}}{r_2}\right)}{2\pi k_{ins} L} + \dfrac{1}{h_{\infty,o} 2\pi r_{ins} L}\right)\right]}{\left[\dfrac{\ln\left(\dfrac{r_{ins}}{r_2}\right)}{2\pi k_{ins} L} + \dfrac{1}{h_{\infty,o} 2\pi r_{ins} L}\right]^2} = 0 \tag{3.59}$$

or

$$\frac{0 - (T_2 - T_{\infty,o})\left[\left(\dfrac{1}{2\pi k_{ins} L r_{ins}} - \dfrac{1}{h_{\infty,o} 2\pi r_{ins}^2 L}\right)\right]}{\left[\dfrac{\ln\left(\dfrac{r_{ins}}{r_2}\right)}{2\pi k_{ins} L} + \dfrac{1}{h_{\infty,o} 2\pi r_{ins} L}\right]^2} = 0 \tag{3.60}$$

$$\frac{1}{2\pi k_{ins} L r_{ins}} - \frac{1}{h_{\infty,o} 2\pi r_{ins}^2 L} = 0 \tag{3.61}$$

$$\frac{1}{k_{ins}} = \frac{1}{h_{\infty,o} r_{ins}} \tag{3.62}$$

The radius of the insulation corresponding to the maximum heat transfer is called the critical radius of insulation.

$$r_{cr,ins} = \frac{k_{ins}}{h_{\infty,o}} \tag{3.63}$$

Similarly let us consider a sphere of inner and outer radii of r_1 and r_2 and covered with an insulation of thickness r_{ins} and thermal conductivity k_{ins} while holding a hot fluid at temperature $T_{\infty,i}$ with heat transfer coefficient $h_{\infty,i}$. If this sphere is exposed to a colder environment, at $T_{\infty,o}$ with a heat transfer coefficient $h_{\infty,o}$, we can write the rate of heat transfer from the outer wall of the sphere to the surrounding air through the insulation as

$$\dot{q} = \frac{(T_2 - T_{\infty,o})}{R_{cond,ins} + R_{conv,ins}} = \frac{(T_2 - T_{\infty,o})}{\dfrac{(r_{ins} - r_2)}{4\pi r_{ins} r_2 k_{ins}} + \dfrac{1}{h_{\infty,o} 4\pi r_{ins}^2}} \tag{3.64}$$

To calculate the insulation radius for maximizing the rate of heat transfer we set $\dfrac{d\dot{q}}{dr_{ins}} = 0$ and solving for r_{ins}, we obtain

$$\frac{d\dot{q}}{dr_{ins}} = \frac{d\left[\dfrac{(T_2 - T_{\infty,o})}{\dfrac{(r_{ins} - r_2)}{4\pi r_{ins} r_2 k_{ins}} + \dfrac{1}{h_{\infty,o} 4\pi r_{ins}^2}}\right]}{dr_{ins}} = 0 \tag{3.65}$$

$$\frac{\left(\dfrac{(r_{ins} - r_2)}{4\pi r_{ins} r_2 k_{ins}} + \dfrac{1}{h_{\infty,o} 4\pi r_{ins}^2}\right)\dfrac{d(T_2 - T_{\infty,o})}{dr_{ins}} - (T_2 - T_{\infty,o})\dfrac{d\left(\dfrac{(r_{ins} - r_2)}{4\pi r_{ins} r_2 k_{ins}} + \dfrac{1}{h_{\infty,o} 4\pi r_{ins}^2}\right)}{dr_{ins}}}{\left[\dfrac{(r_{ins} - r_2)}{4\pi r_{ins} r_2 k_{ins}} + \dfrac{1}{h_{\infty,o} 4\pi r_{ins}^2}\right]^2} = 0 \tag{3.66}$$

$$\frac{0 - (T_2 - T_{\infty,o})\left[\dfrac{1}{4\pi r_{ins}^2 k_{ins}} + \dfrac{-2}{4\pi h_{\infty,o} r_{ins}^3}\right]}{\left[\dfrac{(r_{ins} - r_2)}{4\pi r_{ins} r_2 k_{ins}} + \dfrac{1}{h_{\infty,o} 4\pi r_{ins}^2}\right]^2} = 0 \tag{3.67}$$

$$\frac{1}{4\pi r_{ins}^2 k_{ins}} - \frac{2}{4\pi h_{\infty,o} r_{ins}^3} = 0 \tag{3.68}$$

$$\frac{1}{k_{ins}} = \frac{2}{h_{\infty,o} r_{ins}} \tag{3.69}$$

$$r_{ins} = \frac{2k_{ins}}{h_{\infty,o}} \tag{3.70}$$

As can be seen from the Fig. 3.26, it may appear that the presence of insulation initially increases the heat transfer rates and we may be tempted to think that insulation will not become effective unless the thickness is greater than the critical radius. However, by analyzing the equation for the critical radius considering the practical field conditions, it will be clear that any insulation is better than no insulation. For example, even a simple inexpensive insulating material typically possesses a thermal conductivity of 0.03 $\dfrac{W}{m \cdot {}^\circ C}$, and the average free convective heat transfer coefficient of air starts around 2.5 $\dfrac{W}{m^2 \cdot {}^\circ C}$. Using these values, the critical radius will be

for a cylinder and
$$r_{cr,ins} = \frac{k}{h} = \frac{0.03}{2.5} = 12 \text{ mm}$$

for a sphere
$$r_{cr,ins} = \frac{2k}{h} = \frac{2 \times 0.03}{2.5} = 24 \text{ mm}$$

In addition, in most cases, forced convection conditions apply and the values of h will be greater than 25 $\dfrac{W}{m^2 \cdot {}^\circ C}$. The critical radius will now become as small as 1.2 mm and 2.4 mm for cylinder and sphere, respectively. Commercial insulation materials are typically tens of inches thick and therefore almost always substantially greater than the critical radius. Therefore, for all practical purposes, insulation always decreases the heat transfer rates and it is not necessary to perform the critical radius calculations while selecting insulation.

Example 3.9 (Agricultural Engineering–Safety of an electrical cable):   A 3-m long electrical cable dissipates 150 W of heat energy to the surroundings. The cable consists of 4-mm electric wire that is wrapped with a 2-mm thick plastic insulation with a thermal conductivity of $0.5 \dfrac{W}{m \cdot °C}$. If the cable is exposed to an environment of 25°C with a heat transfer coefficient of $30 \dfrac{W}{m^2 \cdot °C}$, comment on the safety of the cable given the melting point of the plastic is 99°C.

Step 1: Schematic
The problem is depicted in Fig. 3.27.

Step 2: Given data
Length of the cable, $L = 3$ m; Power dissipated, $\dot{q} = 150$ W; Radius of the wire, $r_2 = 2$ mm; Radius of the insulation, $r_{ins} = 4$ mm; Thermal conductivity, $k = 0.5 \dfrac{W}{m \cdot °C}$; Heat transfer coefficient around the cable, $h_{\infty,o} = 30 \dfrac{W}{m^2 \cdot °C}$; and the temperature of the room (surroundings), $T_{\infty,o} = 25°C$

To be determined: Temperature at the interface between the wire and insulation.

Assumptions: Heat transfer occurs only in the radial direction.

Analysis: The system consists of two temperatures, T_2 and $T_{\infty,o}$ separated by two resistances, the conduction resistance of the insulation and convection resistance of the insulation (Fig. 3.27). Using the thermal resistance network, we will first determine the temperature of the wire-insulation interface. If this is greater than 99°C, the plastic insulation will melt and cause unsafe condition for the operator on the floor. Further, we will analyze the problem in light of the critical radius of insulation.

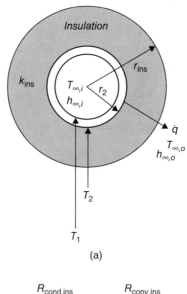

(a)

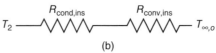

(b)

Figure 3.27 Cross section of the electrical wire with plastic insulation (a) along with the thermal resistance network (b).

Step 3: Calculations

We will first determine the critical radius of insulation for a cylinder as

$$r_{cr,ins} = \frac{k_{ins}}{h_{\infty,ins}} = \frac{0.5}{30} = 0.0167 \text{ m} = 16.67 \text{ mm} \tag{E3.9A}$$

In other words, the heat transfer rate will keep increasing until the radius of insulation reaches 16.67 mm.

The temperature at the interface can be calculated as

$$\dot{q} = \frac{(T_2 - T_{\infty,o})}{\dfrac{\ln\left(\dfrac{r_{ins}}{r_2}\right)}{2\pi k_{ins}L} + \dfrac{1}{h_{\infty,o}2\pi r_{ins}L}} \tag{E3.9B}$$

$$150 = \frac{(T_2 - 25)}{\dfrac{\ln\left(\dfrac{0.004}{0.002}\right)}{2\times3.14\times0.5\times3} + \dfrac{1}{30\times2\times3.14\times0.004\times3}} \tag{E3.9C}$$

Solving for T_2

$$T_2 = 102.38°C$$

Based on the calculations, the interface temperature is slightly greater than 99°C and, therefore, the plastic insulation will likely melt and result in a safety hazard.

Recommendation: Because the radius of the insulation is much less than the critical radius of insulation, we can choose a cable with thicker insulation. Assuming that the heat dissipation rate still remains at 150 W, if we increase the insulation thickness by just 1 mm, the interface temperature becomes

$$150 = \frac{(T_2 - 25)}{\dfrac{\ln\left(\dfrac{0.005}{0.002}\right)}{2\times3.14\times0.5\times3} + \dfrac{1}{30\times2\times3.14\times0.005\times3}} \tag{E3.9D}$$

$$T_{2,(r,ins=0.005\,m)} = 92.6°C$$

which is less than the melting point of the plastic insulation and is therefore a preferred option. As shown in Fig. 3.28, the interface temperature for various insulation thicknesses are plotted.

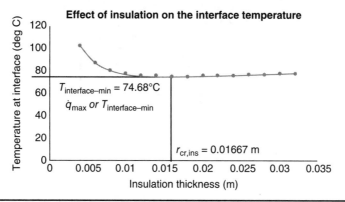

FIGURE 3.28 Plot of the temperatures at the interface between the electrical wire and the insulation for various insulation thicknesses.

3.10   Applications of Numerical Methods in Steady-State Transfer

The equations that we developed in Chap. 2 may be used to solve one-dimensional steady heat transfer problems. While they may be adequate for most cases, heat transfer in two and three dimensions are sometimes encountered by engineers. Solving these equations analytically is highly complicated and sometimes impossible. This is where numerical methods play a great role in simplifying the problems and providing us with quick and reasonably accurate solutions to our problems. However, in this discussion, we will focus only on steady-state two-dimensional heat transfer problems.

Let us consider a rectangular plate (Fig. 3.29a) such that each side of the plate is maintained at different temperatures, T_1, T_2, T_3, and T_4 such that $T_1 \neq T_2 \neq T_3 \neq T_4$. Now due to the temperature gradients, heat will begin to flow within the plate in both x and y-directions. Recalling from Chap. 2, the steady-state heat transfer in two dimensions is expressed as

$$\frac{\partial^2 T}{\partial x^2} + \frac{\partial^2 T}{\partial y^2} + \frac{\dot{e}_{\text{gen}}}{k} = 0 \tag{3.71}$$

To solve Eq. (3.71), we can simply use numerical differentiation such as finite difference method and develop simple algebraic expressions that can be solved by personal calculators. However, we can also use an energy balance method to arrive at the same algebraic expressions that we could have obtained via numerical differentiation and determine the temperatures at any given location on the plate.

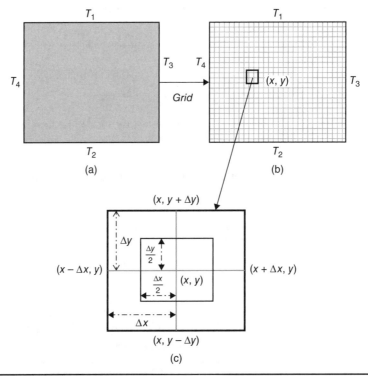

Figure 3.29   Schematic of a plate that is divided into rectangular grid.

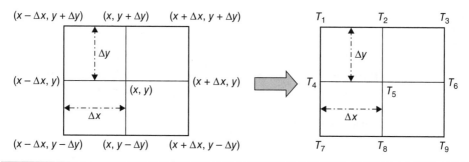

FIGURE 3.30 Simplifying the subscripts will minimize errors due to bookkeeping.

To accomplish that, we will split the plate into numerous rectangular grids (Fig. 3.29b) and simply focus on one grid whose coordinates are presented in Fig. 3.29c. Assuming a unit thickness of the plate ($\Delta z = 1$ *unit*) and performing a heat balance around the plate, we obtain

$$(k)(\Delta y)\left[\frac{T_{x-\Delta x,\,y}-T_{x,y}}{\Delta x}\right]+(k)(\Delta x)\left[\frac{T_{x,\,y+\Delta y}-T_{x,y}}{\Delta y}\right]+(k)(\Delta y)\left[\frac{T_{x+\Delta x,\,y}-T_{x,y}}{\Delta x}\right]$$

$$+(k)(\Delta x)\left[\frac{T_{x,\,y-\Delta y}-T_{x,y}}{\Delta y}\right]+(\Delta x)(\Delta y)(\dot{e}_{gen})=0 \tag{3.72}$$

Dividing the equation by $(k)(\Delta y)(\Delta x)$

$$\left[\frac{T_{x-\Delta x,\,y}-T_{x,y}}{\Delta x^2}\right]+\left[\frac{T_{x,\,y+\Delta y}-T_{x,y}}{\Delta y^2}\right]+\left[\frac{T_{x+\Delta x,\,y}-T_{x,y}}{\Delta x^2}\right]+\left[\frac{T_{x,\,y-\Delta y}-T_{x,y}}{\Delta y^2}\right]+\frac{\dot{e}_{gen}}{(k)}=0 \tag{3.73}$$

Assuming equal grid spacing, that is, $\Delta x = \Delta y$, we obtain

$$T_{x,y}=\frac{1}{4}\left[T_{x-\Delta x,\,y}+T_{x+\Delta x,\,y}+T_{x,\,y+\Delta y}+T_{x,\,y-\Delta y}+\frac{\Delta x^2}{k}\dot{e}_{gen}\right] \tag{3.74}$$

Several biological and agricultural engineering materials do not experience internal energy generation and, therefore, the nodal temperature can be simply expressed as an average temperature of the nodes surrounding the node.

$$T_{x,y}=\frac{1}{4}[T_{x-\Delta x,\,y}+T_{x+\Delta x,\,y}+T_{x,\,y+\Delta y}+T_{x,\,y-\Delta y}] \tag{3.75}$$

Often it will be convenient to avoid the long subscripts and to redraw the grid with simple subscripts as shown in Fig. 3.30 and write the temperature of node 5 as

$$T_5=\frac{1}{4}[T_4+T_6+T_2+T_8] \tag{3.76}$$

Example 3.10 (Agricultural Engineering—Two-dimensional heat transfer in a plate): A small copper plate ($k = 400$ W/m·°C) of dimensions 5 cm × 3 cm is subjected to temperatures as shown in Fig. 3.31. Using an equal grid size of 1 cm, determine the temperature at all interior nodes assuming no energy generation.

Step 1: Schematic
The schematic is shown in Fig. 3.31.

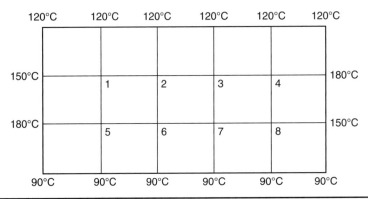

Figure 3.31   Schematic of the copper plate with the given temperatures at the sides.

Step 2: Given data

Thermal conductivity of the copper plate, $k = 400$ W/m·°C; Temperature of the locations on the edges of the plate (Fig. 3.31); Grid size, $\Delta x = \Delta y = 1$ cm; No energy generation, $\dot{e}_{gen} = 0$

To be determined: Temperatures at locations 1 through 9.

Assumptions: Steady-state conditions are maintained, and no energy generation is assumed.

Approach: We will set up eight equations for eight temperatures using the finite difference method and solve them via matrix analysis.

Step 3: Calculations

The temperature of each interior node is the average of temperatures of its surrounding nodes [Eq. (3.76) or (3.77)]. Therefore, for nodes 1–9, we can write the following equations

$$4T_1 = 150 + T_2 + 120 + T_5 \tag{E3.10A}$$

$$4T_2 = T_1 + T_3 + 120 + T_6 \tag{E3.10B}$$

$$4T_3 = T_2 + T_4 + 120 + T_7 \tag{E3.10C}$$

$$4T_4 = T_3 + 180 + 120 + T_8 \tag{E3.10D}$$

$$4T_5 = 180 + T_6 + T_1 + 90 \tag{E3.10E}$$

$$4T_6 = T_5 + T_7 + T_2 + 90 \tag{E3.10F}$$

$$4T_7 = T_6 + T_8 + T_3 + 90 \tag{E3.10G}$$

$$4T_8 = T_7 + 150 + T_4 + 90 \tag{E3.10H}$$

The above equations can be rewritten in the following matrix form

$$
\begin{bmatrix}
4 & -1 & 0 & 0 & -1 & 0 & 0 & 0 \\
-1 & 4 & -1 & 0 & 0 & -1 & 0 & 0 \\
0 & -1 & 4 & -1 & 0 & 0 & -1 & 0 \\
0 & 0 & -1 & 4 & 0 & 0 & 0 & -1 \\
-1 & 0 & 0 & 0 & 4 & -1 & 0 & 0 \\
0 & -1 & 0 & 0 & -1 & 4 & -1 & 0 \\
0 & 0 & -1 & 0 & 0 & -1 & 4 & -1 \\
0 & 0 & 0 & -1 & 0 & 0 & -1 & 4
\end{bmatrix}
\begin{bmatrix}
T_1 \\ T_2 \\ T_3 \\ T_4 \\ T_5 \\ T_6 \\ T_7 \\ T_8
\end{bmatrix}
=
\begin{bmatrix}
270 \\ 120 \\ 120 \\ 300 \\ 270 \\ 90 \\ 90 \\ 240
\end{bmatrix}
$$

After matrix inversion and multiplication, we obtain the temperature of nodes 1–8 as

$$T_1 = 129.8°C,\ T_2 = 121.2°C,\ T_3 = 122.2°C,\ T_4 = 136.0°C,$$

$$T_5 = 128.1°C,\ T_6 = 112.8°C,\ T_7 = 111.7°C,\ T_8 = 121.9°C$$

As shown in the previous example, we can easily compute the temperatures of all the interior nodes by taking an average value of all the surrounding nodes. However, in some cases, we may have to determine the temperatures of the nodes located on the edges (e.g., T_4 or T_3 in Fig. 3.32) and the finite difference equations alone will not be sufficient. In such cases, we need to use the energy balance equation for a given boundary condition.

Boundary conditions: A body can be subjected to several types of boundaries, a few of which will be discussed here. First, let us consider a two-dimensional body of unit thickness ($\Delta z = 1$ unit within which we select an element within the body of dimensions $\frac{\Delta x}{2} \times \Delta y \times 1$ units). Now, we will apply the energy balance to a few commonly found boundary conditions.

Case (1): A body subjected to convection on one side: If the body is subjected to convection heat transfer on the left face, as shown in Fig. 3.32, the energy balance around the element can be written as

$$h\Delta y(T_\infty - T_4) + k\Delta y\frac{(T_5 - T_4)}{\Delta x} + k\frac{\Delta x}{2}\frac{(T_1 - T_4)}{\Delta y} + k\frac{\Delta x}{2}\frac{(T_7 - T_4)}{\Delta y} + \dot{e}_{gen}\Delta y\frac{\Delta x}{2} = 0 \tag{3.77}$$

Further, after assuming $\Delta x = \Delta y$, the above equation becomes

$$h\Delta x T_\infty - h\Delta x T_4 + kT_5 - kT_4 + \frac{k}{2}T_1 - \frac{k}{2}T_4 + \frac{k}{2}T_7 - \frac{k}{2}T_4 + \frac{\Delta x^2}{2}\dot{e}_{gen} = 0 \tag{3.78}$$

After simplifying and rearranging, we obtain

$$T_1 + T_7 + 2T_5 - 2T_4\left(\frac{h\Delta x}{k} + 2\right) + \frac{2h\Delta x T_\infty}{k} + \frac{\Delta x^2}{k}\dot{e}_{gen} = 0 \tag{3.79}$$

or

$$T_4 = \frac{1}{2\left(\frac{h\Delta x}{k} + 2\right)}\left[T_1 + T_7 + 2T_5 + \frac{2h\Delta x T_\infty}{k} + \frac{\Delta x^2}{k}\dot{e}_{gen}\right] \tag{3.80}$$

Case (2): A body subjected to heat flux on one side: Now for the same plate subjected to a uniform heat flux on the left face, as shown in Fig. 3.33, the energy balance around the element becomes

$$\dot{q}''\Delta y + k\Delta y\frac{(T_5 - T_4)}{\Delta x} + k\frac{\Delta x}{2}\frac{(T_1 - T_4)}{\Delta y} + k\frac{\Delta x}{2}\frac{(T_7 - T_4)}{\Delta y} + \dot{e}_{gen}\Delta y\frac{\Delta x}{2} = 0 \tag{3.81}$$

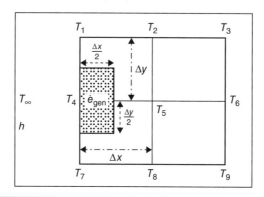

FIGURE 3.32 A rectangular plate subjected to convection (T_∞ and h) on the left face.

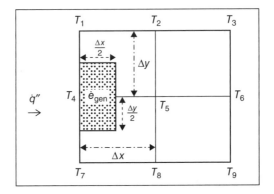

FIGURE 3.33   A rectangular plate subjected to heat flux ($\dot{q}''$) on the left face.

After assuming $\Delta x = \Delta y$ and expanding the above equation and solving for T_4, we obtain

$$2\dot{q}''\frac{\Delta x}{k}+T_1+T_7+2T_5-4T_4+\frac{\Delta x^2}{k}\dot{e}_{gen}=0 \tag{3.82}$$

or

$$T_4=\frac{1}{4}\left[2\dot{q}''\frac{\Delta x}{k}+T_1+T_7+2T_5+\frac{\Delta x^2}{k}\dot{e}_{gen}\right] \tag{3.83}$$

Case (3): A body insulated on one side: In this case, let us insulate the plate on the left face (Fig. 3.34a). We can use the mirror principle to (hypothetically) extend the nodes (dashed lines) to the left as shown in Fig. 3.34b. We can apply the energy balance and determine the temperature at node 4 as below:

$$k\Delta y\frac{(T_5-T_4)}{\Delta x}+k\Delta y\frac{(T_5-T_4)}{\Delta x}+k\Delta x\frac{(T_1-T_4)}{\Delta y}+k\Delta x\frac{(T_7-T_4)}{\Delta y}+\dot{e}_{gen}(\Delta x\Delta y)=0 \tag{3.84}$$

Assuming $\Delta x = \Delta y$ and simplifying, we obtain

$$2T_5+T_1+T_7-4T_4+\frac{\dot{e}_{gen}(\Delta x^2)}{k}=0 \tag{3.85}$$

or

$$T_4=\frac{1}{4}\left[2T_5+T_1+T_7+\frac{\dot{e}_{gen}(\Delta x^2)}{k}\right] \tag{3.86}$$

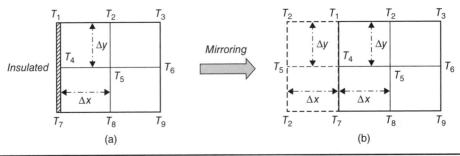

FIGURE 3.34   A rectangular plate insulated on the left face.

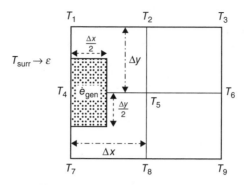

FIGURE 3.35 A rectangular plate subjected to radiation (T_{surr} and ε) on the left face.

Case (4): A body insulated subjected to radiation on one side: In the last case, if the plate is exposed to a radiation heat transfer on the left face (Fig. 3.35), we can write the energy balance around the element as

$$\varepsilon\sigma\Delta y(T_{surr}^4 - T_4^4) + k\Delta y \frac{(T_5 - T_4)}{\Delta x} + k\frac{\Delta x}{2}\frac{(T_1 - T_4)}{\Delta y} + k\frac{\Delta x}{2}\frac{(T_7 - T_4)}{\Delta y} + \dot{e}_{gen}\Delta y\frac{\Delta x}{2} = 0 \qquad (3.87)$$

Example 3.11 (General Engineering—All concentrations): A bottom of a rectangular copper plate ($k = 415$ W/m·K) as shown in Fig. 3.36 is insulated. Determine the temperatures at nodes 5 and 6.

Step 1: Schematic
The schematic is provided in Fig. 3.36.

Step 2: Given data
Thermal conductivity of the copper plate, $k = 415$ W/m·K; Temperature of the locations on the edges of the plate (Fig. 3.35); Grid size, $\Delta x = \Delta y = 0.5$ cm; No energy generation, $\dot{e}_{gen} = 0$

To be determined: Temperatures at nodes 5 and 6.

Assumptions: Steady-state conditions are maintained, and no energy generation is assumed.

Approach: We will set up six equations for eight temperatures using the finite difference method and mirror approach and solve them via matrix analysis.

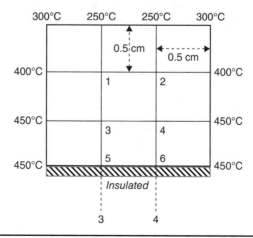

FIGURE 3.36 Schematic of the plate that is insulated at its bottom for Example 3.11.

Step 3: Calculations

The temperature of each interior node is the average of temperatures of its surrounding nodes. Therefore, for nodes 1–6, we can write the following equations

$$4T_1 = 400 + T_2 + 250 + T_3 \tag{E3.11A}$$

$$4T_2 = T_1 + 400 + 250 + T_4 \tag{E3.11B}$$

$$4T_3 = 450 + T_4 + T_1 + T_5 \tag{E3.11C}$$

$$4T_4 = T_3 + 450 + T_2 + T_6 \tag{E3.11D}$$

$$4T_5 = 450 + T_6 + T_3 + T_3 \tag{E3.11E}$$

$$4T_6 = T_5 + 450 + T_4 + T_4 \tag{E3.11F}$$

The above equations can be rewritten in the following matrix form

$$
\begin{bmatrix}
4 & -1 & -1 & 0 & 0 & 0 \\
-1 & 4 & 0 & -1 & 0 & 0 \\
-1 & 0 & 4 & -1 & -1 & 0 \\
0 & -1 & -1 & 4 & 0 & -1 \\
0 & 0 & -2 & 0 & 4 & -1 \\
0 & 0 & 0 & -2 & -1 & 4
\end{bmatrix}
\begin{bmatrix}
T_1 \\ T_2 \\ T_3 \\ T_4 \\ T_5 \\ T_6
\end{bmatrix}
=
\begin{bmatrix}
650 \\ 650 \\ 450 \\ 450 \\ 450 \\ 450
\end{bmatrix}
$$

After matrix inversion and multiplication, we obtain the temperature of nodes 1–8 as

$$T_1 = 352.7°C, \; T_2 = 352.7°C, \; T_3 = 408.3°C,$$
$$T_4 = 408.3°C, \; T_5 = 422.2°C, \; T_6 = 422.2°C$$

Practice Problems for the FE Exam

1. The thermal resistance in a 5-cm thick rectangular wall of area 2 m^2 with thermal conductivity of 10 W/m·K is

   a. 0.0025 K/W.

   b. 0.005 K/W.

   c. 0.1 K/W.

   d. None of the above.

2. The left face and the right face of a 20-cm thick concrete wall ($k = 1.05$ W/m·°C) are subjected to temperatures of 35°C and 12°C, respectively. The heat flux through the wall may be approximated as

   a. 220 W/m^2.

   b. 50 W/m^2.

   c. 120 W/m^2.

   d. None of the above.

3. The same concrete wall described in Problem 2 is now exposed to hot and cold air. The left face of the wall is exposed to cold air at 20°C with a heat transfer coefficient of 25 W/m²·°C while the right face is subjected to hot air flow at 50°C with a heat transfer coefficient of 15 W/m²·°C. The heat flux now becomes approximately

   a. 200 W/m².

   b. 101 W/m².

   c. 170 W/m².

   d. 300 W/m².

4. In Problem 3, the fraction of thermal resistance offered by the wall is close to

   a. 50%.

   b. 75%.

   c. 20%.

   d. 65%.

5. A spherical shell whose inner and outer diameters are 1 m and 1.1 m, respectively, is subjected to a temperature gradient of 50°C. Assuming a thermal conductivity of 40 W/m·K, the conduction heat transfer across the wall of the sphere is closest to

   a. 110 kW.

   b. 100 kW.

   c. 140 kW.

   d. 120 kW.

6. The conduction resistance in a 1-cm thick, 5-cm inner diameter cylindrical copper ($k = 400$ W/m·K) pipe of unit length is approximately

   a. 2 K/m.

   b. 100 K/m.

   c. 0 K/m.

   d. 10 K/m.

7. The conduction heat transfer rate in a 10-m long, 1-cm thick, 5-cm inner diameter cylindrical copper ($k = 400$ W/m·K) pipe subjected to a temperature gradient of 40°C to is approximately

   a. 2.75 MW.

   b. 2.75 kW.

   c. 10 MW.

   d. None of the above.

8. Two 5-cm thick rectangular sections of thermal conductivities 10 and 20 W/m·K, respectively, are joined in series to form a composite wall. The total thermal resistance for this composite wall is

   a. 0.0075 K/m.

   b. 0.005 K/m.

   c. 0.01 K/m.

   d. None of the above.

9. The critical radius of insulation for a cylindrical pipe wrapped with fiberglass ($k = 0.07$ W/m·°C) and subjected to a heat transfer coefficient of 10 W/m²·°C is

   a. 7 mm.

   b. 10 mm.

   c. 15 mm.

   d. None of the above.

10. The critical radius of insulation for a spherical shell wrapped with glasswool ($k = 0.04$ W/m·°C) and subjected to a heat transfer coefficient of 25 W/m²·°C is nearly

    a. 10 mm.

    b. 5 mm.

    c. 3 mm.

    d. None of the above.

Practice Problems for the PE Exam

1. A composite wall in a feed storage facility constructed using two materials as shown below (Fig. 3P1) is subjected to indoor and outdoor temperatures of 27°C and 4°C with respective convective heat transfer coefficients of 20 W/m²·°C and 55 W/m²·°C. Ignoring the effects of radiation, (a) determine the rate of heat loss through the wall and (b) cost of heat loss assuming the cost of electricity as 15 cents per kWh.

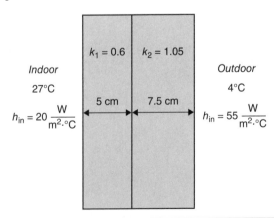

FIGURE 3P1   Schematic for Problem 1 describing heat transfer from inside the feed storage to outside.

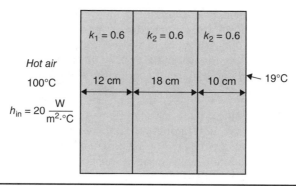

FIGURE 3P2   Schematic for Problem 2 describing heat transfer across the composite wall arranged in series.

2. A composite wall as shown in Fig. 3P2 is exposed to hot air on its left face. Determine the rate of heat transfer and the interface temperatures.

3. A composite wall (5.5 m long × 3 m height × 0.4 m thickness) was built by fusing three carbon-based materials in parallel (Fig. 3P3). The thermal conductivities and areas of cross section for three sections are 0.05, 0.3, and 5 $\frac{W}{m \cdot {}^\circ C}$ and 0.2, 0.5, and 0.5 m², respectively. If the left and right walls are subjected to temperatures of 75°C and 27°C, respectively, determine the heat transfer rate.

4. A shrimp processing facility uses a 1.25 cm-thick stainless steel ($k = 16$ W/m·K) spherical tank of 2 m inner diameter to store hot water at 80°C. Determine the rate of heat transfer to the surroundings maintained at 25°C if the heat transfer coefficients inside and outside the tank are found to be 25 W/m²·°C and 65 W/m²·°C, respectively. Also, calculate the reduction if heat loss rate of a 5-cm thick fiberglass insulation ($k = 0.06$ W/m·K) is wrapped around the tank.

5. A hatchery building maintained at 27°C loses heat to the outside environment at 4°C through its 15-cm thick concrete roof ($k = 1.05$ W/m·°C) with an area 235 m².

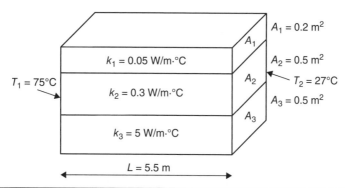

FIGURE 3P3   Schematic for Problem 3 describing heat transfer across the composite wall arranged in parallel.

If the heat transfer coefficients for inside and outside the building are 20 W/m²·°C and 100 W/m²·°C, respectively, determine the rate of heat loss from the building through the roof. Additionally, it was proposed to limit the rate of heat loss to 10 kW by adding insulation to the concrete roof. What should be the R-value of the insulation if the thermal conductivity of the insulation was known to be 0.08 W/m·°C?

6. A meat processing operation uses a 1-cm thick and 3.5-m inner diameter stainless steel ($k = 15$ W/m·°C) spherical container to store hot water at 95°C. If the temperature inside the facility is maintained at 25°C, determine the rate of heat loss from the container assuming the heat transfer coefficients for inside the tank and outside the tank as 10 W/m²·°C and 50 W/m²·°C, respectively. The manager of this facility decides to reduce the rate of heat loss to 5 kW using a fiberglass insulation with a thermal conductivity of 0.08 W/m·°C and approaches you for advice. Perform the necessary calculations and determine the thickness of the insulation required to limit the heat loss to 5 kW.

7. Freshly synthesized biodiesel at 55°C is pumped through a 50-m long copper ($k = 400$ W/m·°C) pipe of 10-cm inner diameter (inside heat transfer coefficient = 100 W/m²·°C). If the heat transfer rate from the biodiesel to the surrounding air at 21°C was measured to be 18,900 W, determine the heat transfer coefficient between the pipe's outer surface and the air. Also determine the temperature of the inner surface of the pipe.

8. It was proposed to insulate a 2-m inner diameter copper ($k = 400$ W/m·°C) spherical shell of 2-cm thickness that holds hot water at 95°C and exposed to air at 5°C. The in-house technician suggested to the manager to procure an insulation material ($k = 0.1$) with an R-value of 1.0. The manager approached you for your opinion. Do you agree with the technician?

9. For Problem 1, what will be the percentage reduction in heat transfer rate if the right face of the wall is insulated with a 4-cm thick fiber glass insulation ($k = 0.07$ W/m·°C)?

10. How important is the concept of critical radius of insulation to regulate heat transfer in (a) a pipe carrying hot water and (b) in a cable carrying electrical current?

CHAPTER **4**

Unsteady-State Conduction

Chapter Objectives

The concepts covered in this chapter will enable students to

- Understand the concept of Biot number and its application in unsteady-state heat transfer.
- Use lumped, approximate, and graphical methods to analyze unsteady-state heat transfer.
- Use superposition principles to solve simple two-dimensional heat transfer problems.

4.1 Motivation

In the previous chapters, we studied the principles of heat transfer under steady-state conditions. In fact, for many engineering problems that we encounter, steady-state assumptions work just fine. However, when certain biological materials are subjected to heating or cooling, the temperature within the body changes with time and the location. To understand this, let us take a simple example. Imagine freezing of fish in a freezer. When subjected to cold environment, the fish will begin to lose heat. If we measure the temperature of the fish, the surface will be colder relative to the inner portions of the fish. As time progresses, the temperature of the fish surface will keep dropping until it reaches an equilibrium. It is also important to note that when any object is exposed to a drastic change in temperature, the heat transfer begins as an unsteady-state process and eventually reaches steady-state equilibrium. To solve such problems, one must employ the principles of unsteady-state heat transfer. We will start this chapter with one-dimensional problems and eventually extend the concepts to two-dimensional objects.

4.2 Solving the Unsteady-State Heat Conduction Problems

There are two approaches in the analysis of unsteady-state heat conduction problems. The first approach, the simpler one, involves the assumption that the temperature only varies with time and not with the location. To illustrate this idea, let us consider a body

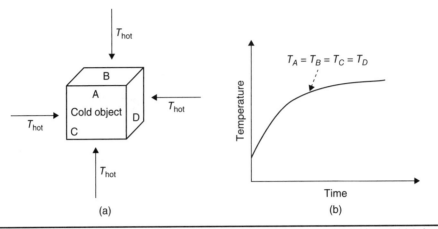

(a)   (b)

FIGURE 4.1 Schematic of a cold body exposed to a hot environment in a lumped model (a) that assumes equal temperatures at all locations within the body (b).

exposed to a heat source (Fig. 4.1a). For clarity, we will choose a cubical object, but the concept is valid for any shape. As time progresses, the body begins to get hotter. However, we will assume that the temperatures at all locations are identical. In other words, we are lumping all the different locations (A, B, C, and D) of the body into one unit (Fig. 4.1b) and analyze only for the variation of temperature as a function of time ($T = f(t)$). This method is commonly called as *the lumped approach*. There are situations where the lumped approach works very well, including heating and cooling of small objects equipped with high thermal conductivities (e.g., metals).

In some cases, our assumption that temperature is only a function of time will not be valid. For example, if we are freezing a large piece of meat in a freezer (Fig. 4.2a), we observe that during the freezing, while the temperature of the meat is continuously decreasing with time, the surface temperature (location C) is much lower than the internal temperature of the meat at two different locations (A and B) (Fig. 4.2b).

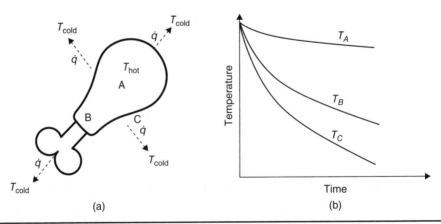

(a)   (b)

FIGURE 4.2 Schematic of a hot body exposed to a cold environment (a) where temperature gradients within the body can exist (b).

In this case, the temperature will be a function of both the time and location within the body $(T = f(x, t))$.

Such problems involve solving of the heat equation to determine the temperature at any location and time. We will begin our discussion with the lumped approach and extend our analysis to cases where temperature varies with time and location.

4.3   The Lumped Approach

To understand the idea of the lumped approach, imagine a body of any random shape is suddenly being cooled (or heated). The lumped approach assumes that every location on (and within) the body is at the same temperature $\left(\dfrac{\partial T}{\partial x} = 0 \right)$ at any given time.

However, the temperature is constantly changing with the time $\left(\dfrac{\partial T}{\partial t} \neq 0 \right)$. Because we are considering the entire body as a single lump, this approach is called "the lumped approach."

4.4   Mathematical Analysis of the Lumped Approach

Consider an object of a mass, m, area, A, volume, V, specific heat, C_p, and density, ρ at an initial temperature, T_i as shown in Fig. 4.3 and it is exposed to an environment that is colder temperature, T_∞ than the body itself. As a result, the body loses heat via convection to its surroundings.

Mathematically,

$$\text{Heat transfer rate via convection: } q_{\text{conv}} = hA(T - T_\infty) \qquad (4.1)$$

$$\text{Rate of energy loss by the body in time } dt = mC_p \frac{dT}{dt} \qquad (4.2)$$

Applying an energy balance around the body will result in

$$-hA(T - T_\infty) = mC_p \frac{dT}{dt} \qquad (4.3)$$

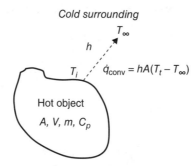

FIGURE 4.3   Heat transfer from a hot body exposed to cold environment via lumped model.

Recalling that $dx = d(x - \text{constant})$, we can rewrite Eq. (4.3) as follows:

$$-hA(T - T_\infty) = mC_p \frac{d(T - T_\infty)}{dt} \tag{4.4}$$

$$-\frac{hA}{mC_p} dt = \frac{d(T - T_\infty)}{(T - T_\infty)} \tag{4.5}$$

Integrating between the limits 0 to t and T_i to T_t, we obtain

$$-\frac{hA}{mC_p} \int_0^t dt = \int_{T_i}^{T_t} \frac{d(T - T_\infty)}{(T - T_\infty)} \tag{4.6}$$

$$-\frac{hA}{mC_p} t = \ln\left[\frac{(T_t - T_\infty)}{(T_i - T_\infty)}\right] \tag{4.7}$$

Therefore, the above equation informs us that the temperature of the body at any given time can be determined from the physical and thermal properties of the body and the environmental temperature the body is subjected to. This is of course under the assumption that the entire body is at the same temperature at any given time.

4.5   The Concept of Biot Number

To continue our discussion on the lumped approach to solve the unsteady-state heat transfer problems, let us first understand how the heat transfer occurs in the system. In the previous example where a body is being cooled (Fig. 4.3), in the first step, heat is lost via conduction from within the body and reaches the outer surface of the body. Subsequently, in the second step, the heat from the surface is lost to the surroundings via convection until a thermal equilibrium is achieved. These steps are mathematically presented as under

$$\textbf{Step 1: } \dot{q}_{\text{cond}} = k\frac{A\Delta T_1}{L}, \text{ where } \Delta T_1 = (T_0 - T_s) \tag{4.8}$$

and

$$\textbf{Step 2: } \dot{q}_{\text{conv}} = hA\Delta T_2, \text{ where } \Delta T_2 = (T_s - T_\infty) \tag{4.9}$$

where T_0 and T_s are the temperatures at the center and at the surface of the body while T_∞ is the temperature of the surrounding medium to which the heat is being lost to. The other parameters have been defined previously.

Using the steady-state equilibrium condition, that is, the rate of conduction is the same as the rate of convection, one can also express the above two steps as follows:

$$hA\Delta T_2 = \frac{kA\Delta T_1}{L} \tag{4.10}$$

$$\frac{hL}{k} = \frac{\Delta T_1}{\Delta T_2} = \frac{\text{Conduction thermal gradient}}{\text{Convection thermal gradient}} \tag{4.11}$$

$$\frac{hL}{k} = \text{Dimensionless parameter} \tag{4.12}$$

The dimensionless ratio $\frac{hL}{k}$ or more generally $\frac{hL_c}{k}$ is defined as the Biot number (N_{Bi}) named after Jean-Baptiste Biot, to acknowledge his significant contributions to the field of heat transfer (ref). Here, L_c is called the characteristic length of the body whose values will depend on the system geometry (slab vs. sphere vs. cylinder that will be discussed later).

In other words, the Biot number is simply a comparison of relative magnitudes of convection and conduction heat transfers the body is subjected to. For a given situation, if the conduction heat transfer rate is significantly higher than the convective heat transfer rate, the value of N_{Bi} will be small and vice versa. Alternatively, it also compares the conduction and convection thermal gradients the body is experiencing. If the convection thermal gradient is significantly higher than the conduction thermal gradient (or very fast conduction relative to convection), the Biot number will assume a small value. Further, if we recall the thermal resistance concepts from Chap. 3, we can also extend those ideas to the Biot number and express the Biot number as

$$N_{Bi} = \frac{hL_c}{k} = \frac{\frac{L_c}{k}}{\frac{1}{h}} \tag{4.13}$$

The ratio $\dfrac{L_c}{k}$ can be interpreted as the internal conduction resistance while $\dfrac{1}{h}$ may be viewed as external convection resistance.

$$N_{Bi} = \frac{hL_c}{k} = \frac{\frac{L_c}{k}}{\frac{1}{h}} = \frac{\text{Internal conduction thermal resistance}}{\text{External convection thermal resistance}} \tag{4.14}$$

4.6 Validity of the Lumped Approach

As can be seen from the definition of the Biot number, large objects with low thermal conductivities experiencing high rates of convection heat transfer possess high values of Biot numbers and vice versa. The lumped approach assumed that there was no thermal gradient within the body. (That is, every location was at the same temperature at any given point of time.) Ideally, for that condition to be satisfied, the conduction gradient in Eq. (4.11) must be zero or the convection thermal gradient must approach infinity. In other words,

$$N_{Bi} = 0 \text{ when } L_c = 0 \text{ or } k = \infty \text{ or } h = 0 \tag{4.15}$$

However, for practical engineering calculations, engineers and scientists came up with critical value of 0.1 for the Biot number to establish its validity in lumped analysis. Therefore, the first step in solving any unsteady-state conduction is to calculate the Biot number using a characteristic length of $\dfrac{\text{Volume}}{\text{Area}}$. If the Biot number is less than 0.1, it may be safe to assume that there is no spatial thermal gradient in the body and the temperature is only a function of time. Mathematically,

$$\forall \ N_{Bi} \leq 0.1 \quad T \neq f(x) \text{ and } T = f(t) \text{ only} \tag{4.16}$$

Example 4.1 (Agricultural Engineering—Heat transfer in a metal sphere): A spherical steel ball of 6 cm diameter at 30°C is placed in a furnace maintained at a temperature of 500°C (Fig. 4.4). If the thermal conductivity, specific heat, and the density of steel are 50 W/m·K, 500 J/kg·K, and 7800 kg/m³, respectively, what will be the temperature at the surface and in the center of the ball after 3 minutes of heating if the heat transfer coefficient between oven and the stainless steel ball is 40 W/m²·°C?

Solution

Step 1: Schematic
The schematic for the problem is shown in Fig. 4.4.

Step 2: Given data
Diameter, $D = 6$ cm; Initial temperature, $T_i = 30°C$; Oven temperature, $T_\infty = 500°C$; Heat transfer coefficient, $h = 40$ W/m²·°C; Thermal conductivity, $k = 50$ W/m·K; Time, $t = 3$ min; Density, $\rho = 7800$ kg/m³; and Specific heat, $C_p = 500$ J/kg·K

Approach: As a first step, we will determine the value of Biot number and verify the validity of the lumped approach. If the Biot number is less than or equal to 0.1, we will use the lumped model to estimate the temperature of the ball after 3 minutes.

Step 3: Calculations
Calculation of the Biot number

$$N_{Bi} = \frac{hL_c}{k} \tag{E4.1A}$$

Recall that the characteristic length, L_c for the verification step is always calculated as $\dfrac{V}{A}$. Therefore,

the volume of the sphere $= \dfrac{4\pi r^3}{3} = \dfrac{4 \times 3.14 \times 0.03^3}{3} = 0.000113$ m³, and the surface area of the sphere $=$

$4\pi r^2 = 4 \times 3.14 \times 0.03^2 = 0.0113$ m² and the Biot number is

$$N_{Bi} = \frac{h\dfrac{V}{A}}{k} = \frac{40 \times \dfrac{0.000113}{0.0113}}{50} \tag{E4.1B}$$

$$N_{Bi} = 0.008 < 0.1$$

Because the value of the Biot number is less than 0.1, the lumped model is applicable.

Calculation of the temperature after 3 minutes of heating
The temperature after 3 minutes of heating can be determined by

$$\ln\left[\frac{(T_t - T_\infty)}{(T_i - T_\infty)}\right] = -\frac{hA}{mC_p}t \tag{E4.1C}$$

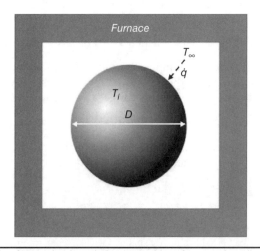

FIGURE 4.4   Schematic of a steel sphere being heated in a furnace described in Example 4.1.

Substituting $m = \rho \times V$

$$\ln\left[\frac{(T_t - T_\infty)}{(T_i - T_\infty)}\right] = -\frac{hA}{\rho V C_p}t = -\frac{h}{\rho C_p}\left(\frac{A}{V}\right)t = -\frac{h}{\rho C_p L_c}t \qquad \text{(E4.1D)}$$

Substituting the relevant values into the equation,

$$\ln\left[\frac{(T_t - 500)}{(30 - 500)}\right] = -\frac{40 \times 3 \times 60}{7800 \times 500 \times 0.01} \qquad \text{(E4.1E)}$$

will result in

$$T_t = 109.2°C$$

The temperature of the surface of the sphere after 3 minutes of heating is 109.2°C. Because the Biot number is less than 0.1, there is no temperature gradient within the sphere (no internal resistance). Therefore, the temperature at the center of the sphere is also 109.2°C.

Example 4.2 (Bioprocess Engineering—Heat transfer in a laboratory reactor): A small thin-walled cylindrical stainless steel reactor is being used to perform a catalytic chemical reaction in an aqueous phase containing sulfuric acid. The catalyst pellets are cylindrical in shape with an average diameter of 1 mm and 5 mm in length (Fig. 4.5). The initial temperature of the reactor and the contents is at 25°C. The reactants will react only when the catalyst pellets reach an internal temperature of 160°C and therefore the operator uses a heating belt to increase the temperature of the reactor inner wall to 200°C. Estimate how many minutes the operator must wait for the reaction to occur and start the data collection process. It is also given that the reactants are well mixed with an overall heat transfer coefficient of 40 W/m²·°C and the thermal properties of the catalyst pellets are $k = 0.11$ W/m·°C, specific heat = 2300 J/kg·K, and density = 500 kg/m³.

Step 1: Schematic
The schematic for the problem is shown in Fig. 4.5.

Step 2: Given data
Catalyst pellet dimensions = 5-mm length and 1-mm diameter; Initial temperature, $T_i = 25°C$; Target temperature, $T_t = 160°C$; Temperature of the environment, $T_\infty = 200°C$; Heat transfer coefficient, $h = 40\ \dfrac{W}{m^2 \cdot °C}$; Thermal conductivity, $k = 0.11\ \dfrac{W}{m \cdot °C}$; Specific heat, $C_p = 2300\ \dfrac{J}{kg \cdot K}$; Density, $\rho = 500\ \dfrac{kg}{m^3}$

To be determined: Time for the reaction to begin to occur (when the target temperature reaches 160°C).

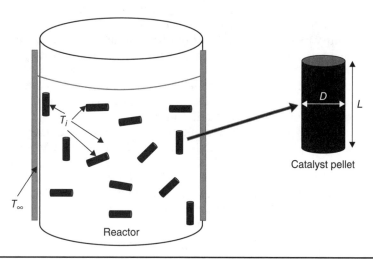

FIGURE 4.5   Schematic of an experimental set up to heat the catalyst pellet to 160°C to initiate the desired chemical reaction.

Assumptions: The wall of the cylinder is thin such that there is no thermal gradient within the wall.

Approach: After confirming that the Biot number is less than 0.1, we will use the lumped approach to determine the heating needed for the pellet to reach the target temperature of 160°C.

Step 3: Calculations
The Biot number is calculated as

$$N_{Bi} = \frac{hL_c}{k} \tag{E4.2A}$$

The characteristic length is determined as

$$L_c = \frac{V}{A} = \frac{\pi R^2 L}{2\pi RL + 2\pi R^2} = \frac{\pi R^2 L}{2\pi R(L+R)} = \frac{RL}{2(L+R)} \tag{E4.2B}$$

Therefore, after substitution of the given data,

$$L_c = \frac{0.0005 \times 0.005}{2(0.005 + 0.0005)} = 2.272 \times 10^{-4} \text{ m}$$

$$N_{Bi} = \frac{40 \times 2.272 \times 10^{-4}}{0.11} = 0.082 < 0.1$$

Therefore, the lumped method will be appropriate here.

$$\ln\left[\frac{(T_t - T_\infty)}{(T_i - T_\infty)}\right] = -\frac{h}{\rho C_p L_c}t \tag{E4.2C}$$

$$\ln\left[\frac{(160 - 200)}{(25 - 200)}\right] = -\frac{40}{500 \times 2300 \times 2.272 \times 10^{-4}}t \tag{E4.2D}$$

$$t = 9.6 \text{ s}$$

The minimum time needed for the pellet to reach 160°C is 9.6 s.

Note: Because the internal resistance is negligible, we can assume that the entire pellet is at 160°C.

4.7   What Happens When the Biot Number Exceeds 0.1?

As we have seen previously, the lumped approach, which we developed in Section 4.3, works very well for small objects possessing high thermal conductivities and exposed to conditions where the convection heat transfer is low, resulting in a Biot number less than or equal to 0.1. However, oftentimes, we are presented with situations in which the temperature varies with time and the position within a given body. Mathematically, this means that the Biot number is greater than 0.1 and the lumped approach will not be appropriate to analyze such unsteady-state conduction problems. Therefore, one has to solve the heat conduction equation to obtain the temperature at any location and time. In this section, we will begin the discussion with a rectangular section and extend the concepts to spherical and cylindrical configurations.

Let us choose an infinitely long plane wall of thickness 2L (Fig. 4.6) with an initial temperature T_i exposed to a hot environment maintained at T_∞. As time progresses, heat is transferred from the surroundings to the wall surface via convection resulting in an increase in the temperature of the wall. Subsequently, heat is carried over into the inner layers of the wall via conduction and eventually reaches T_∞. Recalling from Chap. 2, the one-dimensional heat transfer equation for an isotropic material (k is constant) in Cartesian coordinates is given by the partial differential equation:

$$\frac{\partial^2 T}{\partial x^2} = \frac{1}{\alpha}\frac{\partial T}{\partial t} \tag{4.17}$$

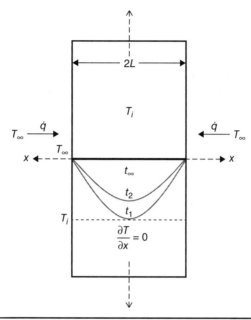

FIGURE 4.6 Schematic of an infinitely long plain wall originally at T_i subjected to a hot environment, T_∞.

where T is the temperature and α is the thermal diffusivity and expressed as a function of thermal conductivity (k), density (ρ), and specific heat (C_p) and mathematically expressed as

$$\alpha = \frac{k}{\rho C_p} \tag{4.18}$$

Such equations are called initial boundary value problems and therefore need the initial and boundary conditions to arrive at an analytical solution. In the context of the problem, at the beginning of the process (initial condition),

$$t = 0,\ T = T_i \tag{4.19}$$

At the outer boundary ($x = L$), between the wall and the surroundings, the convection heat transfer to the wall surface is the same as the conduction heat transfer within the wall (outer boundary condition),

$$hA(T_\infty - T(x = L,\ t)) = -k\frac{\partial T(x = L,\ t)}{\partial x} \tag{4.20}$$

At the inner boundary, that is, the center of the wall ($x = 0$), considering the temperature profile within the wall, the spatial rate of change of the temperature is zero,

$$\frac{\partial T(x = 0,\ t)}{\partial x} = 0 \tag{4.21}$$

> We note that the temperature profiles in the slab are symmetrical. Therefore, it is usually enough to analyze only half of the slab (thickness = L) and yet obtain a complete temperature distribution within the wall section. For analytical solutions, the symmetry may not be significant. However, when numerical techniques are employed, the use of symmetry will significantly reduce the need for computational resources and costs.

Considering that the temperature distribution within the wall is going to depend on eight variables, namely, time, initial and surrounding temperature, thermal conductivity, heat transfer coefficient, the thickness of the wall, density, and specific heat, it may be simpler to transform the heat transfer equation into a nondimensional form. To do this, we define three nondimensional space (ψ), temperature (θ), time (τ) ratios as follows:

$$\psi = \frac{x}{L} \tag{4.22}$$

$$\theta = \frac{T - T_\infty}{T_i - T_\infty} \tag{4.23}$$

$$\tau = \frac{\alpha t}{L^2} \tag{4.24}$$

In other words,

$$\partial x = L\,\partial \psi; \ \partial T = (T_i - T_\infty)\partial\theta; \ \partial t = \frac{L^2}{\alpha}\partial\tau \tag{4.25}$$

The nondimensionalized time, τ is called the Fourier number, N_{Fo}.

Substituting the above equations into the governing equation (Eq. (4.17)) and initial (Eq. (4.19)) and boundary (Eqs. (4.20) and (4.21)) equations and after simplification we obtain the nondimensional heat equation that describes the temperature as a function of only nondimensional time (τ) and space (ψ)

$$\frac{\partial^2 \theta}{\partial \psi^2} = \frac{\partial \theta}{\partial \tau} \tag{4.26}$$

which can be solved by applying the nondimensional initial and boundary conditions as follows:

$$\text{when } \tau = 0; \ \theta = 1 \tag{4.27}$$

$$\text{when } \psi = 0; \ \frac{\partial \theta}{\partial \psi} = 0 \tag{4.28}$$

$$\text{when } \psi = 1; \ \frac{\partial \theta}{\partial \psi} = -\theta N_{\mathrm{Bi}} \tag{4.29}$$

The mathematical solution to the above heat equation for the plain wall result in an infinite series as summarized as follows:

$$\theta = \frac{T_{x=x,t} - T_\infty}{T_i - T_\infty} = \sum_{n=1}^{\infty} C_n \exp(-\xi_n^2 \tau)\cos(\xi_n \psi) \tag{4.30}$$

where

$$C_n = \frac{4\sin\xi_n}{2\xi_n + \sin(2\xi_n)} \tag{4.31}$$

and ξ_n are the roots of Eq. (4.32)

$$\xi_n \tan\xi_n = N_{Bi} \tag{4.32}$$

and the parameter C_n and root of the equation ξ_n are functions of the Biot number of the system subjected to the unsteady-state heat transfer.

However, it may not always be practical to evaluate the infinite series to determine the temperature as a function of time and location. Fortunately, as the time increases the equation converges quickly due to rapid exponential decay in the equation. Usually for $\tau \geq 0.2$, the infinite series can be truncated to a single term with a reasonable degree of accuracy. Hence Eq. (4.30) may be rewritten as

$$\theta = \frac{T_{x=x,t} - T_\infty}{T_i - T_\infty} \approx C_1 \exp(-\xi_1^2 \tau)\cos(\xi_1\psi) \tag{4.33}$$

where

$$C_1 = \frac{4\sin\xi_1}{2\xi_1 + \sin(2\xi_1)} \tag{4.34}$$

$$\xi_1 \tan\xi_1 = N_{Bi} \tag{4.35}$$

For a special case (i.e., $x = 0$), the above equation will reduce to

$$\theta_0 = \frac{T_{x=0,t} - T_\infty}{T_i - T_\infty} \approx C_1 \exp(-\xi_1^2 \tau) \tag{4.36}$$

Further, combining the above equations

$$\theta = \theta_0 \cos(\xi_1\psi) \tag{4.37}$$

Knowing the values of ψ, C_1, and ξ_1 will allow us to solve Eq. (4.33) and determine the temperature and time. Noting that C_1 and ξ_1 are Biot number dependent, their values are calculated for various Biot numbers and tabulated in Tables 4.1A and 4.1B.

Biot Number	Rectangular		Spherical		Cylindrical	
(N_{Bi})	C_1	ξ_1	C_1	ξ_1	C_1	ξ_1
0.1	1.0161	0.3111	1.0298	0.5423	1.0246	0.4417
0.2	1.0311	0.4328	1.0592	0.7593	1.0483	0.6170
0.3	1.0450	0.5218	1.0880	0.9208	1.0712	0.7465
0.5	1.0701	0.6533	1.1441	1.1656	1.1143	0.9408
0.7	1.0918	0.7506	1.1978	1.3525	1.1539	1.0873
0.8	1.1016	0.7910	1.2236	1.4320	1.1724	1.1490
1.0	1.1191	0.8603	1.2732	1.5708	1.2071	1.2558

TABLE 4.1A Values of Parameters C_1 and ξ_1 for One-Term Approximated Analytical Solution ($\tau > 0.2$) of Unsteady-State Heat Transfer for Standard Configurations (Cengel & Ghajar, Heat and Mass Transfer: Fundamentals and Applications, Fifth Edition. McGraw Hill, 2016. Reprinted with permission.)

Biot Number	Rectangular		Spherical		Cylindrical	
(N_{Bi})	C_1	ξ_1	C_1	ξ_1	C_1	ξ_1
2.0	1.1785	1.0769	1.4793	2.0288	1.3384	1.5995
3.0	1.2102	1.1925	1.6227	2.2889	1.4191	1.7887
4.0	1.2287	1.2646	1.7202	2.4556	1.4698	1.9081
5.0	1.2403	1.3138	1.7870	2.5708	1.5029	1.9898
6.0	1.2479	1.3496	1.8338	2.6537	1.5283	2.0490
7.0	1.2532	1.3766	1.8673	2.7165	1.5411	2.0937
8.0	1.2570	1.3978	1.8920	2.7654	1.5526	2.1286
9.0	1.2598	1.4149	1.9106	2.8044	1.5611	2.1566
10.0	1.2620	1.4289	1.9249	2.8363	1.5677	2.1795
20.0	1.2699	1.4961	1.9781	2.9857	1.5919	2.2880
30.0	1.2717	1.5202	1.9898	3.0372	1.5973	2.3261
40.0	1.2723	1.5325	1.9942	3.0632	1.5993	2.3455
50.0	1.2727	1.5400	1.9962	3.0788	1.6002	2.3572
∞	1.2732	1.5708	2.0000	3.1416	1.6021	2.4048

TABLE 4.1A   Values of Parameters C_1 and ξ_1 for One-Term Approximated Analytical Solution ($\tau > 0.2$) of Unsteady-State Heat Transfer for Standard Configurations (Cengel & Ghajar, Heat and Mass Transfer: Fundamentals and Applications, Fifth Edition. McGraw Hill, 2016. Reprinted with permission.) (*Continued*)

	Zeroth Order	First Order
ξ_1	$J_0(\xi_1)$	$J_1(\xi_1)$
0	1	0
0.25	0.984	0.124
0.5	0.938	0.2423
0.75	0.864	0.3492
1	0.765	0.4401
1.25	0.646	0.5106
1.5	0.512	0.5579
1.75	0.369	0.5802
2	0.224	0.5767
2.25	0.083	0.5484
2.5	−0.05	0.4971
2.75	−0.16	0.426
3	−0.26	0.3391
3.25	−0.33	0.2411

TABLE 4.1B   Values of Zeroth and First-Order Bessel Functions for Solving Transient Heat Conduction in a Cylindrical Section Using the One-Term Approximated Analytical Solution ($\tau > 0.2$)

Example 4.3 (Bioprocessing Engineering—Thermal processing of a long and thin meat slab): A long and thin slab of meat of 4-cm thick at 25°C is thermally processed in a hot air environment of 150°C with a heat transfer coefficient of 50 $\frac{W}{m^2 \cdot K}$. The thermophysical properties of the meat are given as thermal conductivity, $k = 0.5 \frac{W}{m \cdot K}$; density, $\rho = 1010 \frac{kg}{m^3}$; specific heat, $C_p = 4000 \frac{J}{kg \cdot K}$. Determine (a) the surface temperature of the slab after 15 minutes of heating and (b) heating time for the meat to reach a well-done temperature at its center (i.e., center temperature ≈ 72°C).

Step 1: Schematic
The problem is depicted in Fig. 4.7.

Step 2: Given data
Thermal conductivity, $k = 0.5 \frac{W}{m \cdot K}$; Heat transfer coefficient, $h = 50 \frac{W}{m^2 \cdot K}$; Density, $\rho = 1010 \frac{kg}{m^3}$;

Specific heat, $C_p = 4000 \frac{J}{kg \cdot K}$; Thickness, $2L = 4$ cm; Initial temperature, $T_i = 25°C$; Temperature of the environment, $T_\infty = 150°C$; Time, $t = 15$ min; Required center temperature, $T_0 = 72°C$

To be determined: (a) Temperature of the surface of the meat slab after 15 minutes of heating and (b) time of processing for the slab to reach a center temperature of 72°C.

Assumptions: The slab is large such that the thickness is negligible relative to width and length.

Step 3: Analysis
As a first step, we will determine the value of the Biot number and verify the validity of the lumped approach. If the Biot number is less than or equal to 0.1, we will directly use the lumped model to estimate the temperature of the slab. If the Biot number is greater than 0.1, we will employ the one-step approximation of the solution to the heat equation.

Step 4: Calculations
(a) Surface temperature of the slab after 15 minutes of heating

The Biot number can be calculated using

$$N_{Bi} = \frac{hL_c}{k} \tag{E4.3A}$$

To assess the validity of the Biot number, the characteristic length, $L_c = \frac{V}{A_S}$

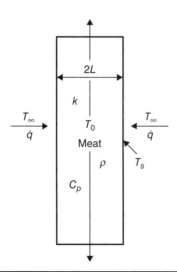

FIGURE 4.7 Schematic of thin and long meat slab being heated via hot air.

Because this is a large slab, $A_S \approx 2 \times$ length $\times$ width (ignoring the area associated with the thickness dimension) and $V =$ length $\times$ width $\times$ thickness.

Therefore,

$$L_c = \frac{\text{length} \times \text{width} \times \text{thickness}}{2 \times \text{length} \times \text{width}} = \frac{\text{length} \times \text{width} \times 2L}{2 \times \text{length} \times \text{width}} = L = 2 \text{ cm} = 0.02 \text{ m}$$

Hence,

$$N_{\text{Bi}} = \frac{hL_c}{k} = \frac{50 \times 0.02}{0.5} = 2$$

Since $N_{\text{Bi}} > 0.1$, the lumped approach is not valid. We will, therefore, use the one-step approximation approach.

The equation for the one-step approximation for a rectangular slab is

$$\theta = \frac{T_{x=x,t} - T_\infty}{T_i - T_\infty} \approx C_1 \exp\left(-\xi_1^2 \tau\right) \cos\left(\xi_1 \psi\right) \qquad \text{(E4.3B)}$$

where

$$C_1 = \frac{4 \sin \xi_1}{2\xi_1 + \sin(2\xi_1)} \qquad \text{(E4.3C)}$$

$$\xi_1 \tan \xi_1 = N_{\text{Bi}} \qquad \text{(E4.3D)}$$

The Biot number is given by

$$N_{\text{Bi}} = \frac{hL_c}{k} \qquad \text{(E4.3E)}$$

The characteristic length, L_c for a slab is half the thickness and therefore 2 cm

$$N_{\text{Bi}} = \frac{hL_c}{k} = \frac{50 \times 0.02}{0.5} = 2$$

For a Biot number of 2, the values of the constants are as follows:

$$\xi_n = 1.07$$
$$C_1 = 1.17$$

We will calculate the thermal diffusivity of the slab as

$$\alpha = \frac{k}{\rho C_p} = \frac{0.5}{(1010 \times 4000)} = 1.23 \times 10^{-7} \frac{\text{m}^2}{\text{s}}$$

From which the dimensionless time parameter, that is, the Fourier number,

$$N_{\text{Fo}} = \tau = \frac{\alpha t}{L_c^2} = \frac{1.23 \times 10^{-7} \times 15 \times 60}{0.02^2} = 0.278$$

And for the surface of the slab, the dimensionless space parameter

$$\psi = \frac{x}{L} = \frac{2}{2} = 1$$

Now,

$$\theta = \frac{T_{x=x,t} - T_\infty}{T_i - T_\infty} \approx 1.17 \times \exp(-1.07^2 \times 0.278)\cos(1.07 \times 1)$$

$$\theta = \frac{T_{x=x,t} - T_\infty}{T_i - T_\infty} \approx 0.40$$

From which, $T_s = T_{x,t} = 0.4 \times (25 - 150) + 150$

The surface temperature of the slab after 15 minutes of thermal processing is approximately 99.6°C.

Configuration	Solution	Constant	Root
Rectangle	$\theta = \sum\limits_{n=1}^{\infty} C_n \exp(-\xi_n^2 \tau) \cos(\xi_n \psi)$	$C_n = \dfrac{4\sin\xi_n}{2\xi_n + \sin(2\xi_n)}$	$N_{Bi} = \xi_n \tan\xi_n$
Sphere	$\theta = \sum\limits_{n=1}^{\infty} C_n \exp(-\xi_n^2 \tau) \dfrac{\sin(\xi_n \psi)}{\xi_n \psi}$	$C_n = \dfrac{4(\sin\xi_n - \xi_n \cos\xi_n)}{2\xi_n - \sin(2\xi_n)}$	$N_{Bi} = 1 - \xi_n \cot\xi_n$
Cylinder	$\theta = \sum\limits_{n=1}^{\infty} C_n \exp(-\xi_n^2 \tau) J_0(\xi_n \psi)$	$C_n = \dfrac{2}{\xi_n}\left[\dfrac{J_1(\xi_n)}{J_0^2(\xi_n) + J_1^2(\xi_n)}\right]$	$N_{Bi} = \xi_n \dfrac{J_1(\xi_n)}{J_0(\xi_n)}$

Note: For spherical and rectangular objects, $\psi = \dfrac{r}{R}$.

TABLE 4.2   Approximate Solution for One-Dimensional Unsteady-State Heat Transfer for Standard Geometries (Cengel & Ghajar, Heat and Mass Transfer: Fundamentals and Applications, Fifth Edition. McGraw Hill, 2016. Reprinted with permission.)

(b) The process time needed for the center temperature to reach 72°C

The temperature at the center of the slab is given by

$$\theta_0 = \frac{T_{0,t} - T_\infty}{T_i - T_\infty} \approx C_1 \exp(-\xi_1^2 \tau) \tag{E4.3F}$$

Substituting the relevant values and solving for the Fourier number, τ

$$\frac{72 - 150}{25 - 150} \approx 1.17 \exp(-(1.07^2)\tau) \tag{E4.3G}$$

$$\tau = 0.55$$

The process time can be calculated as

$$\tau = 0.55 = \frac{\alpha t}{L_c^2} = \frac{1.23 \times 10^{-7} \times t}{0.02^2}$$

The process time is approximately 1794.9 seconds or 29.9 minutes.

The same mathematical treatment can be extended to spherical and cylindrical bodies to obtain the temperature at any point on the surface and within the body at any instant. For completeness, the formulae of all the three standard configurations are summarized in Table 4.2. It is to be noted that the characteristic lengths for an infinite wall, cylinder, and a sphere are defined as half the wall thickness, radius of the cylinder, and the radius of the sphere, respectively.

4.8   Graphical Approach

As seen in the previous section, the approximate analytical solutions provide answers to one-dimensional heat transfer problems under unsteady conditions. Considering the heavy involvement of nondimensional parameters in the analytical approach, it became practical to transform the approximate solutions into graphical plots that can simplify the process further. These plots were prepared by M. P. Heisler in 1947 which included curves relating the Biot number, Fourier number, and temperature ratio to determine the spatial temperature at a given time. In 1961, H. Gröber supplemented these charts with additional plots relating the Fourier and Biot numbers with heat transfer ratios the body is subjected to. These charts are popularly called Heisler–Gröber plots and are

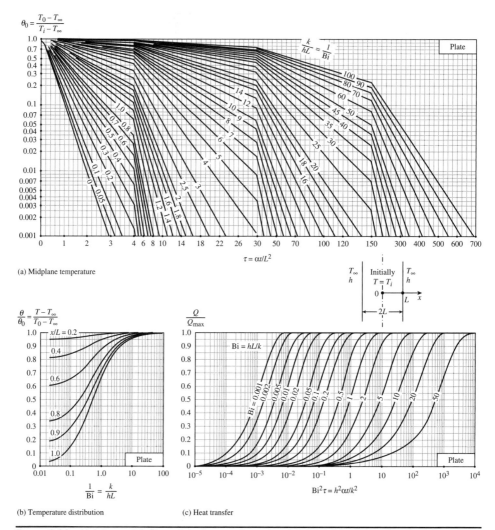

$\theta_0 = \dfrac{T_0 - T_\infty}{T_i - T_\infty}$

(a) Midplane temperature

$\tau = \alpha t / L^2$

$\dfrac{\theta}{\theta_0} = \dfrac{T - T_\infty}{T_0 - T_\infty}$

(b) Temperature distribution

$\dfrac{1}{\text{Bi}} = \dfrac{k}{hL}$

$\dfrac{Q}{Q_{max}}$

(c) Heat transfer

$\text{Bi}^2 \tau = h^2 \alpha t / k^2$

Figure 4.8A Heisler and Gröber plots for a rectangular slab (plate) to determine the (a) center temperature (T_o), (b) temperature distribution (ψ), and (c) the heat transfer (Q). (Image courtesy of McGraw Hill. Cengel & Ghajar, Heat and Mass Transfer: Fundamentals and Applications, Fifth Edition. 2016. Reprinted with permission.)

shown in Figs. 4.8A, B, and C for an infinite wall, cylinder, and sphere, respectively. For the graphical approach also, the characteristic length for an infinite wall, cylinder, and a sphere are defined as half the wall thickness, radius of the cylinder, and the radius of the sphere, respectively.

4.9   Procedure for Using Heisler–Gröber Plots for Solving One-Dimensional Unsteady-State Heat Transfer Problems

To illustrate the practicality and simplicity of Heisler–Gröber plots, let us use an example of a cylindrical-shaped body (e.g., a can of soup) that is to be heated from an initial

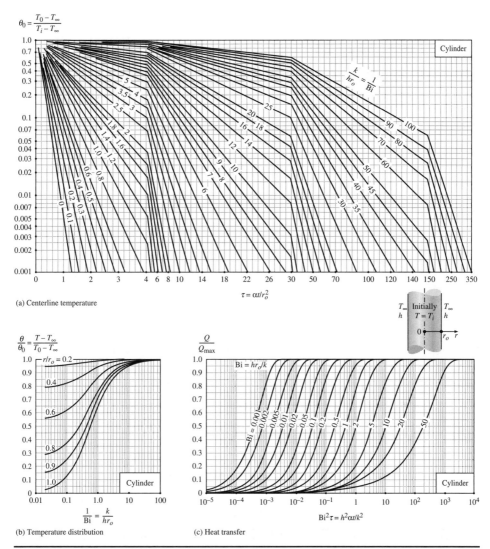

$$\theta_0 = \frac{T_0 - T_\infty}{T_i - T_\infty}$$

(a) Centerline temperature

$$\tau = \alpha t / r_o^2$$

$$\frac{\theta}{\theta_0} = \frac{T - T_\infty}{T_0 - T_\infty}$$

(b) Temperature distribution

$$\frac{1}{Bi} = \frac{k}{hr_o}$$

(c) Heat transfer

$$Bi^2\tau = h^2\alpha t/k^2$$

FIGURE 4.8B Heisler and Gröber plots for a cylinder to determine the (a) center temperature (T_o), (b) temperature distribution (ψ), and (c) the heat transfer (Q). (Image courtesy of McGraw Hill. Cengel & Ghajar, Heat and Mass Transfer: Fundamentals and Applications, Fifth Edition. 2016. Reprinted with permission.)

temperature T_i until the center temperature reaches T_0 by exposing the body to a heating environment T_∞. If the physical properties (dimensions, k and h) are known, we are interested to know how long the heating take should place. The following steps outline the process of calculating the heating time.

Step 1: Calculate the inverse of the Biot number, $1/N_{Bi}$.

Step 2: Estimate the temperature ratio, θ_0.

Step 3: Draw a line that intersects θ_0 and $1/N_{Bi}$ as shown in Fig. 4.9.

Step 4: Read the value of the Fourier number (τ) and calculate the time using $\tau = \frac{\alpha t}{L_c^2}$.

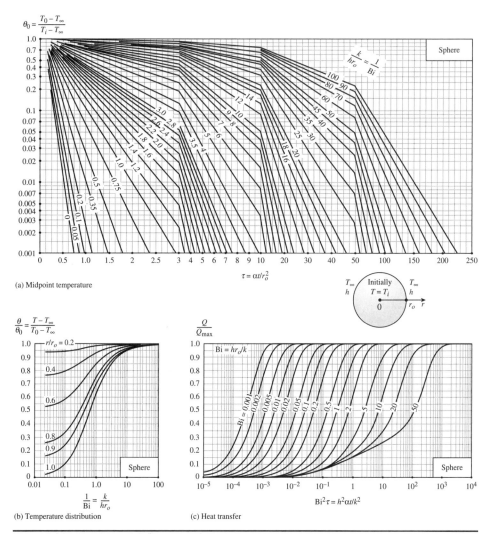

FIGURE 4.8C Heisler and Gröber plots for a sphere to determine the (a) center temperature (T_o), (b) temperature distribution ψ), and (c) the heat transfer (Q). (Image courtesy of McGraw Hill. Cengel & Ghajar, Heat and Mass Transfer: Fundamentals and Applications, Fifth Edition. 2016. Reprinted with permission.)

The following examples demonstrate the applications of graphical approach to solve one-dimensional unsteady-state conduction problems for all three configurations.

Example 4.5 (Bioprocessing Engineering—Heat transfer during chilling of sausage): A sausage considered as a long cylindrical-shaped body of 3-cm radius at 75°C is exposed to a forced stream of cold air at −10°C. What should be the heat transfer coefficient to cool the center of the sausage to 25°C within 60 minutes if the density, thermal conductivity, and specific heat of the sausage were measured to be

$$1070\,\frac{\text{kg}}{\text{m}^3},\ 0.3\,\frac{\text{W}}{\text{m·K}},\ \text{and}\ 3300\,\frac{\text{J}}{\text{kg·K}},\ \text{respectively?}$$

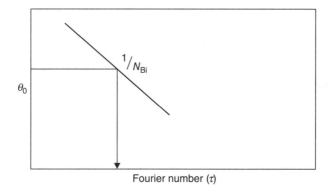

FIGURE 4.9 Illustration of solving unsteady-state conduction problems via graphical (Heisler–Gröber plots) method.

Step 1: Schematic
The problem is described in Fig. 4.10.

Step 2: Given data
Thermal conductivity, $k = 0.3 \ \dfrac{\text{W}}{\text{m·K}}$; Density, $\rho = 1070 \ \dfrac{\text{kg}}{\text{m}^3}$; Specific heat, $C_p = 3300 \ \dfrac{\text{J}}{\text{kg·K}}$; Radius, $R = 3$ cm; Initial temperature, $T_i = 75°C$; Temperature of the environment, $T_\infty = -10°C$; Time, $t = 60$ min; Required center temperature, $T_0 = 25°C$

To be determined: Heat transfer coefficient for cold air flow.

Assumptions: The cylinder is long enough, and heat transfer occurs in one direction (radial) only.

Analysis: We cannot calculate the Biot number because we were not given the value of the heat transfer coefficient. Therefore, we will assume that the Bio number is greater than 0.1 and proceed to determine the Biot number and the heat transfer coefficient. Subsequently, we will verify the validity of our approach by determining the Biot number.

Step 3: Calculations
We will calculate the thermal diffusivity of the sausage as

$$\alpha = \frac{k}{\rho C_p} = \frac{0.3}{(1070 \times 3300)} = 8.49 \times 10^{-8} \ \frac{\text{m}^2}{\text{s}}$$

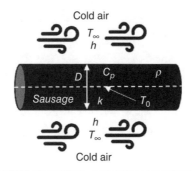

FIGURE 4.10 Schematic for Example 4.5 describing chilling of a sausage.

From which the dimensionless time parameter, that is, the Fourier number is,

$$N_{Fo} = \tau = \frac{\alpha t}{L_c^2} = \frac{8.49 \times 10^{-8} \times 60 \times 60}{0.03^2} = 0.34$$

The temperature ratio at the center of the sausage is given by

$$\theta_0 = \frac{T_{r=0,t} - T_\infty}{T_i - T_\infty} = \frac{25 - (-10)}{75 - (-10)} = 0.41$$

Using the values of $\tau = 0.34$ and $\theta_0 = 0.41$, we can read the value of $\dfrac{1}{N_{Bi}}$ from the chart in Fig. 4.8B (a) as

$$\frac{1}{N_{Bi}} = 0.2$$

or

$$\frac{k}{hL_c} = 0.2 = \frac{0.3}{h \times 0.03}$$

$$h = 50 \ \frac{W}{m^2 \cdot K}$$

The required heat transfer coefficient for the sausage to reach a center temperature of 25°C within 60 minutes is $50 \ \dfrac{W}{m^2 \cdot K}$.

To verify the validity of our approach, we will check if the Biot number is less than 0.1.

$$N_{Bi} = \frac{hL_c}{k} = \frac{50 \times 0.03}{0.3} = 5 \gg 0.1$$

Example 4.6 (Bioprocessing Engineering—Thermal processing of a long and thin meat slab):   In Example 4.3, we used a one-step approximation to determine the time required for the center temperature to reach 72°C. Now, rework the same example problem using the graphical approach.

Step 1: Schematic
The figure described in Example 4.3 (Fig. 4.7) is appropriate to this example problem.

Step 2: Given data
Thermal conductivity, $k = 0.5 \ \dfrac{W}{m \cdot K}$; Heat transfer coefficient, $h = 50 \ \dfrac{W}{m^2 \cdot K}$; Density, $\rho = 1010 \ \dfrac{kg}{m^3}$; Specific heat, $C_p = 4000 \ \dfrac{J}{kg \cdot K}$; Thickness, $2L = 4$ cm; Initial temperature, $T_i = 25°C$; Temperature of the environment, $T_\infty = 150°C$; Time, $t = 15$ min; Required center temperature, $T_0 = 72°C$; Thermal diffusivity, $\alpha = \dfrac{k}{\rho C_p} = \dfrac{0.5}{1010 \times 4000} = 1.23 \times 10^{-7} \ \dfrac{m^2}{s}$

To be determined: (a) Temperature of the surface of the meat slab after 15 minutes of heating and (b) time of processing for the slab to reach a center temperature of 72°C.

Assumptions: The slab is large such that the thickness is negligible relative to width and length.

Step 3: Calculations
The Biot number can be calculated using

$$N_{Bi} = \frac{hL_c}{k} \tag{E4.6A}$$

To assess the validity of the Biot number, the characteristic length,

$$L_c = \frac{V}{A_S} \tag{E4.6B}$$

Because this is a large slab, $A_S \approx 2 \times$ length $\times$ width (ignoring the area associated with the thickness dimension) and $V =$ length $\times$ width $\times$ thickness

Therefore,

$$L_c = \frac{\text{length} \times \text{width} \times \text{thickness}}{2 \times \text{length} \times \text{width}} = \frac{\text{length} \times \text{width} \times 2L}{2 \times \text{length} \times \text{width}} = L = 2 \text{ cm} = 0.02 \text{ m}$$

Hence,

$$N_{Bi} = \frac{hL_c}{k} = \frac{50 \times 0.02}{0.5} = 2$$

Since $N_{Bi} > 0.1$, the lumped approach is not valid. Therefore, we will employ the graphical approach.
We will first determine the temperature ratio at the center of the slab as

$$\theta_0 = \frac{T_{x=0,t} - T_\infty}{T_i - T_\infty} = \frac{(72 - 150)}{(25 - 150)} = 0.62$$

Then we will determine the value of N_{Fo} from the Heisler plot for the slab (Fig. 4.8A (a)) using $\theta_0 = 0.62$ and $\dfrac{1}{N_{Bi}} = 0.5$ as

$$N_{Fo} = 0.55$$

Therefore, the cooling time is calculated as

$$N_{Fo} = 0.55 = \frac{\alpha t}{L_c^2} = \frac{1.23 \times 10^{-7} \times t}{0.02^2}$$

$$t = 1777.6 \text{ s} = 29.6 \text{ min}$$

The minor differences between the answers for Examples 4.3 and 4.6 are due to the visual error in reading the plots.

Example 4.7 (Agricultural Engineering—Cooling of a concrete sphere):   A 10-cm diameter concrete sphere (density = 2400 kg/m³, thermal conductivity = 1.2 W/m·K, and specific heat = 1000 J/kg·°C) at 125°C is taken out of the kiln and cooled in flow of air at 25°C ($h = 100$ W/m²·°C). Estimate the temperature inside the sphere at a location 3 cm from the surface of the sphere after 20 minutes of cooling. In addition, determine the heat transfer from the sphere to the air.

Step 1: Schematic
The problem is depicted in Fig. 4.11.

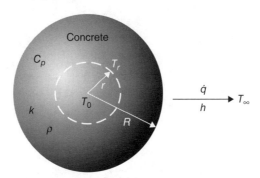

FIGURE 4.11   Schematic of the hot concrete sphere being cooled in air described in Example 4.7.

Step 2: Given data

Thermal conductivity, $k = 1.2 \frac{W}{m \cdot K}$; Heat transfer coefficient, $h = 100 \frac{W}{m^2 \cdot K}$; Density, $\rho = 2400 \frac{kg}{m^3}$;

Specific heat, $C_p = 1000 \frac{J}{kg \cdot K}$; Radius, $r = 5$ cm; Initial temperature, $T_i = 125°C$; Temperature of the environment, $T_\infty = 25°C$; Time, $t = 20$ min

To be determined: (a) Temperature at 3 cm from the outer surface of the sphere after 20 minutes of cooling and (b) heat transfer from sphere to air.

Assumptions: The heat transfer occurs radially.

Step 3: Analysis

As a first step, we will determine the value of the Biot number and verify the validity of the lumped approach. If the Biot number is less than or equal to 0.1, we will use the lumped model to estimate the temperature of the sphere. If the Biot number is greater than 0.1, we will employ the graphical approach to solve for the surface temperature and heat transfer.

Step 4: Calculations

The calculation of the Biot number using the characteristic length, $L_c = \frac{V}{A_s}$ yields

$$L_c = \frac{V}{A_S} = \frac{\frac{4}{3}\pi r^3}{4\pi r^2} = \frac{r}{3} = \frac{0.05}{3} = 0.016 \text{ m}$$

$$N_{Bi} = \frac{hL_c}{k} = \frac{100 \times 0.016}{1.2} = 1.38 > 0.1$$

The lumped approach is not valid, and we will therefore use the graphical method.

(a) Temperature at 3 cm from the outer surface of the sphere after 20 minutes of cooling

The Biot number for the system (using $L_c = r$) is

$$N_{Bi} = \frac{hL_c}{k} = \frac{100 \times 0.05}{1.2} = 4.17$$

And

$$\frac{1}{N_{Bi}} = \frac{1}{4.17} = 0.24$$

The Fourier number,

$$N_{Fo} = \tau = \frac{\alpha t}{L_c^2} = \frac{5 \times 10^{-7} \times 20 \times 60}{0.05^2} = 0.24$$

Because,

$$\alpha = \frac{k}{\rho C_p} = \frac{1.2}{(2400 \times 1000)} = 5 \times 10^{-7} \frac{m^2}{s}$$

Using the values of $\frac{1}{N_{Bi}} = 0.24$ and $\tau = 0.24$, we obtain the value of $\theta_0 = 0.37$ (see Fig. 4.8C (a))

$$\theta_0 = 0.37 = \frac{T_0 - T_\infty}{T_i - T_\infty} = \frac{T_0 - 25}{125 - 25}$$

After solving, we obtain the temperature at the center as

$$T_0 = 62°C$$

However, we are interested in the temperature at a location 3 cm from the surface of the sphere (i.e., $r = 2$ cm from the center). To determine that temperature, we will use the second plot (see Fig. 4.8C (b)) that provides the temperature ratio as a function of location. For the given problem,

$$\frac{r}{R} = \frac{0.02}{0.05} = 0.4$$

From the plot, the value of $\dfrac{\theta}{\theta_0}$ for $\dfrac{r}{R} = 0.4$ and $\dfrac{1}{N_{Bi}} = 0.24$ can be read as

$$\frac{\theta}{\theta_0} = 0.84$$

Hence,

$$\theta_0 \times \frac{\theta}{\theta_0} = 0.37 \times 0.84 = 0.31 = \left[\frac{T_{r=0,t} - T_\infty}{T_i - T_\infty}\right] \times \left[\frac{T_{r,t} - T_\infty}{T_{r=0,t} - T_\infty}\right] = \left[\frac{T_{r,t} - T_\infty}{T_i - T_\infty}\right]$$

$$0.31 = \frac{T_{r=0.02\,m,\,25\,min} - 25}{125 - 25}$$

$$T_{r=0.02\,m,\,25\,min} = 56°C$$

Note: Because we had already determined the center temperature as 62°C, we can directly use $\dfrac{\theta}{\theta_0} = 0.84 = \left[\dfrac{T_{r,t} - T_\infty}{T_{r=0,t} - T_\infty}\right]$ and calculate the temperature at a location 3 cm from the surface of the sphere.

(b) Heat transfer from sphere to air
To determine the amount of heat transfer, we will make use of the third plot (Fig. 4.8C (c)) that provides the heat transfer ratio as a function of $N_{Bi}^2 \tau$ and N_{Bi}. Therefore, for $N_{Bi} = 4.17$ and $N_{Bi}^2 \tau = 4.17^2 \times 0.24 = 4.17$, we obtain

$$\frac{q}{q_{max}} \approx 0.7$$

We calculate $q_{max} = mC_p\Delta T = (\rho V C_p \Delta T) = \left(2400 \times \dfrac{4}{3} \times 3.14 \times 0.05^3\right) \times 1000 \times (125 - 25) = 125{,}600$ J
Heat transfer from the hot sphere to the air,

$$q = 0.7 \times 125{,}600 = 87{,}920 \text{ J}$$

4.10   One-Dimensional Unsteady-State Heat Transfer in Semi-Infinite Bodies

Sometimes, we would like to determine the temperature near the surface of a semi-infinite body. In this context, a semi-infinite body is assumed to have an infinitely large plane surface and infinite depth (e.g., a football field). As usual, assuming constant thermal properties and no heat generation within the body, we can write the one-dimensional heat equation in the familiar form as

$$\frac{\partial^2 T}{\partial x^2} = \frac{1}{\alpha}\frac{\partial T}{\partial t} \tag{4.38}$$

With the initial and boundary conditions as follows:

Initial condition: At the beginning of the heat transfer process ($t = 0$), the temperature at any point in the x-direction, the temperature is T_i

$$T_{x,0} = T_i \tag{4.39}$$

Boundary condition #1: At the surface of the body at any time t, the surface temperature will remain constant at T_s

$$T_{0,t} = T_s \tag{4.40}$$

Boundary condition #2: The temperature far away from the surface ($x = \infty$), at any time t, the temperature will remain constant at T_i

$$T_{\infty,t} = T_i \tag{4.41}$$

The above partial differential equation along with the initial and boundary conditions can be solved by converting the partial differential equation into an ordinary differential equation by defining a single special variable called similarity variable that describes both x and t as follows:

$$\zeta = \frac{x}{2\sqrt{\alpha t}} \tag{4.42}$$

After substituting the expression for ζ in Eq. (4.38) and employing the separation of variables approach, the solution to the differential equation will result in

$$\frac{T_{x,t} - T_s}{T_i - T_s} = \text{erf}(\zeta) \tag{4.43}$$

or in an expanded form

$$\frac{T_{x,t} - T_s}{T_i - T_s} = \text{erf}\left(\frac{x}{2\sqrt{\alpha t}}\right) \tag{4.44}$$

Sometimes, the above equation is also expressed as

$$\frac{T_{x,t} - T_s}{T_i - T_s} = 1 - \text{erfc}\left(\frac{x}{2\sqrt{\alpha t}}\right) \tag{4.45}$$

The values of the error function and the complementary error function are summarized in Fig. 4.12 and Table 4.3.

Example 4.8 (Bioprocessing Engineering): A large rectangular carton of meat at 5°C is being frozen in a plate freezer. The plate on the top of the slab is maintained at a constant temperature of –25°C. The processing time is considered adequate when the temperature reaches –2°C at a section 2 cm from the surface of the slab. As the floor manager, you just received a report that the slabs from the previous shift were processed for 4 hours by maintaining a constant surface temperature of –25°C. Would you accept the product to be shipped for consumption? For your analysis, assume that the properties of the meat are the same as those of water and the packaging material offers negligible resistance to heat transfer.

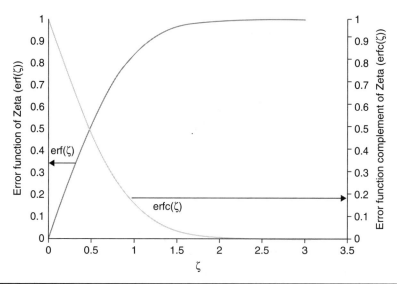

FIGURE 4.12   Plot of error function and its complement for the similarity variable, ζ.

Zeta	erf (ζ)	erfc (ζ)
0	0	1
0.1	0.11246292	0.88753708
0.2	0.22270259	0.77729741
0.3	0.32862676	0.67137324
0.4	0.42839236	0.57160764
0.5	0.52049988	0.47950012
0.6	0.60385609	0.39614391
0.7	0.67780119	0.32219881
0.8	0.74210096	0.25789904
0.9	0.79690821	0.20309179
1	0.84270079	0.15729921
1.1	0.88020507	0.11979493
1.2	0.91031398	0.08968602
1.3	0.93400794	0.06599206
1.4	0.95228512	0.04771488
1.5	0.96610515	0.03389485
1.6	0.97634838	0.02365162
1.7	0.98379046	0.01620954
1.8	0.9890905	0.0109095
1.9	0.99279043	0.00720957

TABLE 4.3   Numerical Values of Error Function and Its Complement for the Similarity Variable, ζ

Zeta	erf (ζ)	erfc (ζ)
2	0.99532227	0.00467773
2.1	0.99702053	0.00297947
2.2	0.99813715	0.00186285
2.3	0.99885682	0.00114318
2.4	0.99931149	0.00068851
2.5	0.99959305	0.00040695
2.6	0.99976397	0.00023603
2.7	0.99986567	0.00013433
2.8	0.99992499	7.5013E-05
2.9	0.9999589	4.1098E-05
3	0.99997791	2.209E-05

TABLE 4.3 Numerical Values of Error Function and Its Complement for the Similarity Variable, ζ (*Continued*)

Step 1: Schematic
The problem is described in Fig. 4.13.

Step 2: Given data
Thermal conductivity, $k = 0.6 \ \dfrac{W}{m \cdot K}$; Heat transfer coefficient, $h = 100 \ \dfrac{W}{m^2 \cdot K}$; Density, $\rho = 1000 \ \dfrac{kg}{m^3}$;

Specific heat, $C_p = 4200 \ \dfrac{J}{kg \cdot K}$; Depth, $x = 2$ cm; Initial temperature, $T_i = 5°C$; Surface temperature of the carton, $T_s = -25°C$; Time, $t = 4$ h.

To be determined: Freezing time needed for reaching a temperature of $-2°C$ at a depth of 2 cm from the surface.

Assumptions: The properties of the meat are the same as water, and carton offers not resistance to heat transfer.

Step 3: Calculations
Thermal diffusivity, $\alpha = \dfrac{k}{\rho C_p} = \dfrac{0.6}{1000 \times 4200} = 1.42 \times 10^{-7} \ \dfrac{m^2}{s}$

The temperature ratio,

$$\frac{T_{x,t} - T_s}{T_i - T_s} = \text{erf}\left(\frac{x}{2\sqrt{\alpha t}}\right) \qquad \text{(E4.8A)}$$

$$\frac{-2 - (-25)}{5 - (-25)} = \text{erf}\left(\frac{0.02}{2\sqrt{1.42 \times 10^{-7} t}}\right) \qquad \text{(E4.8B)}$$

$$0.77 = \text{erf}\left(\frac{0.02}{2\sqrt{1.42 \times 10^{-7} t}}\right) \qquad \text{(E4.8C)}$$

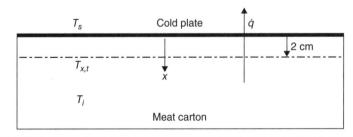

FIGURE 4.13 Schematic of meat carton being frozen.

From Table 4.3, the value of the similarity variable for an error function for 0.77 is 0.85. Therefore,

$$\frac{0.02}{2\sqrt{1.42 \times 10^{-7} t}} = 0.85 \qquad \text{(E4.8D)}$$

$$t = 968 \text{ s}$$

The freezing time should be at least 968 seconds. Therefore, the product is suitable for shipping.

4.11   Two-Dimensional Unsteady-State Heat Transfer in Finite-Sized Bodies

In some cases, we may have to deal with objects that are of finite shape such as short cylinders and short slabs. The principles covered in the treatment of one-dimensional problems can be extended to one-dimensional problems using the concept of superposition. To illustrate this idea, let us consider an infinitely long cylinder and an infinitely long slab of the same material of a known thermal conductivity (k). Now the perpendicular intersection of the infinitely long cylinder (radius = R) and an infinitely long slab (thickness = $2L$) can be visualized as a short cylinder of length $2L$ and radius R (Fig. 4.14).

In its current form, this short cylinder is now subjected to two-dimensional heat transfer: along the length direction and the radial direction. If we assume that there is no heat generation in the short cylinder and is exposed to the same temperature (T_∞) and heat transfer coefficient (h), the nondimensional temperature ratio for the short

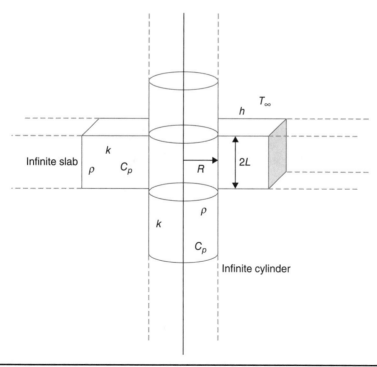

Figure 4.14   Heat transfer from a short cylinder may be visualized as a combination of an infinite cylinder perpendicularly interesting with an infinite slab.

cylinder can be obtained by superimposing the equations for long slab and cylinder. Mathematically,

$$\theta_{\text{short cyl}} = \theta_{\text{long cyl}} \times \theta_{\text{long slab}} \tag{4.46}$$

$$\left[\frac{T_{x,r,t} - T_\infty}{T_i - T_\infty}\right]_{\text{short cyl}} = \left[\frac{T_{r,t} - T_\infty}{T_i - T_\infty}\right]_{\text{long cyl}} \times \left[\frac{T_{x,t} - T_\infty}{T_i - T_\infty}\right]_{\text{long slab}} \tag{4.47}$$

Similarly, the solution to a finite-size rectangular section may be expressed as a perpendicular superposition of two infinitely long rectangular slabs (Fig. 4.14).

$$\left[\frac{T_{x,t} - T_\infty}{T_i - T_\infty}\right]_{\text{short slab}} = \left[\frac{T_{x,t} - T_\infty}{T_i - T_\infty}\right]_{\text{long slab}} \times \left[\frac{T_{y,t} - T_\infty}{T_i - T_\infty}\right]_{\text{long slab}} \tag{4.48}$$

Example 4.9 (Bioprocessing Engineering—Freezing of biocomposite):   A short cylindrical biocomposite (4-cm diameter and 10-cm length) of thermal conductivity of $0.7 \dfrac{\text{W}}{\text{m·°C}}$ at 15°C is cooled for 28 minutes using a cold stream of a fluid (−10°C) with a heat transfer coefficient of $35 \dfrac{\text{W}}{\text{m}^2\text{·°C}}$. Determine the temperature at the center and the surface of the cylindrical biocomposite if the density and specific heat are given as $1100 \dfrac{\text{kg}}{\text{m}^3}$ and $2700 \dfrac{\text{J}}{\text{kg·°C}}$, respectively.

Step 1: Schematic
The problem is depicted in Fig. 4.15.

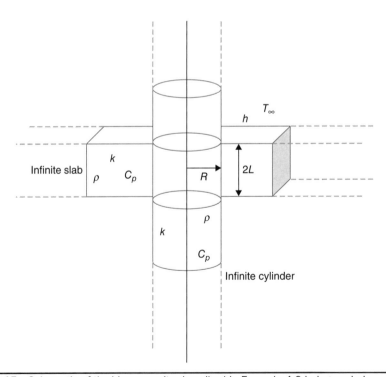

Figure 4.15   Schematic of the biocomposite described in Example 4.9 being cooled.

Step 2: Given data

Short cylinder (Length, $2L = 10$ cm and Radius, $R = 2$ cm); Thermal conductivity, $k = 0.7 \dfrac{W}{m \cdot K}$;

Heat transfer coefficient, $h = 35 \dfrac{W}{m^2 \cdot K}$; Density, $\rho = 1100 \dfrac{kg}{m^3}$; Specific heat, $C_p = 2700 \dfrac{J}{kg \cdot K}$; Initial

temperature, $T_i = 15°C$; Temperature of the environment, $T_\infty = -10°C$; Cooling time, $t = 28$ min

To be determined: Temperature at the center and the surface of the cylinder after 28 minutes of cooling.

Assumptions: There is no heat generation, and the short cylinder has constant thermal conductivity and is exposed to the same temperature and heat transfer coefficient so that the superposition principle is applicable.

Step 3: Analysis

The short cylinder may be viewed as a combination of long cylinder of radius, $R = 2$ cm and along rectangular slab of thickness, $2L = 10$ cm. Therefore, the temperature at the center of the short cylinder is the same as the center of the long slab and long cylinder. Similarly, the temperature of the top surface of the cylinder is the same as the temperature of the surface of the wall and at the center of the cylinder (Fig. 4.15).

Step 4: Calculations
Long cylinder
The Biot number for a cylindrical-shaped body is given by

$$N_{Bi} = \frac{hL_c}{k} = \frac{35 \times 0.02}{0.7} = 1.0$$

The Fourier number is calculated as

$$\tau = \frac{\alpha t}{L_c^2} = \frac{2.35 \times 10^{-7} \times 28 \times 60}{0.02^2} = 0.98$$

Since

$$\alpha = \frac{k}{\rho C_p} = \frac{0.7}{1100 \times 2700} = 2.35 \times 10^{-7} \frac{m^2}{s}$$

From the Heisler plot for the cylinder, the temperature ratio for $\dfrac{1}{N_{Bi}} = 1$ and $\tau = 0.98$ can be approximated as 0.25.

$$\theta_{cyl,center} = \frac{T_0 - T_\infty}{T_i - T_\infty} = 0.25$$

Similarly, for the long slab,

$$N_{Bi} = \frac{hL_c}{k} = \frac{35 \times 0.05}{0.7} = 2.5$$

$$\tau = \frac{\alpha t}{L_c^2} = \frac{2.35 \times 10^{-7} \times 28 \times 60}{0.05^2} = 0.158$$

From the Heisler plot for a rectangular slab, the temperature ratio for $\dfrac{1}{N_{Bi}} = 0.4$ and $\tau = 0.158$ can be approximated as 0.97

$$\theta_{slab,center} = \frac{T_0 - T_\infty}{T_i - T_\infty} \approx 0.97$$

Now, the temperature at the center of the short cylinder can be determined via the superposition principle

$$\theta_{short\ cyl,center} = \theta_{cyl,center} \times \theta_{slab,center}$$

$$\left[\frac{T_{0,0,t}-T_\infty}{T_i-T_\infty}\right]_{\text{short cyl}} = \left[\frac{T_{0,t}-T_\infty}{T_i-T_\infty}\right]_{\text{long cyl}} \times \left[\frac{T_{0,t}-T_\infty}{T_i-T_\infty}\right]_{\text{long slab}} \tag{E4.9A}$$

$$\left[\frac{T_{0,0,t}-T_\infty}{T_i-T_\infty}\right]_{\text{short cyl}} \approx 0.25 \times 0.97 = 0.246 \tag{E4.9B}$$

$$\left[\frac{T_{0,0,t}-(-10)}{15-(-10)}\right]_{\text{short cyl}} = 0.246 \tag{E4.9C}$$

$$T_{0,0,t} = -3.8°C$$

The temperature at the center of the biocomposite cylinder after 28 minutes is −3.8°C.

The temperature at the upper surface of the cylinder is the same as that of the top surface ($x = L$) of the slab ($\psi = \dfrac{x}{L} = 1$). Therefore, the temperature ratio for $\dfrac{1}{N_{\text{Bi}}} = 0.4$, $\tau = 0.158$, and $\dfrac{x}{L} = 1$ is about 0.4. That is,

$$\theta_{\text{slab}(L,t)} = \frac{T_{L,t}-T_\infty}{T_0-T_\infty} \approx 0.40 \tag{E4.9D}$$

And by superposition the temperature ratio at the top of the short cylinder is given by

$$\left[\frac{T_{L,0,t}-T_\infty}{T_i-T_\infty}\right]_{\text{short cyl}} = \theta_{\text{cyl,center}} \times \theta_{\text{slab,surface}} \tag{E4.9E}$$

$$\left[\frac{T_{L,0,t}-T_\infty}{T_i-T_\infty}\right]_{\text{short cyl}} = 0.25 \times 0.40 = 0.1 \tag{E4.9F}$$

Hence,

$$\frac{T_{L,0,t}-(-10)}{15-(-10)} = 0.1$$

The temperature at the center and on the upper surface of the short cylinder is −3.8°C and −7.5°C, respectively.

Practice Problems for the FE Exam

1. For a given thermal conductivity k, heat transfer coefficient h, and the characteristic length L_c, the mathematical expression of the Biot number is:

   a. $\dfrac{hL_c}{k}$.

   b. $\dfrac{kL_c}{h}$.

   c. $\dfrac{hk}{L_c}$.

   d. $\dfrac{L_c}{hk}$.

2. The thermal diffusivity of biodiesel whose thermal conductivity, density, and specific heat are 0.16 W/m·°C, 890 kg/m³, and 1700 J/kg·°C, respectively is nearly:

   a. $1.06 \times 10^{-6} \dfrac{m^2}{s}$.

   b. $2 \times 10^{-7} \dfrac{m^2}{s}$.

   c. $1.06 \times 10^{-7} \dfrac{m^2}{s}$.

   d. $1.60 \times 10^{-7} \dfrac{m^2}{s}$.

3. A small Biot number ($N_{Bi} \leq 0.1$) indicates that:

   a. Convection resistance is negligible.

   b. Conduction resistance is negligible.

   c. Both conduction and convection resistances are negligible.

   d. None of the above.

4. A small copper (density = 8950 kg/m³ and specific heat = 380 J/kg·°C) cube of dimensions 1 cm × 1 cm × 1 cm initially at a room temperature of 27°C is exposed to a hot environment maintained at 200°C with a heat transfer coefficient of 25 W/m²·°C for 4 minutes. Assuming a small Biot number, the temperature of the cube at its center will be:

   a. 117°C.

   b. 200°C.

   c. 110°C.

   d. 98°C.

5. A body with a characteristic length of 0.05 m and a thermal conductivity of 50 W/m·K is subjected to convection heat transfer. The limiting heat transfer coefficient required for the lumped system model to be applicable is:

   a. $27 \dfrac{W}{m^2 \cdot °C}$.

   b. $35 \dfrac{W}{m^2 \cdot °C}$.

   c. $100 \dfrac{W}{m^2 \cdot °C}$.

   d. None of the above.

6. A 1-cm diameter copper ($k = 405 \dfrac{W}{m \cdot K}$) sphere at 110°C is cooled suddenly in ice water at 0°C with a heat transfer coefficient of 35 $\dfrac{W}{m^2 \cdot °C}$. Given that the density and specific heat of copper as 9000 $\dfrac{kg}{m^3}$ and 390 $\dfrac{J}{kg \cdot °C}$, respectively, the temperature of the center of the sphere after 8 minutes of cooling is closest to:

    a. 17 $\dfrac{W}{m^2 \cdot °C}$.

    b. 47 $\dfrac{W}{m^2 \cdot °C}$.

    c. 33 $\dfrac{W}{m^2 \cdot °C}$.

    d. None of the above.

7. Potato granules (density = 700 $\dfrac{kg}{m^3}$ and specific heat = 3200 $\dfrac{J}{kg \cdot °C}$) assumed to be 5-mm cubes of thermal conductivity of 0.6 W/m·K at 25°C is heated in a hot air environment ($h = 10 \dfrac{W}{m^2 \cdot °C}$) maintained at 135°C. The time needed for the surface of the cubes to reach a temperature of 80°C is approximately:

    a. 819 s.

    b. 780 s.

    c. 121 s.

    d. None of the above.

8. The Biot number for a cylindrical 2.5-cm diameter and 15-cm long hot dog (initial temperature = 27°C) with thermal conductivity of 0.7 W/m·K placed in a cold-water bath (5°C) with a heat transfer coefficient of 60 $\dfrac{W}{m^2 \cdot °C}$ is:

    a. 0.05.

    b. 1.7.

    c. 0.53.

    d. None of the above.

9. A solid aluminum cylinder ($k = 100 \dfrac{W}{m \cdot K}$; $\rho = 2700 \dfrac{kg}{m^3}$; $C_p = 850 \dfrac{J}{kg \cdot °C}$) of dimensions, diameter = 2.5 cm and length 20 cm at 27°C is heated in a furnace maintained at 410°C for 4 minutes. If the heat transfer coefficient inside the furnace is 55 $\dfrac{W}{m^2 \cdot °C}$, the temperature of the cylinder at a location 1 cm from the center is nearly:

    a. 257°C.

    b. 315°C.

    c. 351°C.

    d. None of the above.

10. A large fish (assumed to be flat surface) initially at 15°C was covered with a layer of ice at –2°C for 25 minutes. If the properties of fish are assumed to be similar to that of water (density = 1000 kg/m^3, specific heat = 4200 J/kg·°C, thermal conductivity = 0.6 W/m·°C), the temperature of the fish 1 cm from the surface is nearly:

   a. 7°C.

   b. 0°C.

   c. –2°C.

   d. 5°C.

Practice Problems for the PE Exam

1. It is proposed to convert poultry litter ($k = 0.21\dfrac{\text{W}}{\text{m·°C}}$, $C_p = 1800\dfrac{\text{J}}{\text{kg}}°\text{C}$, $\rho = 700\dfrac{\text{kg}}{\text{m}^3}$) into biochar. Pelletized poultry litter samples of 1-cm diameter and 5 cm in length at an initial temperature of 21°C are placed in a furnace with a convective heat transfer coefficient of 7 W/m^2·°C. The carbonization of the poultry litter is considered complete if the center of the pellet is exposed to a temperature of 400°C for 12 minutes. Considering the physical properties of the poultry litter, at what temperature should the furnace operate?

2. A slab of meat (10 cm × 5 cm × 1 cm) initially at 7°C is placed in a freezer that was maintained at –30°C. If the convective heat transfer coefficient within the freezer is 5 W/m^2·°C, determine the temperature of the surface and the center of the meat slab after 60 minutes of cooling assuming the properties of the meat are the same as those of the water.

3. A hot iron (density = 7900 kg/m^3, thermal conductivity = 80 W/m·°C, and specific heat = 450 J/kg·°C) cylinder at 200°C is suddenly dropped in a tank of cold water at 15°C. If the temperature of the cylinder surface was measured to be 50°C after 4 minutes, determine the heat transfer coefficient between the cylinder and the water.

4. Wood chips approximated as 6-cm diameter spheres are being gasified at 800°C in a reactor in which the heat transfer coefficient is calculated as 55 W/m^2·°C. If the initial temperature of the wood chips is 25°C, determine the temperature of the wood chip at its surface, center, and 1 cm from the surface after 22 minutes of gasification assuming the thermal conductivity, density, and specific heat are given as 0.15 W/m·°C, 400 kg/m^3, and 2200 J/kg·°C, respectively.

5. A long and thin slab of meat of 7.5-cm thick at 8°C is placed in an environment of 180°C with a heat transfer coefficient of 95 $\dfrac{\text{W}}{\text{m}^2\text{·K}}$. The thermophysical properties of the meat may be assumed to be the same as those of the water. Using the one-term approximated analytical method, determine (a) the surface temperature of the slab after 20 minutes of heating and (b) heating time for the meat to reach a center temperature of 80°C.

6. Repeat Problem 4 using the graphical method and determine (a) the surface temperature of the slab after 20 minutes of heating and (b) heating time for the meat to reach a center temperature of 80°C. Additionally, determine the corresponding amount of energy transferred per kilogram of meat after 20 minutes of heating.

7. Repeat Problem 2 using the graphical method and compare the answers.

8. A long sausage (3-cm diameter and 30-cm long) at 95°C is placed in an agitated cold-water tank maintained at 10°C ($h = 300 \text{ W/m}^2 \cdot °\text{C}$) for 8 minutes. The thermal conductivity, specific heat, and the density of the sausage are given as $0.35 \dfrac{\text{W}}{\text{m} \cdot \text{K}}$, $3400 \dfrac{\text{J}}{\text{kg} \cdot \text{K}}$, and $1100 \dfrac{\text{kg}}{\text{m}^3}$, respectively. Determine the surface temperature of the sausage.

9. A carbon composite initially at 40°C was covered with a layer of ice at 0°C for 17 minutes. What will be the temperature of the composite 0.5 cm from the surface if the density, thermal conductivity, and specific heat are given as 900 kg/m^3, $0.68 \text{ W/m} \cdot °\text{C}$, and $710 \text{ J/kg} \cdot °\text{C}$, respectively.

10. A short cylindrical biocomposite (3-cm diameter and 5-cm length) of thermal conductivity of $0.55 \dfrac{\text{W}}{\text{m} \cdot °\text{C}}$ is cooled for 30 minutes using a stream of a fluid at −10°C with a heat transfer coefficient of $50 \dfrac{\text{W}}{\text{m}^2 \cdot °\text{C}}$. Determine the temperature at the center and the surface of the cylindrical biocomposite if the density and specific heat are given as $1200 \dfrac{\text{kg}}{\text{m}^3}$ and $2900 \dfrac{\text{J}}{\text{kg} \cdot °\text{C}}$, respectively.

CHAPTER 5

Fundamentals of Convection Heat Transfer

Chapter Objectives

After studying this chapter, students will be able to

- Understand the physical concept of convection heat transfer.
- Interpret the meaning of various dimensional parameters used in convection heat transfer.
- Use the correlations to determine the heat transfer coefficients for free and forced convection.

5.1 Motivation

Until now, we dealt with heat transfer within the solids that occurred via conduction. However, several practical situations involve conduction heat transfer via the bulk motion of a fluid around a solid. For example, imagine freezing of freshly harvested meat in a cold storage. Due to the circulation of the cold air, the heat is transferred out of the slab, and the slab eventually freezes. Our goal, in all such cases, is to determine the rate of heat transfer by the cold air from the slab by combining the principles of fluid mechanics with conduction heat transfer. Therefore, this chapter will provide physical and mathematical background required to solve most common problems encountered in convection heat transfer.

5.2 The Concept of Convection Heat Transfer

In Chapter 1, we learned that an ice-cold soda can, when placed in front of a running fan, will become warm quickly (Fig. 5.1a). Now, we will try to understand the processes involved in the warming of the soda can. As a first step, the flowing air (at higher temperature) molecules will encounter the wall of the (cold) soda can. Subsequently, conduction heat transfer occurs between the hot air molecules and cold metal surface

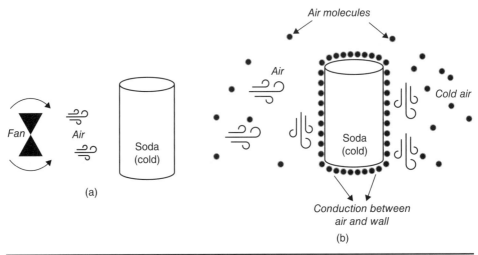

FIGURE 5.1   Illustration of convection heat transfer between a cold soda can and the air (a) due to combined effects of conduction and bulk motion of air (b).

resulting in soda can's surface getting slightly warmer and air molecules getting slightly colder, and the heat from the wall of the soda can is transferred to the cold soda. Next, the air molecules move away due to the flow and the next batch of molecules will repeat the process and until a thermal equilibrium is reached (Fig. 5.1b). This type of heat transfer that involves conduction accompanied by a flow of a fluid is called *convection heat transfer*.

5.3   Quantifying Convection Heat Transfer

In our day-to-day experiences, we are familiar that increasing the airflow rate around a hot object will cool it faster. Experimental investigations indicated that the rate of heat transfer via convection was found to depend on several variables, including the temperature gradient, the properties of the fluid involved in heat transfer such as viscosity, velocity, density, specific heat and the thermal conductivity, geometry, and even the configuration of the solid surface including its exposed surface area, roughness, and its orientation in relation to the fluid flow. In other words,

$$\dot{q}_{conv} = f(\Delta T, \mu, v, \rho, C_p, k, A, \text{roughness, geometry, configuration, fluid flow pattern})$$

By now you would have figured out the mathematical complexity of convection heat transfer due to the dependence of several variables. In engineering practice, however, convection heat transfer is expressed in its simplest form

$$\dot{q}_{conv} = hA\Delta T \tag{5.1}$$

where h is defined as the heat transfer coefficient $\left(\dfrac{W}{m^2 \cdot °C}\right)$ and is a strong function of several variables.

$$h = f(\mu, \rho, v, C_p, k, \text{roughness, orientation, fluid flow pattern}) \tag{5.2}$$

Equation (5.1) is called Newton's law of cooling because of the similarity of the concept that was originally proposed by Issac Newton in early 1700s.

The temperature gradient and the area of the solid body are easy to measure and Eq. (5.1) becomes simple, only if we know the value of h. However, we must realize that determining the values of h analytically is extremely difficult. Therefore, engineers rely on experimental data to determine the values of h for standard situations and extend them to solve problems that mimic the standard situations.

5.4   Nusselt Number

One straightforward way to quantify the extent of convection heat transfer is use a nondimensional parameter called the Nusselt number, named after Prof. Wilhelm Nusselt, an outstanding mechanical engineer from Germany, who developed the concept of dimensionless heat transfer. It will be entirely appropriate to use an example to understand the concept of Nusselt number and its versatility. Let us imagine two long parallel metal plates separated by stagnant air of conductivity k and thickness L (Fig. 5.2a). Now we will heat the left plate and cool the right plate such that the left plate is constantly maintained at T_h (say 90°C) while the right plate is maintained at T_c (say 5°C). We can easily see that the heat will transfer from the left plate to the right plate through the stagnant layer of air. Additionally, because the air is completely stagnant (assuming that a stagnant air layer behaves like a solid), the heat transfer is via conduction (Fig. 5.2a). Therefore, the conduction flux (when area = 1 m²) is expressed as

$$\dot{q}''_{\text{cond}} = \frac{k(T_h - T_c)}{L} \tag{5.3}$$

Now, we disturb the air column such that the air begins to mix vigorously. At this stage, convection heat transfer sets in due to the bulk motion of the air (Fig. 5.2b) whose convection flux (again assuming an area of 1 m²) is written as

$$\dot{q}''_{\text{conv}} = h(T_h - T_c) \tag{5.4}$$

The **Nusselt number**, thus, is defined as the ratio of convection heat transfer to conduction heat transfer. Hence,

$$N_{Nu} = \frac{\dot{q}''_{\text{conv}}}{\dot{q}''_{\text{cond}}} = \frac{h(T_h - T_c)}{\left(\dfrac{k(T_h - T_c)}{L}\right)} = \frac{hL}{k} \tag{5.5}$$

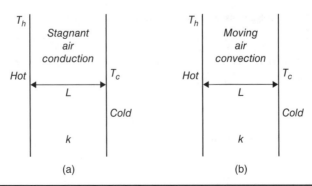

$$\text{(a)} \qquad\qquad\qquad \text{(b)}$$

FIGURE 5.2   Heat transfer between two plates separated by stagnant air occurs via conduction (a) while by moving air occurs via convection (b).

Typically, we use a characteristic length L_c to represent the dimension of any given body undergoing heat transfer. Therefore, the general expression for Nusselt number is

$$N_{Nu} = \frac{hL_c}{k} \tag{5.6}$$

5.5 Physical Meaning of Nusselt Number

The Nusselt number is an indicator of the extent of increase in the heat transfer due to the flow of the fluid (i.e., convection) when compared to stagnant fluid (i.e., conduction). For example, if we determine that the Nusselt number to be 125, we interpret that the bulk flow of the fluid had increased the heat transfer rate by 125 times (compared to no fluid flow). Therefore, if we want to cool the birds in a poultry barn in peak summer, we prefer a higher Nusselt number (airflow rates). On the other hand, if we want to conserve the heat, lower Nusselt numbers are desired.

5.6 Nusselt Number Versus Biot Number

At first glance, the mathematical form of Nusselt number (Eq. 5.6) looks identical to the Biot number that we defined, discussed, and used in Chapter 4. If we recall, we defined Biot number as

$$N_{Bi} = \frac{hL_c}{k} \tag{5.7}$$

which looks very similar to the Nusselt number that was defined in the previous section

$$N_{Nu} = \frac{hL_c}{k} \tag{5.8}$$

However, we must note that the Biot number calculation involves the thermal conductivity of the solid undergoing the heat transfer, whereas the Nusselt number uses the thermal conductivity of the fluid involved in the convection heat transfer process. Therefore, despite appearing to have similar formulation, both of these numbers describe different phenomena and hence should be interpreted differently. The following example will clarify the difference.

Example 5.1 (General Engineering–All concentrations): A hot metal sphere of 2-cm diameter at 90°C is immersed in a bucket of cold water at 4°C with a heat transfer coefficient of $120 \frac{W}{m^2 \cdot °C}$. If the thermal conductivities of the metal cylinder and the water are $50 \frac{W}{m \cdot °C}$ and $0.6 \frac{W}{m \cdot °C}$, respectively, determine the Biot and Nusselt numbers and explain their physical significance.

$$N_{Bi} = \frac{hL_c}{k} = \frac{hL_c}{k_{metal}} = \frac{120 \times 0.02}{50} = 0.048 \tag{E5.1A}$$

Physical significance: The Biot number of 0.048 (which is less than 0.1) suggests that the temperature gradient within the metal sphere is negligible and the entire sphere is at the same temperature at any given time.

$$N_{Nu} = \frac{hL_c}{k} = \frac{hL_c}{k_{water}} = \frac{120 \times 0.02}{0.6} = 4 \tag{E5.1B}$$

Physical significance: The Nusselt number of 4 informs us that the movement of cold water around the metal ball increased the heat transfer rate by 400% relative to pure conduction alone (stagnant water).

5.7   Relationship with Fluid Mechanics

Because convection always involves fluid flow, we will study the convection heat transfer in light of fluid mechanics. From fluid mechanics, we recall that when the fluid is moving, the fluid molecules can either move in an orderly fashion (like a marching of army soldiers) or completely in a random and chaotic motion. When the molecules travel in a smooth and orderly manner, we can visualize the fluid element to be consisting of layers (or lamina) of fluid moving one over the other. Such motion of the fluid is called laminar flow. On the other hand, when the molecules are sufficiently disturbed, the layers are broken and the fluid undergoes intense and random mixing resulting in what is called as turbulent flow (we will discuss these concepts in detail in Chapter 8). The intermediate stage, that is, the stage between the laminar and turbulent flow, during which the fluid erratically fluctuates between laminar and turbulent flow is called the transition flow and is generally avoided by engineers.

Through systematic experimentation, Osborne Reynolds, a famous Irish engineer, demonstrated that the velocity of the flow, density and viscosity of the fluid, and the characteristic length of the system (although Reynolds used diameter as the characteristic length for a circular pipe) together will determine whether laminar or transition or turbulent flow condition prevails in a given system. Subsequently, these observations resulted into a nondimensional parameter that we famously call Reynolds number.

$$N_{Re} = \frac{\text{Inertial forces}}{\text{Viscous forces}} \tag{5.9}$$

Mathematically,

$$N_{Re} = \frac{\rho v L_c}{\mu} \tag{5.10}$$

where $\rho \left(\dfrac{\text{kg}}{\text{m}^3} \right)$ is the density of the fluid, $v \left(\dfrac{\text{m}}{\text{s}} \right)$ is the mean velocity of the fluid, $L_c (\text{m})$ is the allowable travel distance of the molecules or characteristic length of the conduit, and $\mu \, (\text{Pa} \cdot \text{s})$ is the viscosity of the fluid.

5.8   Physical Meaning of Reynolds Number

Reynolds number is an indicator of the extent of chaos or randomness present in the fluid. The inertial force (the numerator, $\rho V L_c$) pushes the fluid forward toward the state of randomness while the viscous forces (denominator, μ) resists the movement. Fluids with low densities, travelling in smaller diameter pipes at lower velocities cannot create randomness strong enough to overcome the viscous effects of the fluid. Such situations will result in a smaller Reynolds number and such flow is called laminar flow. However, fluids flowing at high velocities in large diameter pipes will easily overcome resistance offered by the viscosity of the fluid and still create random motion among the fluid molecules resulting in high Reynolds number and is therefore aptly termed as turbulent flow.

Configuration	Laminar	Turbulent
Flow around a sphere	<10	>200,000
Flow past a cylinder	<400,000	>400,000
Flow within a cylinder	<2100	>4000
Flow in a packed bed cylinder	<10	>2000
Stirred cylinder	<10	>10,000
Open channel flow	500	>2000
Flow around a flat plate	<300,000	>500,000

*Note:** While calculating the Reynolds number one must exercise caution to choose an appropriate characteristic length.

TABLE 5.1   Range of Reynolds Numbers for Various Geometries*

In the original Reynolds experiments, when the Reynolds number was less than about 2100, the fluid behaved in an orderly manner and therefore was considered as the upper limit for the laminar flow. However, beyond a value of about 4000 the fluid began to experience randomness and hence became the minimum value to attain turbulent flow.

For circular pipes, the range of Reynolds numbers is given by

$$N_{Re} < 2100, \text{ the flow is laminar.} \tag{5.11}$$

$$2100 < N_{Re} < 4000, \text{ the flow is transitional.} \tag{5.12}$$

$$N_{Re} > 4000, \text{ the flow is turbulent.} \tag{5.13}$$

Similarly, the range of Reynolds numbers for other geometries is summarized in Table 5.1.

5.9   Boundary Layer Formation in Convection Heat Transfer

Continuing our discussion on fluid flow, we will now focus on the boundary layers which form when a fluid flows over a solid surface. Let us imagine a hot fluid at a temperature, T_∞ with a viscosity, μ flowing over a cold solid plate maintained at temperature of 0°C with a velocity, v_∞. From our knowledge of fluid mechanics, we know that the viscosity causes the velocity of the fluid very close to the plate to decrease significantly relative to v_∞. In fact, the fluid comes to a complete rest (no slip) at the surface of the plate ($y = 0$) and the velocity gradually increases to almost free stream velocity, $v_\infty (y = \delta_v)$ (Fig. 5.3a). It is this layer of the fluid, wherein the velocity of the fluid is affected by the viscosity that is called the velocity boundary layer (VBL). The thickness of the VBL (δ_v) is defined as the vertical distance from the surface of the plate to a specific location in the fluid layer at which the fluid velocity attains 99% of the free stream velocity. In other words, within the VBL, the viscosity dominates causing velocity gradient. Beyond the boundary, however, the effect of viscosity is negligible, and no velocity gradient exists. The concept of boundary later was originally proposed by Ludwig Prandtl, the famous German scientist who pioneered research in fluid dynamics.

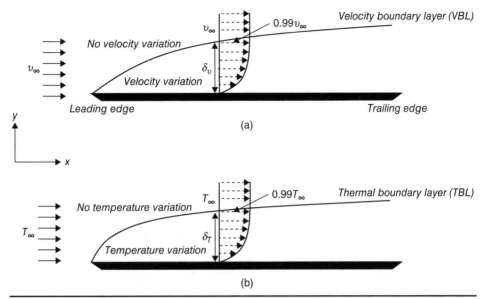

FIGURE 5.3 Illustration of velocity boundary and thermal boundary layers over a flat plate.

Similarly, due to the temperature gradient between the hot fluid and the cold plate ($T_\infty - 0°C$), another boundary layer called thermal boundary layer (TBL) also develops (Fig. 5.3b). The fluid's temperature at the plate's surface will be 0°C ($y = 0$) and will gradually increase almost to T_∞ ($y = \delta_T$). The TBL is the layer of the fluid within which a temperature gradient can exist. Beyond the TBL, the temperature of the fluid is identical to T_∞ (the free stream temperature). Similar to the VBL, the thickness of the TBL (δ_T) is the distance between the surface of the plate where the temperature is 0°C and the location along the y-axis where the temperature reaches 99% of the free stream temperature.

5.10 Prandtl Number (N_{Pr})

The thickness of the VBL relative to TBL is a dimensionless parameter called Prandtl number, named in honor of Ludwig Prandtl.

Mathematically,

$$N_{Pr} = \frac{\delta_v}{\delta_T} = \frac{\text{Momentum diffusivity}}{\text{Thermal diffusivity}} \tag{5.14}$$

$$N_{Pr} = \frac{v}{\alpha} \tag{5.15}$$

Sometimes, the Prandtl number described above is simplified as

$$N_{Pr} = \frac{\dfrac{\mu}{\rho}}{\dfrac{k}{\rho C_p}} = \frac{\mu C_p}{k} \tag{5.16}$$

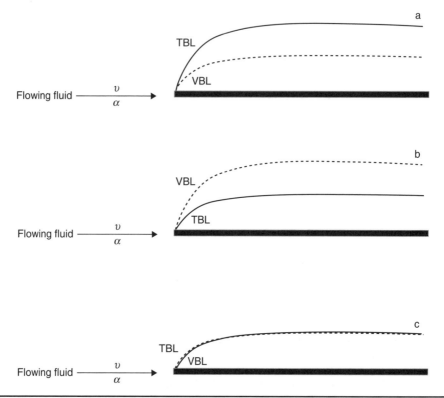

Figure 5.4   Illustration of relative thicknesses of velocity and thermal boundary layers during the flow of hot fluid over a cold plate.

5.11   Physical Meaning of Prandtl Number

The Prandtl number is to be viewed as a measure of how fast will heat diffuse through the fluid when compared to the momentum. When we encounter low Prandtl numbers (<0.10) (e.g., liquid metals), the heat will diffuse quickly when compared to the momentum (Fig. 5.4a).

On the other extreme, when the VBL is significantly thicker than the TBL, the dissipation of momentum is faster than the temperature resulting in larger values of Prandtl number (>100) (e.g., oils) (Fig. 5.4b). In the intermediate range (1–20), lie the fluids such as water and gases (Fig. 5.4c) that are commonly encountered in biological and agricultural engineering where both the VBL and TBL are of similar orders of magnitude.

5.12   Free and Forced Convection

Convection heat transfer can be broadly categorized into two types depending on the type of the fluid flow. To explain this, let us revisit the soda can example that we began with. If you recall, we introduced a flow of air to warm the can faster. Because we are forcing the air to flow over the soda can's surface, we call this type of heat transfer as forced convection heat transfer or simply *forced convection*. We can readily see examples of forced convection all around us. Our personal computer has a fan inside to transfer the heat from the electronics via forced convection. Other common examples include ceiling fans in our homes, convection ovens, air conditioners, to name a few.

We realize that even if we had not provided any flow around the soda can, heat transfer would still have occurred (albeit at a lower rate) and the cold soda can would have eventually become warmer. The air molecules on the surface of the can transfer the heat to the surface via conduction. This will result in the air molecules becoming colder and denser and will eventually flow down. The next batch of warm air molecules will continue with the heat transfer process causing a subtle downward airflow around the can's surface. Because heat transfer is occurring due to the natural movement of air, we term this as *free or natural convection*. Some of the examples of natural convection include sea breeze, cooling of our cell phones, and melting of ice cream.

5.13 Grashof Number (N_{Gr})

Similar to the Reynolds number in forced flow conditions, the Grashof number serves as an indicator of the flow regime in the context of free convection. To explain this further, we will consider a simple case of free convection transfer in which hot surface is exchanging heat with cold fluid (Fig. 5.5).

When a molecule of a cold fluid is in contact with the hot surface, the molecule will heat up and as a result its density decreases and will move upward due to the buoyancy. However, as we can imagine, the viscous forces of the fluid resist the flow. The Grashof number compares the extent of these opposing forces and provides us with a mathematical metric about the flow regime. Therefore,

$$N_{Gr} = \frac{\text{Buoyancy forces}}{\text{Viscous forces}} \tag{5.17}$$

Mathematically,

$$N_{Gr} = \frac{\rho^2 L_c^3 g \beta \Delta T}{\mu^2} \tag{5.18}$$

where $\rho \left(\frac{\text{kg}}{\text{m}^3} \right)$ is the density of the fluid, L_c (m) is the characteristic length of the system, $g \left(\frac{\text{m}}{\text{s}^2} \right)$ is the acceleration due to gravity, $\beta \left(\frac{1}{\text{°C}} \right)$ is the coefficient of volume expansion, ΔT (°C) is the temperature gradient, and μ (Pa · s) is the viscosity of the fluid.

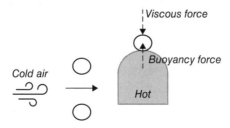

Figure 5.5 Schematic of movement of cold air around a hot body via free convection.

The coefficient of volume expansion, β is mathematically defined as

| β | $\dfrac{1}{T\,(\text{K})}$ for ideal gases | $\dfrac{(\rho_\infty - \rho_s)(T_s - T_\infty)}{\rho_s}$ for other fluids | (5.19) |

where T is the absolute temperature (K), ρ_s and ρ_∞ are the densities of the fluid at the interface and at a location far away from the surface $\left(\dfrac{\text{kg}}{\text{m}^3}\right)$, and T_s and T_∞ are the absolute temperature of the air at the surface and at a location far away from the surface (K).

We must note that forced and free convection can occur simultaneously, especially, during low flow rates (blowing of air to cool coffee). One way to separate out the effects is to use nondimensional ratio as below:

$$\frac{N_{Gr}}{N_{Re}^2} \cong 1 \qquad \text{Mixed convection (both effects are similar)} \qquad (5.20)$$

$$\frac{N_{Gr}}{N_{Re}^2} \gg 1 \qquad \text{Natural convection (inertial forces are small)} \qquad (5.21)$$

$$\frac{N_{Gr}}{N_{Re}^2} \ll 1 \qquad \text{Forced convection (Buoyancy forces are small)} \qquad (5.22)$$

5.14   Determining Heat Transfer Coefficients

One of the major tasks of a practicing engineer in solving convection heat transfer problems is to determine the value of the convection heat transfer coefficient. Once this is accomplished, solving Newton's law of cooling is fairly straightforward. However, we realize that determining the expression analytically is complicated. Therefore, we will employ simpler approaches such as dimensional analysis, very similar to what Nusselt originally proposed to find mathematical relations between the main nondimensional parameters involved in heat transfer. We will begin our discussion with free convection and extended the same ideas to forced convection followed by relevant example problems.

5.15   Heat Transfer Coefficient for Free Convection

The heat transfer coefficient in free convection is dependent on several variables and expressed mathematically as below:

$$h = f(\mu, L_c, k, \beta, g, C_p, \rho, \Delta T) \qquad (5.23)$$

We see that there are a total of nine variables with dimensions (Table 5.2) and four primary dimensions (i.e., Mass (M), Length (L), Time (T), and temperature (θ)). Hence, the nondimensional π terms are $9 - 4 = 5$.

By choosing μ, L_c, k, and β as repeating variables and h, ΔT, ρ, C_p, and g as nonrepeating variables, the nondimensional π terms are expressed as

$$\pi_1 = \mu^a L_c^b k^c \beta^d h \qquad (5.24)$$

$$\pi_2 = \mu^a L_c^b k^c \beta^d \Delta T \qquad (5.25)$$

Variable	Unit	Dimension
Heat transfer coefficient (h)	$\dfrac{W}{m^2 \cdot {}^{\circ}C}$	$MT^{-3}\theta^{-1}$
Density (ρ)	$\dfrac{kg}{m^3}$	ML^{-3}
Viscosity (μ)	$Pa \cdot s$	$ML^{-1}T^{-1}$
Thermal conductivity (k)	$\dfrac{W}{m \cdot {}^{\circ}C}$	$MLT^{-3}\theta^{-1}$
Characteristic length (L_c)	m	L
Specific heat (C_p)	$\dfrac{J}{kg \cdot {}^{\circ}C}$	$L^2T^{-2}\theta^{-1}$
Coefficient of expansion (β)	$\dfrac{1}{{}^{\circ}C}$	θ^{-1}
Gravity (g)	$\dfrac{m}{s^2}$	LT^{-2}
Temperature gradient (ΔT)	${}^{\circ}C$	θ

TABLE 5.2 Variables Affecting Free Convection and Their Dimensions

$$\pi_3 = \mu^a L_c^b k^c \beta^d \rho \tag{5.26}$$

$$\pi_4 = \mu^a L_c^b k^c \beta^d C_p \tag{5.27}$$

$$\pi_5 = \mu^a L_c^b k^c \beta^d g \tag{5.28}$$

As the next step, we will express each π term in its primary dimension

$$\pi_1 = M^0 L^0 T^0 \theta^0 = (ML^{-1}T^{-1})^a (L)^b (MLT^{-3}\theta^{-1})^c (\theta^{-1})^d (MT^{-3}\theta^{-1}) \tag{5.29}$$

Now, equating the exponents will yield the following equations:

$$0 = a + c + 1 \tag{5.30}$$

$$0 = -a + b + c \tag{5.31}$$

$$0 = -a - 3c - 3 \tag{5.32}$$

$$0 = -c - d - 1 \tag{5.33}$$

which after solving will result in

$$a = 0,\, b = 1,\, c = -1,\, \text{and } d = 0 \tag{5.34}$$

After substituting the values of a, b, c, and d in the expression for π_1 we will obtain:

$$\pi_1 = \mu^0 L_c^1 k^{-1} \beta^0 h \tag{5.35}$$

$$\pi_1 = \frac{hL_c}{k} = N_{Nu} \tag{5.36}$$

For the second π term,

$$\pi_2 = \mu^a L_c^b k^c \beta^d \Delta T \tag{5.37}$$

$$\pi_2 = M^0 L^0 T^0 \theta^0 = (ML^{-1}T^{-1})^a (L)^b (MLT^{-3}\theta^{-1})^c (\theta^{-1})^d (\theta) \tag{5.38}$$

Equating the exponents,

$$0 = a + c \tag{5.39}$$

$$0 = -a + b + c \tag{5.40}$$

$$0 = -a - 3c \tag{5.41}$$

$$0 = -c - d + 1 \tag{5.42}$$

And solving them will yield,

$$a = 0, b = 0, c = 0, \text{ and } d = 1 \tag{5.43}$$

After substituting in the expression for π_2

$$\pi_2 = \mu^0 L_c^0 k^0 \beta \Delta T \tag{5.44}$$

Or

$$\pi_2 = \beta \Delta T \tag{5.45}$$

We can repeat these steps for π_3 and obtain π_3 as

$$\pi_3 = \frac{\rho L_c k^{(1/2)}}{\beta^{(1/2)} \mu^{(3/2)}} \tag{5.46}$$

Similarly, for

$$\pi_4 = \mu^a L_c^b k^c \beta^d C_p \tag{5.47}$$

$$\pi_4 = M^0 L^0 T^0 \theta^0 = (ML^{-1}T^{-1})^a (L)^b (MLT^{-3}\theta^{-1})^c (\theta^{-1})^d (L^2 T^{-2}\theta^{-1}) \tag{5.48}$$

Equating the exponents and solving for $a, b, c,$ and d,

$$a = 1, b = 0, c = -1, \text{ and } d = 0 \tag{5.49}$$

and substituting into the expression for π_4

$$\pi_4 = \mu^1 L_c^0 k^{-1} \beta^0 C_p \tag{5.50}$$

$$\pi_4 = \frac{\mu C_p}{k} = \left(\frac{\mu}{\rho}\right)\left(\frac{\rho C_p}{k}\right) = \frac{\nu}{\alpha} = N_{Pr} \tag{5.51}$$

Finally, the fifth π term ($\pi_5 = \mu^a L_c^b k^c \beta^d g$) can also be evaluated using the same procedure outlined in the previous steps and can be expressed as

$$\pi_5 = \mu^1 L_c^1 k^{-1} \beta^1 g \tag{5.52}$$

$$\pi_5 = \frac{\mu L_c \beta g}{k} \tag{5.53}$$

At this time, we can create a new nondimensional, π_6 parameter such that

$$\pi_6 = \pi_2 \pi_3^2 \pi_5 \tag{5.54}$$

Or

$$\pi_6 = (\beta \Delta T) \left(\frac{\rho L_c k^{(1/2)}}{\beta^{(1/2)} \mu^{(3/2)}} \right)^2 \frac{\mu L_c \beta g}{k} \tag{5.55}$$

Or

$$\pi_6 = \frac{\rho^2 \beta L_c^3 \Delta T g}{\mu^2} = N_{Gr} \tag{5.56}$$

Therefore, the π terms can be written as

$$\pi_1 = f(\pi_6, \pi_4) \tag{5.57}$$

And the heat transfer coefficient for free convection is written as

$$N_{Nu} = f(N_{Gr}, N_{Pr}) \tag{5.58}$$

In general terms,

$$N_{Nu} = K N_{Gr}^{n_1} N_{Pr}^{n_2} \tag{5.59}$$

where N, n_1, and n_2 are dimensionless constants to account for the shape and the flow regime within the system.

The product of N_{Gr} and N_{Pr} is called Rayleigh number, N_{Ra} and is expressed as

$$N_{Ra} = \frac{\rho^2 \beta L_c^3 \Delta T g}{\mu^2} \times \frac{\mu}{\rho \alpha} = \frac{\rho \beta L_c^3 \Delta T g}{\mu \alpha} \tag{5.60}$$

Several researchers and engineers have performed numerous experiments with different geometries and determined the values for the constants, N, n_1, and n_2; a few of them will be presented here. For special configurations, the students are encouraged to look up the literature for the correct values of N, n_1, and n_2 for the appropriate flow regime.

5.16 Heat Transfer Coefficients for Free Convection for Flow over a Vertical Plate and Cylinder (Characteristic Length = Length of the Cylinder)

For a plate of length L in a vertical direction, the correlation recommended by Churchill and Chu (1975) can be used.

$$N_{Nu} = \left[0.825 + \frac{0.387 N_{Ra}^{\left(\frac{1}{6}\right)}}{\left(1 + \left(\frac{0.492}{N_{Pr}} \right)^{\left(\frac{9}{16}\right)} \right)^{\frac{8}{27}}} \right]^2 \quad \forall N_{Ra} \tag{5.61}$$

$L_c = L$

Similarly, for a vertical cylinder of length L, the same equation can be used if

$$N_{Nu} = \left[0.825 + \frac{0.387 N_{Ra}^{\left(\frac{1}{6}\right)}}{\left(1+\left(\frac{0.492}{N_{Pr}}\right)^{\left(\frac{9}{16}\right)}\right)^{\frac{8}{27}}} \right]^2 \quad \text{when } D \geq \frac{35L}{N_{Gr}^{0.25}} \qquad (5.62)$$

5.17   Heat Transfer Coefficients for Flow over a Sphere and Cylinder

For a sphere of diameter D, the correlation provided by Churchill is appropriate

$$N_{Nu} = 2 + \frac{0.589(N_{Ra})^{0.25}}{\left\{1+\left[\frac{0.469}{N_{Pr}}\right]^{\left(\frac{9}{16}\right)}\right\}^{\left(\frac{4}{9}\right)}} \qquad (5.63)$$

when $N_{Ra} \leq 10^{11}$ and $N_{Pr} > 0.7$

Similarly, for a cylinder of length L and diameter D, the Nusselt number can be estimated via the following correlation[2]:

$$N_{Nu} = \left[0.6 + \frac{0.387 N_{Ra}^{\left(\frac{1}{6}\right)}}{\left\{1+\left[\frac{0.559}{N_{Pr}}\right]^{\left(\frac{9}{16}\right)}\right\}^{\left(\frac{8}{27}\right)}} \right]^2 \qquad (5.64)$$

when $N_{Ra} \leq 10^{12}$

The horizontal cylinder can also be analyzed via the following correlation by choosing the diameter as the characteristic length (Henderson et al., 1997).

$$N_{Nu} = 0.56 N_{Ra}^{0.25} \qquad \text{when } 10^4 < N_{Ra} < 10^8 \text{ (laminar)} \qquad (5.65)$$

$$N_{Nu} = 0.13 N_{Ra}^{0.33} \qquad \text{when } 10^8 < N_{Ra} < 10^{12} \text{ (turbulent)} \qquad (5.66)$$

Several other correlations are available for various shapes and configurations under laminar and turbulent flow regimes in references 3–5. The following examples will illustrate the use of appropriate correlations in free convection conditions.

Example 5.2 (Agricultural Engineering—Heat loss from a spherical water tank):   A spherical hot water storage tank of 2 m diameter whose surface was measured to be at 95°C was exposed to air at 25°C. What will be the rate of heat loss experienced by the tank and how much will it cost per day if the price of heating is $10 cents/kWh?

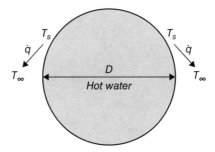

FIGURE 5.6 Schematic of the spherical hot water tank losing heat to its surrounding air.

Step 1: Schematic
The problem is depicted in Fig. 5.6.

Step 2: Given data
Diameter (L_c) of the spherical tank, $D = 2$ m; Temperature of the tank surface, $T_s = 95°C$; Temperature of the surrounding air, $T_\infty = 25°C$; Heating cost $= \dfrac{0.1\$}{\text{kWh}}$

To be determined: Rate of heat loss and its cost per day.

Assumptions: One-dimensional steady-state heat transfer occurs (in radial direction) and thermal conductivities remain constant throughout. The air in the room is not actively circulated and therefore free convection prevails.

Analysis: The heat transfer occurs via free convection whose flow regime will be governed by the Grashof number (and Rayleigh number). If the Rayleigh number is less than 10^{11} and the Prandtl number is equal to or greater than 0.7, we can use the correlation for the sphere for the calculation of the Nusselt number, followed by the heat transfer coefficient, rate of heat transfer, and the cost of this heat loss.

Step 3: Calculations
The temperature of the interface between the air and the tank surface is the average temperature, T_{avg}.

$$T_{avg} = \frac{95 + 25}{2} = 60°C \tag{E5.2A}$$

The physical and thermal properties of air at 60°C are Viscosity, $\mu = 0.00002$ Pa·s, Density, $\rho = 1.06\ \dfrac{\text{kg}}{\text{m}^3}$, Thermal conductivity, $k = 0.028\ \dfrac{\text{W}}{\text{m·°C}}$, Coefficient of volume expansion, $\beta = \dfrac{1}{(273 + 60)} = \dfrac{0.003}{°C}$, Thermal diffusivity, $\alpha = 0.00002674\ \dfrac{\text{m}^2}{\text{s}}$, and Prandtl number, $N_{Pr} = 0.7$.

The Grashof number is determined as below:

$$N_{Gr} = \frac{\rho^2 L_c^3 g \beta \Delta T}{\mu^2} = \frac{1.06^2 \times 2^3 \times 9.81 \times 0.003 \times (95 - 25)}{0.00002^2} \approx 0.46 \times 10^{11} \tag{E5.2B}$$

Rayleigh number will be

$$N_{Ra} = N_{Gr} \times N_{Pr} = 0.46 \times 10^{11} \times 0.7 \approx 0.32 \times 10^{11} \tag{E5.2C}$$

Because $N_{Ra} = 0.32 \times 10^{11}$ and $N_{Pr} = 0.7$, we use the following correlation to determine the Nusselt number

$$N_{Nu} = 2 + \frac{0.589(N_{Ra})^{0.25}}{\left\{1 + \left[\dfrac{0.469}{N_{Pr}}\right]^{\left(\frac{9}{16}\right)}\right\}^{\left(\frac{4}{9}\right)}} \tag{E5.2D}$$

Substituting the values of Rayleigh and Prandtl numbers into the correlation

$$N_{Nu} = 2 + \frac{0.589(0.32 \times 10^{11})^{0.25}}{\left\{ 1 + \left[\frac{0.469}{0.7} \right]^{\left(\frac{9}{16}\right)} \right\}^{\left(\frac{4}{9}\right)}} = 194.58 \tag{E5.2E}$$

The heat transfer coefficient can be determined as

$$N_{Nu} = 194.58 = \frac{hL_c}{k} = \frac{h \times 2}{0.028} \Rightarrow h = 2.72 \frac{W}{m^2 \cdot {}^{\circ}C} \tag{E5.2F}$$

The heat transfer rate can now be easily determined.

$$\dot{q} = hA(T_s - T_\infty) = 2.72 \times 4 \times 3.14 \times 1^2 \times (95 - 25) = 2395 \ W = 2.395 \ kW \tag{E5.2G}$$

The cost of the loss per day is $2.395 \ kW \times 24 \ h \times 0.1 \frac{\$}{kWh} \approx \$5.75$.

Example 5.3 (Agricultural Engineering—Heat loss from a water heater): A water heater of dimensions 0.51-m diameter and 1.5-m long stores hot water for preprocessing sweet potatoes. If the surface temperature of the tank was measured to be 90°C, determine the rate of heat loss to a room maintained at 27°C.

Step 1: Schematic
The problem is depicted in Fig. 5.7.

Step 2: Given data
Length (L_c) of the tank, $L = 1.5$ m; Diameter of the tank, $D = 0.51$ m; Temperature of the tank surface, $T_s = 90$°C; Temperature of the surrounding air, $T_\infty = 27$°C. Note that for a vertical cylinder, the characteristic length is the same as the length of the cylinder ($L_c = L$).

To be determined: Rate of heat transferred to the air.

Assumptions: One-dimensional steady-state heat transfer occurs (in radial direction) and thermal conductivities remain constant throughout. The air in the room is not actively circulated and therefore free convection prevails.

Analysis: The heat transfer occurs via free convection whose flow regime will be governed by the Grashof number (and Rayleigh number). The water tank is a vertical cylinder with a characteristic length of 1.5 m and same correlation for the vertical plate is applicable provided that $D \geq \frac{35L}{N_{Gr}^{0.25}}$.

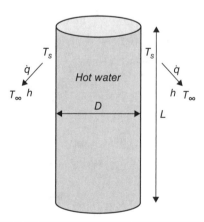

Figure 5.7 Schematic of the hot water tank as described in Example 5.3 for washing sweet potatoes.

Once the Nusselt number is determined the heat transfer can be calculated via the convection governing equation.

Step 3: Calculations

The interface (average) temperature is

$$T_{avg} = \frac{90 + 27}{2} = 58.5°C \qquad \text{(E5.3A)}$$

At 58.5°C, the relevant properties of air are Viscosity, $\mu = 0.00002$ Pa·s, Density, $\rho = 1.07 \frac{kg}{m^3}$, Thermal conductivity, $k = 0.028 \frac{W}{m·°C}$, Coefficient of volume expansion, $\beta = \frac{1}{(273 + 58.5)} = \frac{0.003}{°C}$, Thermal diffusivity, $\alpha = 0.0000267 \frac{m^2}{s}$, and Prandtl number, $N_{Pr} = 0.7$.

With this data, the Grashof number can be calculated as

$$N_{Gr} = \frac{\rho^2 L_c^3 g \beta \Delta T}{\mu^2} = \frac{1.07^2 \times 1.5^3 \times 9.81 \times 0.003 \times (90 - 27)}{0.00002^2} \approx 1.79 \times 10^{10} \qquad \text{(E5.3B)}$$

Now, the Rayleigh number becomes,

$$N_{Ra} = N_{Gr} \times N_{Pr} = 1.8 \times 10^{10} \times 0.7 \approx 1.25 \times 10^{10} \qquad \text{(E5.3C)}$$

We can directly calculate the Rayleigh number as

$$N_{Ra} = \frac{\rho \beta L_c^3 \Delta T g}{\mu \alpha} = \frac{1.07 \times 0.003 \times 1.5^3 \times (90 - 27) \times 9.81}{0.00002 \times 0.0000267} \approx 1.26 \times 10^{10} \qquad \text{(E5.3D)}$$

Now to verify that the diameter is large enough,

$$D \geq \frac{35L}{N_{Gr}^{0.25}}$$

$$0.51 \geq \frac{35 \times 1.5}{(1.8 \times 10^{10})^{0.25}} \text{ (Valid)} \qquad \text{(E5.3E)}$$

The value of the Nusselt number can now be determined by

$$N_{Nu} = \left[0.825 + \frac{0.387 N_{Ra}^{\left(\frac{1}{6}\right)}}{\left(1 + \left(\frac{0.492}{N_{Pr}} \right)^{\left(\frac{9}{16}\right)} \right)^{\frac{8}{27}}} \right]^2 \qquad \text{(E5.3F)}$$

Substituting the relevant values,

$$N_{Nu} = \left[0.825 + \frac{0.387 \times (1.26 \times 10^{10})^{\frac{1}{6}}}{\left(1 + \left(\frac{0.492}{0.7} \right)^{\left(\frac{9}{16}\right)} \right)^{\frac{8}{27}}} \right]^2 = 270.91 \qquad \text{(E5.3G)}$$

The heat transfer coefficient, h can be calculated as

$$N_{Nu} = 270.91 = \frac{h L_c}{k} = \frac{h \times 1.5}{0.028} \qquad \text{(E5.3H)}$$

or,

$$h = 5.05 \frac{W}{m^2 · °C}$$

Therefore, the rate of free convection heat transfer from the tank to the room is

$$\dot{q} = hA(T_s - T_\infty) = 5.05 \times 2 \times 3.14 \times \frac{0.51}{2} \times 1.5 \times (90 - 27) = 765.28 \text{ W} \qquad \text{(E5.3I)}$$

Example 5.4 (Bioprocessing Engineering–Heat transfer to a sausage): A 120-mm long frozen sausage of 40-mm diameter at –2°C is thawed by water at 32°C. Determine the amount of heat transferred in 30 minutes assuming that the sausage was placed horizontally.

Step 1: Schematic
The problem is depicted in Fig. 5.8.

Step 2: Given data
Length of the sausage, L = 120 mm; Diameter (L_c) of the sausage, D = 40 mm; Temperature of the sausage surface, T_s = –2°C; Temperature of the surrounding water, T_∞ = 32°C; Heat transfer time, ΔT = 30 min

To be determined: Amount of heat transferred to sausage in 30 minutes.

Assumptions: One-dimensional heat transfer occurs (in radial direction), the volume of water is high enough such that the temperature of the water remains constant, and thermal conductivities remain constant throughout.

Analysis: Because the water is still, heat transfers via free convection. Therefore, we will determine the flow regime based on the Grashof number. Subsequently, we will use the appropriate correlation to determine the Nusselt number followed by the heat transfer and the amount of heat transfer.

Step 3: Calculations
The temperature at the interface is given by the average temperature, which is

$$T_{avg} = \frac{-2 + 32}{2} = 15°C \qquad \text{(E5.4A)}$$

The properties of water at 15°C are Density, $\rho = 999 \ \frac{\text{kg}}{\text{m}^3}$, Viscosity, $\mu = 0.001139 \text{ Pa} \cdot \text{s}$, Thermal conductivity, $k = 0.58 \ \frac{\text{W}}{\text{m} \cdot °\text{C}}$, and $N_{Pr} = 8.19$.

We can calculate the Grashof number as

$$N_{Gr} = \frac{\rho^2 L_c^3 g \beta \Delta T}{\mu^2} = \frac{999^2 \times 0.04^3 \times 9.81 \times 0.00012 \times (32 - (-2))}{0.001139^2} = 1.97 \times 10^6 \qquad \text{(E5.4B)}$$

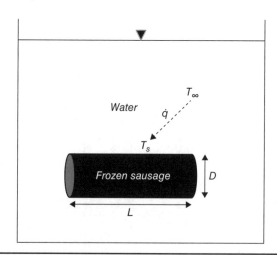

Figure 5.8 Schematic of the sausage experiencing a convection heat transfer during thawing in (still) water.

Now,

$$N_{Ra} = N_{Gr}N_{Pr} = 2,107,696 \times 8.19 = 0.16 \times 10^8 \text{ (laminar)} \qquad \text{(E5.4C)}$$

Therefore,

$$N_{Nu} = 0.56 N_{Ra}^{0.25} = 0.56 \times (0.16 \times 10^8)^{0.25} = 35.49$$

$$N_{Nu} = 35.49 = \frac{hL_c}{k} = \frac{h \times 0.04}{0.58}$$

$$h = 514.7 \frac{W}{m^2 \cdot °C} \qquad \text{(E5.4D)}$$

Heat transfer rate,

$$\dot{q} = hA(T_\infty - T_s) = 514.7 \times 2 \times 3.14 * \left(\frac{0.04}{2}\right) \times 0.12 \times (32 - (-2)) = 263.73 \text{ W} \qquad \text{(E5.4E)}$$

Heat transferred to the sausage in 30 minutes

$$q = 234.95 \times 30 \times 60 = 474.7 \text{ kJ} \qquad \text{(E5.4F)}$$

5.18 Heat Transfer Coefficient for Forced Convection

In a forced convection heat transfer, there will be an active flow of fluid that will cause the thermal and velocity boundary layers to develop. Therefore, our final mathematical relation will involve Nusselt number (to describe the heat transfer coefficient), Reynolds number (to describe the flow of the fluid), and Prandtl number (to describe the extent of TBL and VBL).

Mathematically, the heat transfer coefficient for forced convection can be expressed as

$$h = f(\rho, \mu, k, L_c C_p, v) \qquad (5.67)$$

Because, we have 7 variables with the dimensions (Table 5.3) and 4 primary dimensions, the number of nondimensional π terms are $7 - 4 = 3$.

Variable	Unit	Dimension
Heat transfer coefficient (h)	$\dfrac{W}{m^2 \cdot °C}$	$MT^{-3}\theta^{-1}$
Density (ρ)	$\dfrac{kg}{m^3}$	ML^{-3}
Viscosity (μ)	$Pa \cdot s$	$ML^{-1}T^{-1}$
Thermal conductivity (k)	$\dfrac{W}{m \cdot °C}$	$MLT^{-3}\theta^{-1}$
Characteristic length (L_c)	m	L
Specific heat (C_p)	$\dfrac{J}{kg \cdot °C}$	$L^2T^{-2}\theta^{-1}$
Velocity (v)	$\dfrac{m}{s}$	LT^{-1}

TABLE 5.3 Variables Affecting Free Convection and Their Dimensions

Let us select the repeating variables as ρ, μ, k and L_c and h, C_p, and v as nonrepeating variables and expressed the π terms as

$$\pi_1 = \rho^a \mu^b k^c L_c^d h \tag{5.68}$$

$$\pi_2 = \rho^a \mu^b k^c L_c^d C_p \tag{5.69}$$

$$\pi_3 = \rho^a \mu^b k^c L_c^d v \tag{5.70}$$

Expressing each π term in its fundamental dimensions and equating the exponents for each of the fundamental dimensions will result in

$$\pi_1 = M^0 L^0 T^0 \theta^0 = (ML^{-3})^a (ML^{-1}T^{-1})^b (MLT^{-3}\theta^{-1})^c (L)^d (MT^{-3}\theta^{-1}) \tag{5.71}$$

$$0 = a + b + c + 1 \tag{5.72}$$

$$0 = -3a - b + c + d \tag{5.73}$$

$$0 = -b - 3c - 3 \tag{5.74}$$

$$0 = -c - 1 \tag{5.75}$$

Solving the above equations will yield,

$$a = 0,\ b = 0,\ c = -1,\ \text{and}\ d = 1 \tag{5.76}$$

After substituting the exponents in the expression for π_1, we will obtain

$$\pi_1 = \rho^0 \mu^0 k^{-1} L_c^1 h \tag{5.77}$$

Or

$$\pi_1 = \frac{hL_c}{k} = N_{Nu} \tag{5.78}$$

Similarly,

$$\pi_2 = M^0 L^0 T^0 \theta^0 = (ML^{-3})^a (ML^{-1}T^{-1})^b (MLT^{-3}\theta^{-1})^c (L)^d (L^2 T^{-2}\theta^{-1}) \tag{5.79}$$

$$0 = a + b + c \tag{5.80}$$

$$0 = -3a - b + c + d + 2 \tag{5.81}$$

$$0 = -b - 3c - 2 \tag{5.82}$$

$$0 = -c - 1 \tag{5.83}$$

Solving the above equations, we obtain

$$a = 0,\ b = 1,\ c = -1,\ \text{and}\ d = 0 \tag{5.84}$$

Now after substituting the values of a, b, c, and d in the expression for π_2, we obtain

$$\pi_2 = \rho^0 \mu^1 k^{-1} L_c^0 C_p \tag{5.85}$$

Or

$$\pi_2 = \frac{\mu C_p}{k} = \left(\frac{\mu}{\rho}\right)\left(\frac{\rho C_p}{k}\right) = \frac{v}{\alpha} = N_{Pr} \tag{5.86}$$

Finally, we can repeat the same exercise with π_3

$$\pi_3 = M^0 L^0 T^0 \theta^0 = (ML^{-3})^a (ML^{-1}T^{-1})^b (MLT^{-3}\theta^{-1})^c (L)^d (LT^{-1}) \tag{5.87}$$

$$0 = a + b + c \tag{5.88}$$

$$0 = -3a - b + c + d + 1 \tag{5.89}$$

$$0 = -b - 3c - 1 \tag{5.90}$$

$$0 = -c \tag{5.91}$$

After solving, we obtain

$$a = 1, b = -1, c = 0, \text{ and } d = 1 \tag{5.92}$$

Which after substitution into the expression for π_3 will result in

$$\pi_3 = \rho^1 \mu^{-1} k^0 L_c^1 v \tag{5.93}$$

$$\pi_3 = \frac{\rho v L_c}{\mu} = N_{Re} \tag{5.94}$$

Putting all the π terms together will yield the mathematical relation for the heat transfer coefficient for forced convection.

$$\pi_1 = f(\pi_2, \pi_3) \tag{5.95}$$

Or

$$N_{Nu} = f(N_{Pr}, N_{Re}) \tag{5.96}$$

In general terms, the relation between the Nusselt number, the Prandtl number, and the Reynolds number may be expressed as

$$N_{Nu} = K N_{Pr}^{n_1} N_{Re}^{n_2} \tag{5.97}$$

where N, n_1, and n_2 are dimensionless constants to account for the shape and the flow regime within the system.

Numerous experimental investigations have yielded several correlations for various geometries. A few simple geometries are listed here while references are provided for others. Students are as usual encouraged to look up the literature for other available correlations for the appropriate flow regime.

5.19 Internal Flow in Circular and Noncircular Pipes

For fluids flowing through pipes and ducts, correlations provided by Dittus-Boelter and Cengel are appropriate as below

$$N_{Nu} = 3.66 \qquad N_{Re} < 2100 \text{ (laminar)} \tag{5.98}$$

$$N_{Nu} = 0.023 N_{Re}^{0.8} N_{Pr}^{n_2} \qquad N_{Re} > 10{,}000 \text{ (turbulent)} \tag{5.99}$$

$$n_2 = 0.4 \text{ (heating) and } n_2 = 0.3 \text{ (cooling)} \tag{5.100}$$

5.20   External Flow over Flat Plates

For fluids flowing over the flat plates, correlations provided by Cengel (Eqs. (5.101) and (5.102)) may be used

$$N_{Nu} = 0.664 N_{Re}^{0.5} N_{Pr}^{0.333} \qquad \text{when } N_{Re} < 500,000 \text{ (laminar) and } N_{Pr} > 0.6 \qquad (5.101)$$

$$N_{Nu} = 0.037 N_{Re}^{0.8} N_{Pr}^{0.33} \qquad \text{when } N_{Re} = 500,000\text{--}10,000,000 \text{ (turbulent) and } N_{Pr} = 0.6\text{--}60 \qquad (5.102)$$

5.21   Flow over a Sphere

Using the experimental data, Whitaker (1972) related the Nusselt number to Reynolds and Prandtl numbers for a sphere of diameter D as below:

$$N_{Nu} - 2 = \left(0.4 N_{Re}^{0.5} + 0.06 N_{Re}^{0.667} \right) N_{Pr}^{0.4} \left[\frac{\mu_{\infty}}{\mu_s} \right]^{0.25} \qquad \begin{aligned} N_{Re} &= 3.5 - 8 \times 10^4 \\ N_{Pr} &= 0.7 - 380 \\ \frac{\mu_{\infty}}{\mu_s} &= 1 - 3.2 \end{aligned} \qquad (5.103)$$

The fluid properties to determine the Reynolds, Nusselt, and Prandtl numbers correspond to the medium temperature (T_{∞}).

5.22   Flow over a Cylinder

The heat transfer coefficient for a fluid over a cylinder can be determined via the correlation developed by Churchill and Bernstein (1977) as below:

$$N_{Nu} = 0.3 + \frac{0.62 N_{Re}^{0.5} N_{Pr}^{0.333}}{\left[1 + \left(\frac{0.4}{N_{Pr}} \right)^{0.667} \right]^{0.25}} \left\{ 1 + \left[\frac{N_{Re}}{282,000} \right]^{0.625} \right\}^{0.8} \qquad N_{Re} N_{Pr} > 0.2 \qquad (5.104)$$

The applications of some of the above correlations are described via the following numerical examples.

Example 5.5 (Bioprocessing Engineering—Heat transfer in a hydrochar reactor):   A 50-cm diameter stainless steel spherical shell at a temperature of 30°C is being heated to facilitate a hydrothermal carbonization reaction of manure slurry. The air around the shell was measured to be 210°C and anemometer reading indicated that the hot airflow rate around the shell is $5 \frac{m}{s}$. Determine the rate of heat transfer from the environment to the shell.

Step 1: Schematic
The given problem is depicted in Fig. 5.9.

Step 2: Given data
Temperature of the shell's surface, $T_s = 30°C$; Temperature of the surrounding environment, $T_{\infty} = 210°C$; Diameter (L_c) of the spherical shell, $D = 0.5$ m; Velocity of the airflow, $v = 5 \frac{m}{s}$
To be determined: Rate of heat loss from the hot air to the shell.

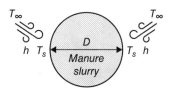

Figure 5.9 Schematic of the spherical reactor for hydrothermal carbonization of manure slurry undergoing forced convection heat transfer.

Assumptions: One-dimensional steady-state heat transfer occurs (in radial direction) and thermal conductivities remain constant throughout. The air around the shell is actively moved via a fan and therefore forced convection heat transfer occurs.

Analysis: The heat transfer occurs via forced convection whose flow regime will be governed by the Reynolds number. If the Reynolds number is less than 80,000 and the Prandtl number is between 0.7 and 380, and the ratio of viscosity of the air around the sphere and on the surface is between 1 and 3.2, we can directly use the Whitaker's correlation for the sphere for the calculation of the Nusselt number, followed by the heat transfer coefficient, and rate of heat transfer.

Step 3: Calculations

The properties of air (heating medium) at 210°C are Viscosity, $\mu_\infty = 0.000026$ Pa·s, Density, $\rho = 0.73 \dfrac{\text{kg}}{\text{m}^3}$, Thermal conductivity, $k = 0.038 \dfrac{\text{W}}{\text{m}\cdot °\text{C}}$, and Prandtl number, $N_{Pr} = 0.7$.

The viscosity of the air at the surface of the shell (at 30°C), $\mu_s = 0.000018$ Pa·s

The Reynolds number, N_{Re} will be

$$N_{Re} = \frac{\rho v L_c}{\mu} = \frac{0.73 \times 5 \times 0.5}{0.000026} = 70,192.3 < 80,000 \tag{E5.5A}$$

$$N_{Pr} = 0.7 \text{ (between 0.7 and 380)}$$

$$\frac{\mu_\infty}{\mu_s} = \frac{0.000026}{0.000018} = 1.44 \text{ (between 1 and 3.2)}$$

Therefore, we can use the Whitaker's correlation to determine the Nusselt number as below:

$$N_{Nu} = 2 + \left(0.4 N_{Re}^{0.5} + 0.06 N_{Re}^{0.667}\right) N_{Pr}^{0.4} \left[\frac{\mu_\infty}{\mu_s}\right]^{0.25} \tag{E5.5B}$$

Substituting the relevant data will result in

$$N_{Nu} = 2 + (0.4 \times (70,192.3)^{0.5} + 0.06 \times (70,192.3)^{0.667}) \times 0.7^{0.4} \times \left[\frac{0.000026}{0.000018}\right]^{0.25} \tag{E5.5C}$$

$$N_{Nu} = 200.1$$

The heat transfer coefficient, h

$$h = \frac{N_{Nu} \times k}{L_c} = \frac{200.1 \times 0.038}{0.5} = 15.2 \frac{\text{W}}{\text{m}^2 \cdot °\text{C}} \tag{E5.5D}$$

The heat transfer rate, $\dot{q}$

$$\dot{q} = hA(T_\infty - T_s) = 15.2 \times 4 \times 3.14 \times \left(\frac{0.5}{2}\right)^2 \times (210 - 30) = 2,147.8 \text{ W} \tag{E5.5E}$$

Example 5.6 (Bioprocessing Engineering—Cooling of a gasifier): An experimental steel cylindrical gasifier of 0.3-m diameter and 1 m in length at 320°C is being cooled with a fan that moves the air at 20°C at a rate of $30 \dfrac{\text{km}}{\text{h}}$. Determine the heat transfer coefficient and the rate of heat loss to the surroundings.

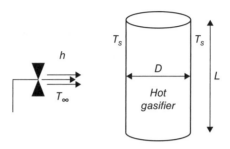

Figure 5.10 Schematic for Example 5.6 describing the cooling of the stainless steel gasifier via forced convection in air.

Step 1: Schematic
The given problem is depicted in Fig. 5.10.

Step 2: Given data
Dimensions of the cylinder, $L = 1$ m and $D = 0.3$ m ($L_c = D = 0.3$ m); Temperature of the cylinder's surface, $T_s = 320°C$; Temperature of the surrounding (air), $T_\infty = 20°C$; Velocity of the airflow,

$$v = 30 \ \frac{km}{h} = \frac{30,000}{60 \times 60} = 8.33 \ \frac{m}{s}$$

To be determined: Heat transfer coefficient and the rate of heat loss.

Assumptions: One-dimensional steady-state heat transfer occurs (in radial direction) and thermal conductivities remain constant throughout. The air around the hot cylinder is actively moved via a fan and therefore forced convection heat transfer occurs.

Analysis: The heat transfer occurs via forced convection whose flow regime will be governed by the Reynolds number. If the product of Reynolds number and Prandtl number is greater than 0.2, we can consider that the Churchill and Bernstein will be valid and heat transfer coefficient and the rate of heat transfer can be calculated via the given correlation.

Step 3: Calculations
The average temperature of the fluid is

$$T_{avg} = \frac{320 + 20}{2} = 170°C \tag{E5.6A}$$

The properties of air (coolant) at 170°C are Viscosity, $\mu = 0.000024$ Pa $\cdot$ s, Density, $\rho = 0.79 \ \frac{kg}{m^3}$, Thermal conductivity, $k = 0.035 \ \frac{W}{m \cdot °C}$, and Prandtl number, $N_{Pr} = 0.70$.

Using these data, the Reynolds number, N_{Re} will be

$$N_{Re} = \frac{\rho v L_c}{\mu} = \frac{0.79 \times 8.33 \times 0.3}{0.000024} = 82,258.75 \tag{E5.6B}$$

$$N_{Re} N_{Pr} = 82,258.75 \times 0.7 = 57,581.12 \gg 0.2$$

We calculate the Nusselt number as

$$N_{Nu} = 0.3 + \frac{0.62 \times 82,258.75^{0.5} \times 0.7^{0.333}}{\left[1 + \left(\frac{0.4}{0.7}\right)^{0.667}\right]^{0.25}} \left\{1 + \left[\frac{82,258.75}{282,000}\right]^{0.625}\right\}^{0.8} \tag{E5.6C}$$

$$N_{Nu} = 188.1$$

The heat transfer coefficient, h

$$h = \frac{N_{Nu} \times k}{L_c} = \frac{188.1 \times 0.035}{0.3} = 21.94 \ \frac{W}{m^2 \cdot °C} \tag{E5.6D}$$

The heat transfer rate, $\dot{q}$

$$\dot{q} = hA(T_\infty - T_s) = 21.94 \times 2 \times 3.14 \times \frac{0.3}{2} \times 1 \times (320 - 20) \approx 6202 \ W \tag{E5.6E}$$

We can easily increase the rate of heat transfer (by increasing value of the heat transfer coefficient) by increasing the fan speed.

Practice Problems for the FE Exam

1. The amount of heat transferred between a hot object of an area of 2 m² at 70°C and the surrounding air at 25°C with a heat transfer coefficient of 50 $\frac{W}{m^2 \cdot °C}$ in 20 minutes is

   a. 5.4 MJ.

   b. 4 kW.

   c. 8500 kJ.

   d. 10,000 J.

2. Which of the following variables do not affect the Reynolds number in a circular pipe?

   a. Diameter

   b. Roughness

   c. Density

   d. Pipe length

3. The flow of water in a pipe of 30 cm diameter at 100 LPM at 20°C (density = 1000 kg/m³ and viscosity = 0.001 Pa·s) is considered

   a. laminar.

   b. transitional.

   c. turbulent.

   d. critical.

4. For a given temperature gradient, the convection heat transfer can be increased by typically

   a. increasing the N_{Re}.

   b. increasing the surface area.

   c. a only.

   d. both a and b.

5. The value of Prandtl number for water at 20°C with a viscosity of 0.001 Pa·s, density of 1000 kg/m³, thermal conductivity of 0.6 W/m·°C, and a specific heat of 4.2 kJ/kg·°C is

   a. 25.

   b. 5.5.

   c. 7.

   d. 10.

6. Which of the following statements is false?

   a. Grashof number in free convection is akin to Reynolds number in forced convection.

   b. Low Prandtl number indicates slower rate of heat diffusion relative to rate of momentum diffusion.

   c. Nusselt number of 100 suggests that movement of fluid increases the conduction heat transfer by 100 times.

   d. Free and forced convection can occur simultaneously.

7. A 1.5-m vertical cylinder at 25°C is exposed to hot air at 75°C. The air properties at the interface (i.e., at 50°C) are viscosity, $\mu = 0.000019$ Pa·s, specific heat, $C_p =$ 1.007 $\dfrac{\text{kJ}}{\text{kg}\cdot°\text{C}}$, density, $\rho = 1.09$ $\dfrac{\text{kg}}{\text{m}^3}$, thermal conductivity, $k = 0.027$ $\dfrac{\text{W}}{\text{m}\cdot°\text{C}}$. The Grashof number for this system is closest to

   a. 1.7×10^{12}.

   b. 3×10^8.

   c. 1.7×10^{10}.

   d. 2×10^{11}.

8. The Rayleigh number for the above problem is nearly

   a. 1.3×10^{10}.

   b. 500,000.

   c. 1.7×10^{10}.

   d. 2.1×10^9.

9. The heat transfer coefficient in $\dfrac{\text{W}}{\text{m}^2\cdot°\text{C}}$ in a 1-cm diameter cylindrical pipe carrying air at 50°C at 3 m/s and exposed to constant air temperature at 25°C with air properties of density = 1.09 kg/m³, viscosity = 0.0000196 Pa·s, specific heat = 1.007 kJ/kg·°C, and thermal conductivity = 0.02 W/m·°C is

   a. 75.

   b. 14.5.

   c. 3.66.

   d. None of the above.

10. Which of the following relations is true?

 a. $N_{Pr} = \dfrac{\text{Thermal boundary layer thickness}}{\text{Velocity boundary layer thickness}}$

 b. $N_{Nu} = \dfrac{hk}{L_c}$

 c. For forced convection, $N_{Nu} = f(N_{Pr}, N_{Re})$

 d. $N_{Ra} = \dfrac{\rho^2 \beta L_c^3 \Delta T g}{\mu^2}$

Practice Problems for the PE Exam

1. A biochar pellet (diameter = 1 cm, length = 5 cm) at 200°C is immersed suddenly in water at 25°C that was constantly agitated. Assuming a heat transfer coefficient of 100 W/m²·°C, determine the heat transfer flux from biochar to the water.

2. Calculate the Prandtl number for air and water for temperatures between 0°C and 100°C (with 10°C increments) and plot N_{Pr} versus temperature curves.

3. A hot copper sphere of 5-cm diameter at an initial temperature of 100°C is taken out of a furnace is exposed to cold air at 2°C flowing at 30 km/h. Determine the heat transfer coefficient between the sphere and the air. Also determine the rate of convective heat transfer.

4. Repeat Problem 3 without the forced airflow and compare the result.

5. A 5-cm diameter stainless steel cylinder at 20°C is placed horizontally in a hot air environment of 150°C at 5 m/s. Calculate the rate of heat transfer between the hot air and the cylinder.

6. Nusselt number and Biot number have similar mathematical forms. Explain the differences between N_{Nu} and N_{Bi}. Also, describe the physical significance of the Reynolds, Prandtl, Grashof, and Nusselt numbers in the context of convection heat transfer.

7. A cylindrical container of dimensions 1-m diameter and 2-m long stores hot water in a poultry processing facility. If the surface temperature of the tank was measured to be 95°C, determine the rate of heat loss to a room maintained at 20°C.

8. Biomass is being processed via pyrolysis in a 30-cm diameter stainless steel spherical shell (reactor). The initial temperature of the reactor is 25°C. During the pyrolysis process, the airflow around the reactor was measured to be at 20 km/h and 250°C. Determine the rate of heat transfer from the air to the reactor.

9. A hot cylinder of 0.3 m diameter and 1 m in length at 400°C is being cooled with a fan that moves the air at 25°C at a rate of 20 $\dfrac{\text{km}}{\text{h}}$. Determine the heat transfer coefficient and the rate of heat loss to the air.

CHAPTER **6**

Design and Analysis of Heat Exchangers

Chapter Objectives

After the completion of this chapter, students will be able to

- Design and size heat exchangers for given situations.
- Determine the rate of heat transfer in a heat exchanger.
- Use analytical and graphical methods to analyze heat exchangers.

6.1 Motivation

Several engineering operations such as chemical processing generate large amounts of heat that needs to be regularly expelled for ensuring the safety of the equipment. Similarly, in HVAC systems, heat generated by the furnace is to be distributed to the entire dwelling to maintain an optimum level of comfort. Even in our automobiles, the engine block must be continuously cooled to prevent overheating. All the aforementioned processes require equipment capable to transferring the excess heat effectively. Such equipments are called heat exchangers which are the focus of this chapter.

6.2 The Principles of Heat Exchanger

As the name suggests, the main task of the heat exchanger (HX) is to transfer heat from the hot stream of fluid to the cold stream of fluid. From thermodynamics, we recall that when the hot and cold fluids are brought into contact, the heat energy from the hot stream will be transferred to the cold stream resulting in an equilibrium temperature (Fig. 6.1).

Depending on the type of contact between the fluids, heat exchangers may be classified as direct contact, regenerative, or recuperative heat exchangers. In its simplest form, a direct contact heat exchanger (Fig. 6.2) involves the intimate mixing of two usually immiscible fluids at the same pressure (or a gas and a liquid) at different temperatures.

During mixing, the heat is rapidly transferred mainly via convection. A popular example of a direct contact heat exchanger is a cooling tower extensively used in the power industry to reject excess heat (Fig. 6.3).

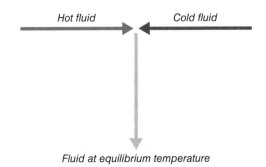

FIGURE 6.1 Contact between hot and cold fluid streams will result in an equilibrium.

FIGURE 6.2 A simple schematic of a direct contact heat exchanger in which a hot gas is cooled by flowing (bubbling) through a cold fluid.

FIGURE 6.3 An image of a cooling tower used in power plants for heat transfer. (Photo courtesy of Tim Reckmann ©. Reprinted with permission.)

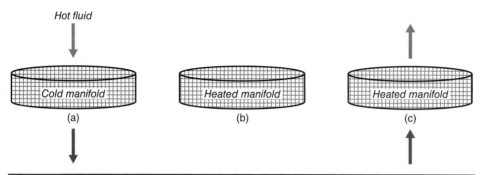

FIGURE 6.4 A circular regenerator absorbs the heat from the hot stream (a) and stores the heat (b) which is absorbed away by the subsequent cold stream moving in the opposite direction (c).

On the other hand, a regenerative heat exchanger (or a regenerator) involves a heat sink/source that facilitates the transfer of heat in a stepwise fashion (Fig. 6.4). In the first step, the hot fluid passes through a high-surface medium (usually a solid) and heats up the medium that acts as a heat sink. Subsequently, in the second step, the cold fluid is passed through the same heated medium which now acts as a heat source to remove the heat. A common example of a regenerator is shown in Fig. 6.5.

Finally, a recuperative heat exchanger involves the simultaneous flow of hot and cold fluids separated by a thin but highly conductive (usually metal) surface (Fig. 6.6). During the heat transfer process, heat is transferred from the hot fluid to the separator. In the next step, the hot separator transfers the heat to cold fluid and increases the

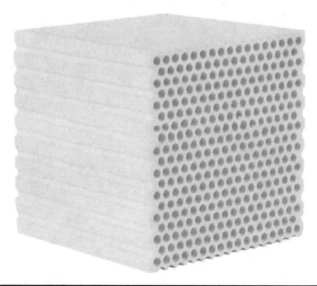

FIGURE 6.5 An image of a cubic monolithic regenerator heat exchanger. (Image courtesy of Ping Yin, Beihai Kaite Chemical Packing Co., Ltd. Reprinted with permission.)

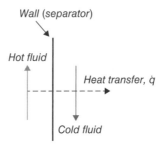

FIGURE 6.6 A general schematic of a recuperative heat exchanger in which heat is transferred from the hot fluid to the cold fluid via the wall of the heat exchanger.

temperature of the cold fluid. In most industrial situations, several tubes are combined and placed in large baffle-equipped shell. The tubes and shell carry fluids at different temperatures resulting in heat transfer from the hot fluid to the cold fluid. This type of heat exchanger is called shell-and-tube heat exchanger. In some cases, the residence time for heat transfer is increased by rerouting tubes through the shell. Such heat exchangers are called multi-pass shell-and-tube heat exchanger.

Recuperative heat exchangers are extremely common in the industry and come in several different configurations (Fig. 6.7) and therefore will be the focus of this chapter.

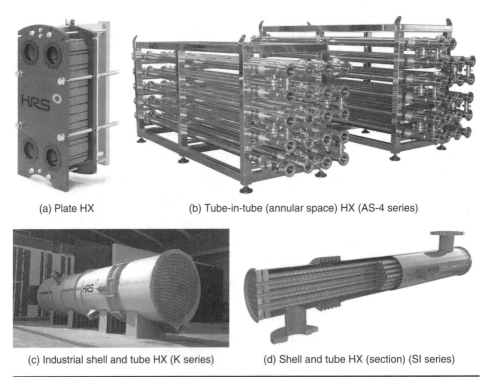

(a) Plate HX   (b) Tube-in-tube (annular space) HX (AS-4 series)

(c) Industrial shell and tube HX (K series)   (d) Shell and tube HX (section) (SI series)

FIGURE 6.7 Images of popular and commonly used plate (a), tube-in-tube (b), and shell-and-tube heat exchangers (c and d). (Images courtesy of HRS Heat Exchangers, UK. Reprinted with permission.)

6.3   Common Recuperative Heat Exchanger Configurations Counter-Flow Heat Exchanger (CFHXs)

As one of the most ubiquitous configurations, a counter-flow heat exchanger involves the simultaneous movement of hot and cold fluids in the opposite directions (Fig. 6.8a). Because the fluids flow in the opposite directions, the cold fluid may exit the system at a temperature higher than the outlet temperature of the hot fluid (Fig. 6.8b).

6.3.1   Parallel-Flow Heat Exchangers (PFHXs)

In a parallel-flow heat exchanger, the hot and the cold fluids flow in the same direction (Fig. 6.9a). Although the heat transfer rates offered by PFHXs are not as high as CFHXs, they still find applications in a few situations where overheating and over-cooling of the heat exchanger walls needs to be avoided to minimize the effects of wall fouling. In a PFHX, the outlet temperatures of the cold and hot fluids approach each other (Fig. 6.9b).

6.3.2   Cross-Flow Heat Exchangers (CRFHXs)

In CRFHX, the hot and cold fluids travel in perpendicular directions (Fig. 6.10a and b). It is very commonly used when one fluid is a gas, and another fluid is a liquid. Generally, CRFHXs are more effective than CFHXs and PFHXs and therefore can provide high compactness that are suitable in space-constrained applications such as radiators in automobiles (Fig. 6.11).

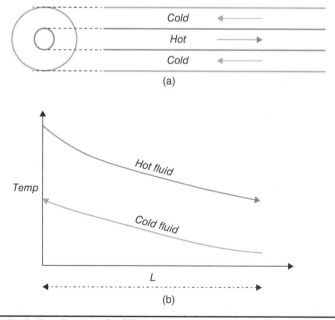

FIGURE 6.8   A simple line diagram of a CFHX (a) with temperature profiles of the hot and cold fluids flowing in the opposite directions (b).

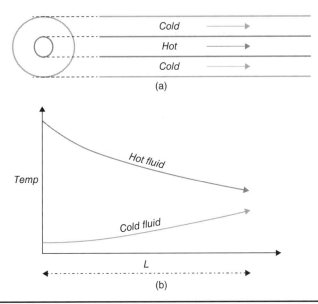

FIGURE 6.9   A simple line diagram of a PFHX (a) with temperature profiles of the hot and cold fluids flowing in the same directions (b).

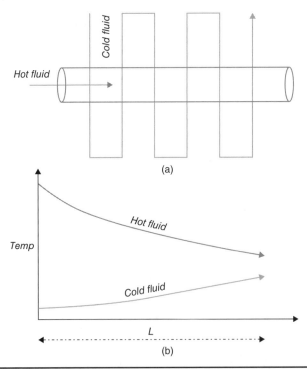

FIGURE 6.10   A simple line diagram of a CRFHX (a) with temperature profiles of the hot and cold fluids flowing in the perpendicular directions (b). **Note:** The actual temperature profile depends on the configuration.

FIGURE 6.11 An image of a car radiator that functions as a CRFHX. (Image courtesy of U.S. Radiator, USA. Reprinted with permission.)

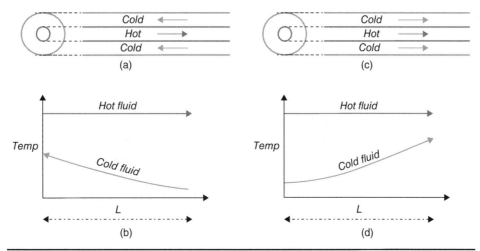

FIGURE 6.12 Line diagrams of phase-change heat exchangers and their temperature profiles subjected to counter-flow (a and b) and parallel-flow (c and d) configurations.

6.3.3 Phase-Change Heat Exchangers (PCHXs)

These types of HXs are very common in food, petrochemical, and energy industries. Typically, high-pressure steam is contacted with a cold liquid (Fig. 6.12a and c) during which the cold liquid absorbs the latent heat of the steam (Fig. 6.12b and d).

6.4 Overall Heat Transfer Coefficient

The transfer of heat in a recuperative HX is usually via a combination of conduction and convection (convection may also include the radiation effects if any). To explain this idea, let us consider a simple double-pipe HX as an example (Fig. 6.13a).

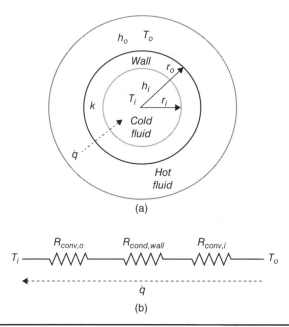

(a)

(b)

FIGURE 6.13 A schematic of a double-pipe HX carrying cold fluid in the inner pipe and the hot fluid in the annulus (a) with the corresponding thermal resistances in series (b).

Let us also imagine that the hot fluid at a temperature T_o, flows through the outer pipe with a heat transfer coefficient, h_o, while the cold fluid at temperature T_i, flows through the inner pipe of outer radius, r_i with thermal conductivity, k, and a heat transfer coefficient, h_i. Recalling the concept of thermal resistance from Chapter 3, we can express the rate of heat transfer from the hot fluid (outer pipe) to the cold fluid (inner pipe) through the separator (wall of the inner pipe) as

$$\dot{q} = \frac{\Delta T}{R_{conv,o} + R_{cond,wall} + R_{conv,i}}$$

where ΔT is the temperature gradient, and $R_{conv,o}$, $R_{cond,wall}$, and $R_{conv,i}$ are the thermal resistances associated with the convection within the outer pipe, conduction through the wall of the inner pipe, and the convection within the inner pipe (Fig. 6.13b).

After expressing the thermal resistance in their mathematical forms, we arrive at

$$\dot{q} = \frac{(T_o - T_i)}{\dfrac{1}{h_o A_o} + \dfrac{\ln \dfrac{r_o}{r_i}}{2\pi L k} + \dfrac{1}{h_i A_i}} \tag{6.1}$$

Because the heat transfer through the HX occurs via conduction and convection, we can express the heat transfer rate similar to Newton's law of cooling as in Eq. (6.2)

$$\dot{q} = UA(T_o - T_i) \tag{6.2}$$

where U is the overall heat transfer coefficient $\left(\dfrac{W}{m^2 \cdot {}^{\circ}C}\right)$ and A is the area of cross section
(m^2) through which heat transfer occurs due to the temperature gradient (°C) of $(T_o - T_i)$.
Combining Eqs. (6.1) and (6.2), we obtain Eq. (6.3)

$$\frac{1}{UA} = \frac{1}{h_o A_o} + \frac{\ln \frac{r_o}{r_i}}{2\pi L k} + \frac{1}{h_i A_i} \tag{6.3}$$

We must note that the HX consists of two areas, A_o and A_i and hence, the HX will
also have two overall heat transfer coefficients, U_o and U_i. Therefore, for the given heat
transfer rate and the temperature gradient, we can write as

$$U_o A_o = U_i A_i \tag{6.4}$$

and

$$\frac{1}{U_o A_o} = \frac{1}{h_o A_o} + \frac{\ln \frac{r_o}{r_i}}{2\pi L k} + \frac{1}{h_i A_i} \tag{6.5}$$

$$\frac{1}{U_i A_i} = \frac{1}{h_o A_o} + \frac{\ln \frac{r_o}{r_i}}{2\pi L k} + \frac{1}{h_i A_i} \tag{6.6}$$

Further, one must understand that the overall heat transfer coefficient is always
determined along with the corresponding area of the system that was employed in the
calculation.

In certain special cases, when wall thickness is negligible ($D_i \approx D_0$) or the thermal
conductivity of the wall is very high $\left(k \geq 100 \dfrac{W}{mK} \right)$ or both, we obtain

$$\frac{1}{U_o} \approx \frac{1}{U_i} \approx \frac{1}{U} \approx \left[\frac{1}{h_o} + \frac{1}{h_i} \right] \tag{6.7}$$

For plate heat exchangers that employ parallel plates (Fig. 6.14), the analysis is
straight forward. Assuming the heat transfer area (A) of the plate (m^2) is the same for
hot and cold streams, the overall heat transfer coefficient, $U \left(\dfrac{W}{m^2 \, {}^{\circ}C} \right)$ can now be
expressed as

$$\frac{1}{U} = \frac{1}{h_{hot}} + \frac{L}{k} + \frac{1}{h_{cold}} \tag{6.8}$$

where h_{hot} and h_{cold} are the heat transfer coefficients $\left(\dfrac{W}{m^2 \, {}^{\circ}C} \right)$ associated with hot and
cold streams, L is the thickness of the plate (m), and k is the thermal conductivity of the
plate (W/m·°C).

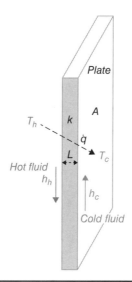

Figure 6.14 A general schematic of a plate heat exchanger in which heat is transferred from the hot fluid to the cold fluid via the wall of the heat exchanger.

6.5 Governing Equations

The fundamental equation that governs the transfer of heat from one fluid stream to the other is the convection equation, whose general form we are very familiar with

$$\dot{q} = UA\Delta T_{\text{rep}} \tag{6.9}$$

where U is the overall heat transfer coefficient $\left(\dfrac{\text{W}}{\text{m}^2 \cdot {}^\circ\text{C}}\right)$ that includes the effects of convection, conduction, and radiation, A is the area of the cross section of the tube (m²), and ΔT_{rep} is the representative temperature gradient between the cooling and heating fluids taken together.

During the process of the heat exchange, the rate of heat lost by hot fluid is given by

$$\dot{q}_h = \dot{m}_h C_{P,h} \Delta T_h = \dot{m}_h C_{P,h}\left(T_{h,i} - T_{h,o}\right) \tag{6.10}$$

Similarly, the rate of heat gained by cold fluid is given by

$$\dot{q}_c = \dot{m}_c C_{P,c} \Delta T_c = \dot{m}_c C_{P,c}\left(T_{c,o} - T_{c,i}\right) \tag{6.11}$$

In the above equations, $\dot{m}$, $\dot{q}$, C_P, and T are the mass flow rates of the fluid, heat transfer rates, specific heats, and temperatures of the cold stream (c) and hot stream (h), respectively (Table 6.1).

Finally, the maximum rate of heat transfer is given by

$$\dot{q}_{\text{max}} = C_{\text{min}}\left(T_{h,i} - T_{c,i}\right) \tag{6.12}$$

where $\dot{q}_{\text{max}}$ is the maximum possible rate of heat transfer (W) and C_{min} is the smaller of C_c and C_h and $C_c = m_c C_{P,c}$ and $C_h = m_h C_{P,h}$.

Variable	Cold Stream	Hot Stream
Heat transfer rate (W)	$\dot{q}_c$	$\dot{q}_h$
Mass flow rate (kg/s)	m_c	m_h
Specific heat (J/kg·°C)	$C_{P,c}$	$C_{P,h}$
Inlet temperature (°C)	$T_{c,i}$	$T_{h,i}$
Outlet temperature (°C)	$T_{c,o}$	$T_{h,o}$

TABLE 6.1 Variables Employed (with Units) in the Design and Analysis of Heat Exchangers

Since the heat exchanger transfers the heat from the hot stream to the cold stream, Eqs. (6.9)–(6.12) can be combined to solve for a given heat exchanger. However, the most important question that comes to our mind is that what does ΔT_{rep} in Eq. (6.9) represent? Let us consider a simple parallel-flow heat exchanger as shown in Fig. 6.15a.

Recalling that both hot stream and the cold stream have separate temperature gradients, $\Delta T_h = (T_{h,i} - T_{h,o})$ and $\Delta T_c = (T_{c,o} - T_{c,i})$, respectively, and that the temperatures and the gradients are continuously changing along the length of the heat exchanger, which temperature gradient do we use as ΔT_{rep}? Often, in lieu of ΔT_{rep}, we are tempted to use an average of the temperature gradients at the inlet and the outlet of the heat exchanger (Fig. 6.15b).

$$\Delta T_{ave} = \frac{\Delta T_1 + \Delta T_2}{2} \qquad (6.13)$$

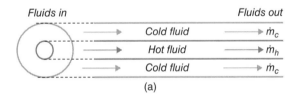

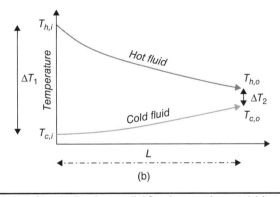

FIGURE 6.15 Schematic of a tube-in-tube parallel-flow heat exchanger (a) in which the temperature gradient between the hot and cold streams decreases along the length of the heat exchanger (b).

This may work if the temperature difference between hot and cold fluids is small. However, in most situations this is not the case and therefore will lead to erroneous results. For a heat exchanger to perform effectively, the ΔT_{rep} should represent a temperature gradient that accurately captures the overall temperature difference between the hot fluid and the cold fluid across the entire length of the heat exchanger. The following section will focus on determining the most accurate temperature gradient using the log mean temperature difference (LMTD).

6.6 Approach 1—The Log Mean Temperature Difference (LMTD) Method

Because most of the tubular heat exchangers are of counter-flow type, we will begin our analysis using a counter-flow double tube heat exchanger (Fig. 6.16) whose temperature profiles are also sketched. The nomenclature for the heat exchanger is summarized in Table 6.2.

$$d\dot{q}_h = \dot{m}_h C_{p,h}(-dT_h) \tag{6.14}$$

$$d\dot{q}_c = \dot{m}_c C_{p,c}(-dT_c) \tag{6.15}$$

It is to be noted that the negative signs for the temperatures are to accommodate the negative temperature gradient for the hot stream and the reverse direction (right to left) for the cold stream.

Recall that

$$d\dot{q}_h = d\dot{q}_c = d\dot{q} \tag{6.16}$$

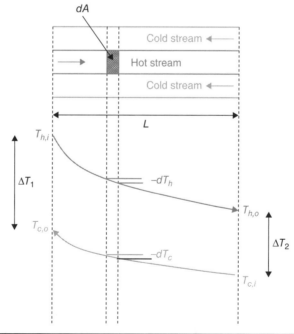

Figure 6.16 Schematic of a tube-in-tube counter-flow heat exchanger along with the temperature profiles of the hot and cold streams.

Variable	Cold Stream	Hot Stream
Heat transfer rate (W)	$\dot{q}_c$	$\dot{q}_h$
Mass flow rate (kg/s)	$\dot{m}_c$	$\dot{m}_h$
Specific heat (J/kg·°C)	$C_{P,c}$	$C_{P,h}$
Inlet temperature (°C)	$T_{c,i}$	$T_{h,i}$
Outlet temperature (°C)	$T_{c,o}$	$T_{h,o}$

TABLE 6.2 Variables Employed (with Units) in the Design and Analysis of Counter-Flow Heat Exchangers

After expressing $\dot{m}_h C_{p,h}$ as C_h and $\dot{m}_c C_{p,c}$ as C_c, where C_h and C_c are called heat capacity rates $\left(\dfrac{W}{°C}\right)$, we can combine Eqs. (6.14) and (6.15) with Eq. (6.16) as below

$$dT_h = \frac{-d\dot{q}}{\dot{m}_h C_{p,h}} = \frac{-d\dot{q}}{C_h} \tag{6.17}$$

$$dT_c = \frac{-d\dot{q}}{\dot{m}_c C_{p,c}} = \frac{-d\dot{q}}{C_c} \tag{6.18}$$

Now,

$$d\left(T_h - T_c\right) = -d\dot{q}\left(\frac{1}{C_h} - \frac{1}{C_c}\right) \tag{6.19}$$

We also know that

$$\dot{q} = \dot{m}_h C_{p,h}(T_{h,i} - T_{h,o}) = C_h(T_{h,i} - T_{h,o}) \tag{6.20}$$

$$\dot{q} = \dot{m}_c C_{p,c}(T_{c,o} - T_{c,i}) = C_c(T_{c,o} - T_{c,i}) \tag{6.21}$$

and therefore

$$C_h = \frac{\dot{q}}{(T_{h,i} - T_{h,o})} \tag{6.22}$$

$$C_c = \frac{\dot{q}}{(T_{c,o} - T_{c,i})} \tag{6.23}$$

After substituting Eqs. (6.22) and (6.23) into Eq. (6.19), we obtain

$$d(T_h - T_c) = -d\dot{q}\left\{\frac{1}{\dfrac{\dot{q}}{(T_{h,i} - T_{h,o})}} - \frac{1}{\dfrac{\dot{q}}{(T_{c,o} - T_{c,i})}}\right\} \tag{6.24}$$

$$d(T_h - T_c) = \frac{-d\dot{q}}{\dot{q}}\left\{(T_{h,i} - T_{h,o}) - (T_{c,o} - T_{c,i})\right\} \tag{6.25}$$

$$d(T_h - T_c) = \frac{-d\dot{q}}{\dot{q}}\left\{(T_{h,i} - T_{c,o}) - (T_{h,o} - T_{c,i})\right\} \tag{6.26}$$

$$d(T_h - T_c) = \frac{-d\dot{q}}{\dot{q}}\left\{\Delta T_1 - \Delta T_2\right\} \tag{6.27}$$

Because,

$$(T_{h,i} - T_{c,o}) = \Delta T_1 \qquad \text{(Fig. 6.16)} \tag{6.28}$$

$$(T_{h,o} - T_{c,i}) = \Delta T_2 \qquad \text{(Fig. 6.16)} \tag{6.29}$$

Now, using the relation

$$d\dot{q} = U(dA)(T_h - T_c) \tag{6.30}$$

and substituting Eq. (6.30) in Eq. (6.27), we obtain

$$d(T_h - T_c) = \frac{-U(dA)(T_h - T_c)}{\dot{q}}\{\Delta T_1 - \Delta T_2\} \tag{6.31}$$

Rearranging will result in

$$\frac{d(T_h - T_c)}{(T_h - T_c)} = -\frac{U(dA)}{\dot{q}}\{\Delta T_1 - \Delta T_2\} \tag{6.32}$$

Integrating between the inlet and the outlet temperature conditions

At the inlet of the heat exchanger $L = 0$, $A = 0$, $T_h = T_{h,i}$, and $T_c = T_{c,o}$

At the outlet of the heat exchanger $L = L$ $A = A$, $T_h = T_{h,o}$, and $T_c = T_{c,\text{in}}$

$$\ln\left[\frac{T_{h,o} - T_{c,i}}{T_{h,i} - T_{c,o}}\right] = -\frac{UA}{\dot{q}}\{\Delta T_1 - \Delta T_2\} \tag{6.33}$$

$$\ln\left[\frac{\Delta T_2}{\Delta T_1}\right] = -\frac{UA}{\dot{q}}\{\Delta T_1 - \Delta T_2\} \tag{6.34}$$

or

$$\ln\left[\frac{\Delta T_1}{\Delta T_2}\right] = \frac{UA}{\dot{q}}\{\Delta T_1 - \Delta T_2\} \tag{6.35}$$

With rearrangement, we obtain the design equation for the CFHX as below:

$$\dot{q}_{\text{CFHX}} = UA\frac{\{\Delta T_1 - \Delta T_2\}}{\ln\left[\dfrac{\Delta T_1}{\Delta T_2}\right]} \tag{6.36}$$

Comparing Eq. (6.36) with Eq. (6.9), we can see that

$$\Delta T_{\text{rep}} = \frac{\{\Delta T_1 - \Delta T_2\}}{\ln\left[\dfrac{\Delta T_1}{\Delta T_2}\right]} \tag{6.37}$$

and the mathematical expression $\dfrac{\{\Delta T_1 - \Delta T_2\}}{\ln\left[\dfrac{\Delta T_1}{\Delta T_2}\right]}$ is also called the log mean temperature difference (LMTD) of the temperatures of the hot and cold streams and hence called the LMTD approach. It is noted that our approach for LMTD was based on

$\Delta T_1 > \Delta T_2$. However, the LMTD approach is equally valid when $\Delta T_1 < \Delta T_2$ and the heat transfer rate is now expressed as

$$\dot{q}_{\text{CFHX}} = UA \frac{\{\Delta T_2 - \Delta T_1\}}{\ln\left[\dfrac{\Delta T_2}{\Delta T_1}\right]} \tag{6.38}$$

In general terms, however, the expressions for LMTD are written as

$$\Delta T_{\text{rep}} = \frac{\{\Delta T_L - \Delta T_S\}}{\ln\left[\dfrac{\Delta T_L}{\Delta T_S}\right]} \tag{6.39}$$

and

$$\dot{q} = UA \frac{\{\Delta T_L - \Delta T_S\}}{\ln\left[\dfrac{\Delta T_L}{\Delta T_S}\right]} \tag{6.40}$$

where ΔT_L and ΔT_S are the larger and smaller temperature gradients, respectively while the other variables are defined as before.

6.7 The LMTD Method for Parallel-Flow Heat Exchangers

The same mathematical treatment described in counter-flow heat exchanger can as well be extended to a parallel-flow heat exchanger (Fig. 6.17) to obtain the representative temperature gradient that is identical to what we obtained in the counter-flow heat exchanger.

$$\Delta T_{\text{rep}} = \frac{\{\Delta T_1 - \Delta T_2\}}{\ln\left[\dfrac{\Delta T_1}{\Delta T_2}\right]} \tag{6.41}$$

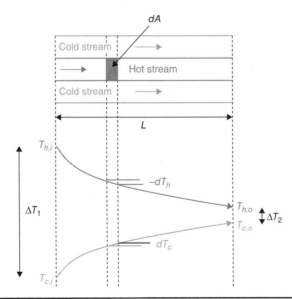

Figure 6.17 Schematic of a tube-in-tube parallel-flow heat exchanger along with the temperature profiles of the hot and cold streams.

and

$$\dot{q}_{\text{PFHX}} = UA \frac{\{\Delta T_1 - \Delta T_2\}}{\ln\left[\dfrac{\Delta T_1}{\Delta T_2}\right]} \tag{6.42}$$

Therefore, for LMTD approach, the rate of heat transfer is independent of the flow direction

$$\dot{q}_{\text{CFHX}} = UA \frac{\{\Delta T_L - \Delta T_S\}}{\ln\left[\dfrac{\Delta T_L}{\Delta T_S}\right]} = \dot{q}_{\text{PFHX}} \tag{6.43}$$

Example 6.1 (Environmental Engineering—Heating of water in aquaculture):   An aquacultural facility is planning to heat the water from 10°C to 25°C using hot water from a nearby waste combustion facility using thin-walled double-tube heat exchanger with a counter-flow configuration (Fig. 6.18). The hot water enters the inner tube of a 5-cm diameter made of copper at 85°C at 5 kg/s and exits the tube at 65°C. Determine the heat transfer rate and sizing of the heat exchanger needed if the overall heat transfer coefficient was estimated to be 750 W/m².°C. Disregard the effects of temperature on the specific heat of the water.

Step 1: Schematic
The problem is described in Fig. 6.18.

Step 2: Given data
Inlet temperature of the cold stream, $T_{c,i} = 10°C$; Outlet temperature of the cold stream, $T_{c,o} = 25°C$; Inlet temperature of the hot stream, $T_{h,i} = 85°C$; Outlet temperature of the hot stream, $T_{h,o} = 65°C$;

Mass flow rate of the hot stream, $\dot{m}_h = 5 \dfrac{\text{kg}}{\text{s}}$; Diameter of the inner tube, $D = 5$ cm; and The overall

heat transfer coefficient, $U = 750 \dfrac{\text{W}}{\text{m}^2 \cdot °\text{C}}$

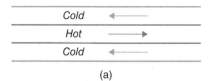

(a)

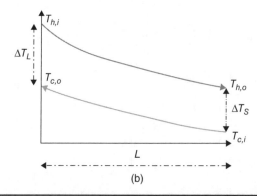

(b)

FIGURE 6.18   The schematic for the counter-flow heat exchanger for Example 6.1 (a) and the temperature profiles for hot and cold streams (b).

To be determined: Rate of heat transfer, q, and size (length) of the heat exchanger.

Assumptions: Steady-state flow, no loss of heat to the external surroundings, and the effects of temperature on the change in the specific heat of water is negligible, that is, the specific heat of water,

$$C_p = 4200 \ \frac{J}{kg \cdot °C} = C_{p,h} = C_{p,c}.$$

Approach: Noting that the rate of heat transfer out of the hot stream is the same as the heat transferred into the cold stream, we will first calculate the rate of heat transfer from the available data for the hot stream. Subsequently, we will use the governing equation for the CFHX to determine the area and the length of the HX needed.

Step 3: Calculations
Rate of heat transfer from the hot stream, $\dot{q}_h = \dot{m}_h C_{p,h} (T_{h,i} - T_{h,o})$ (E6.1A)

Substituting the data into the above equation, we obtain

$$\dot{q}_h = 5 \times 4200 \times (85 - 65) = 420,000 \ W$$

Now, this is the same rate of heat transfer into the cold stream. Therefore,

$$\dot{q}_h = \dot{q}_c = 420,000 \ W = \dot{m}_c C_{p,c} (T_{c,o} - T_{c,i}) \quad\quad\quad (E6.1B)$$

Substituting the data into the above equation, we obtain

$$420,000 \ W = \dot{m}_c \times 4200 \times (25 - 10)$$

$$\dot{m}_c = 6.67 \ \frac{kg}{s}$$

HX Design
The governing equation for the CFHX is given by

$$\dot{q} = UA \frac{\{\Delta T_L - \Delta T_S\}}{\ln\left[\frac{\Delta T_L}{\Delta T_S}\right]} \quad\quad\quad (E6.1C)$$

From the given inlet and outlet temperatures of hot and cold streams, we can calculate T_L and T_S as

$$T_L = 85°C - 25°C = 60°C \quad \text{and} \quad T_S = 65°C - 10°C = 55°C$$

$$\dot{q} = 420,000 = U(\pi DL) \frac{\{\Delta T_L - \Delta T_S\}}{\ln\left[\frac{\Delta T_L}{\Delta T_S}\right]} \quad\quad\quad (E6.1D)$$

Substituting the values into the above equation, we obtain the length of the CFHX as

$$420,000 = 750 \times 3.14 \times 0.05 \times L \frac{\{60 - 55\}}{\ln\left[\frac{60}{55}\right]} \quad\quad\quad (E6.1E)$$

$$L = 62 \ m$$

Therefore, a 62.07-m long 5-cm diameter tube carrying cold water at 6.67 kg/s should be adequate to absorb 420,000 W of energy.

Example 6.2 (Agricultural Engineering—Heating of air in animal systems):   Air in a small experimental pig house is being heated in winter via a solar heating system (Fig. 6.19). The air is heated by the solar heater and flows into a 10-cm diameter, thin-walled parallel-flow heat exchanger system ($U = 45\ \text{W}/\text{m}^2\cdot{}^\circ\text{C}$) at 60°C and exits the heat exchanger at 30°C at 0.3 kg/s. The cold air has an inlet temperature of 12°C at 1 kg/s. What will be the outlet temperature of the cold air? How big will the heat exchanger be? In addition, what is the cost of heat harvested by the heat exchanger in a 24-hour period assuming an efficiency of 75% and an electricity price of 15 cents/kWh?

Step 1: Schematic
The problem is depicted in Fig. 6.19.

Step 2: Given data
Inlet temperature of the cold stream, $T_{c,i} = 12°C$; Mass flow rate of the cold stream, $\dot{m}_c = 1\ \dfrac{\text{kg}}{\text{s}}$; Inlet temperature of the hot stream, $T_{h,i} = 60°C$; Outlet temperature of the hot stream, $T_{h,o} = 30°C$; Mass flow rate of the hot stream, $\dot{m}_h = 0.3\ \dfrac{\text{kg}}{\text{s}}$; Diameter of the inner tube, $D_i = 10$ cm; The overall heat transfer coefficient for the inner tube, $U_i = 45\ \dfrac{\text{W}}{\text{m}^2\cdot{}^\circ\text{C}}$; Efficiency of the HX, $\eta = 75\%$; and cost of energy = 15 cents/kWh

To be determined: Outlet temperature of the cold stream, $T_{c,o}$, size (length) of the heat exchanger, and the daily cost of energy harvested by the solar heating system.

Assumptions: Steady-state flow, no loss of heat to the external surroundings, and the effects of temperature on the change in the specific heat of air are negligible, that is, the specific heat of air, $C_p = 1060\ \dfrac{\text{J}}{\text{kg}\cdot{}^\circ\text{C}} = C_{p,h} = C_{p,c}$.

Approach: We will first calculate the rate of heat transfer from the available data for the hot stream. Subsequently, we will determine the temperature of the outlet of the cold stream and use the governing equation for the PFHX to determine the length of the HX needed. Finally, using the heat transfer rate, we will determine the cost of the harvested heat.

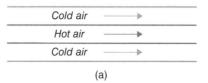

(a)

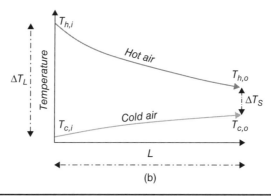

(b)

Figure 6.19   The schematic for the parallel-flow heat exchanger for Example 6.2 (a) and the temperature profiles for hot and cold streams (b).

Step 3: Calculations

Rate of heat transfer from the hot stream, $\dot{q}_h = \dot{m}_h C_{p,h}(T_{h,i} - T_{h,o})$ (E6.2A)

Substituting the data into the above equation, we obtain

$$\dot{q}_h = 0.3 \times 1060 \times (60 - 30) = 9540 \ \text{W}$$

Now, this is the same rate of heat transfer into the cold stream.

$$\dot{q}_h = \dot{q}_c = 9540 \ \text{W} = \dot{m}_c C_{p,c}(T_{c,o} - T_{c,i})$$ (E6.2B)

Therefore, using the given data, we obtain

$$9540 \ \text{W} = 1 \times 1060 \times (T_{c,o} - 12)$$ (E6.2C)

The outlet temperature of the cold stream, $T_{c,o} = 21°C$

HX Design

The governing equation for the PFHX is given by

$$\dot{q} = U_i A_i \frac{\{\Delta T_L - \Delta T_S\}}{\ln\left[\dfrac{\Delta T_L}{\Delta T_S}\right]}$$ (E6.2D)

From the known inlet and outlet temperatures of hot and cold streams, we can calculate T_L and T_S as

$$T_L = 60°C - 12°C = 48°C \quad \text{and}$$

$$T_S = 30°C - 21°C = 9°C$$

$$\dot{q} = 9540 = U_i(\pi D_i L)\frac{\{\Delta T_L - \Delta T_S\}}{\ln\left[\dfrac{\Delta T_L}{\Delta T_S}\right]}$$ (E6.2E)

After substituting the values of q, U, D, ΔT_L, and ΔT_S into the above equation, we obtain the length of the CFHX as

$$9540 = 45 \times 3.14 \times 0.1 \times L \times \frac{\{48 - 9\}}{\ln\left[\dfrac{48}{9}\right]}$$ (E6.2F)

$$L = 29 \ \text{m}$$

The cost of the heat harvested from the solar heating system is determined as

$$\text{Cost} = \frac{9540 \times 24 \times 0.75 \times 0.15}{1000} \approx \$25.76$$

Example 6.3 (Bioprocessing Engineering—Cooling of oil in a plate heat exchanger): It was proposed to use a 1-mm thick copper ($k = 405/\text{m}\cdot°C$) plate heat exchanger cool hot oil (specific heat = 2900 J/kg·°C) at 120°C using cold water (specific heat = 4200 J/kg·°C) available at 5°C (Fig. 6.20). The hot oil enters at 0.3 kg/s while the cold water enters the heat exchanger in the counterclockwise direction at 0.9 kg/s and exits at 15°C. If the heat transfer coefficients associated with the hot and cold plates are 75 and 175 W/m²·°C, respectively, determine the area of the plates needed for this project.

Step 1: Schematic

The problem is illustrated in Fig. 6.20.

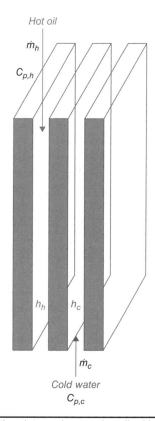

Hot oil

$\dot{m}_h$

$C_{p,h}$

h_h h_c

$\dot{m}_c$

Cold water

$C_{p,c}$

Figure 6.20 The schematic for the plate exchanger described in Example 6.3.

Step 2: Given data

Thermal conductivity of copper plate, $k = 405 \frac{W}{m}°C$; Thickness of the plate, $L = 1$ mm; Inlet temperature of the hot stream (oil) stream, $T_{h,i} = 120°C$; Mass flow rate of the hot stream, $\dot{m}_h = 0.3 \frac{kg}{s}$; Specific heat of the hot stream, $C_{p,h} = 2900 \frac{J}{kg·°C}$; Inlet temperature of the cold stream, $T_{c,i} = 5°C$; Outlet temperature of the cold stream, $T_{h,o} = 15°C$; Mass flow rate of the cold stream, $\dot{m}_c = 0.9 \frac{kg}{s}$; Specific heat of the cold stream, $C_{p,c} = 4200 \frac{J}{kg·°C}$; The heat transfer coefficient for the hot plate, $h_h = 75 \frac{W}{m^2·°C}$; and The heat transfer coefficient for the cold plate, $h_c = 175 \frac{W}{m^2·°C}$

To be determined: The size of the heat exchanger.

Assumptions: Steady-state flow, no loss of heat to the external surroundings, and the effects of temperature on the change in the specific heats of water and oil are negligible, that is, $C_{p,h}$ and $C_{p,c} \neq f$ (temperature).

Approach: We will first determine the overall heat transfer coefficient, U and subsequently determine area of the heat exchanger needed to cool the oil from 120°C to 80°C.

Step 3: Calculations

The overall heat transfer coefficient is given by

$$\frac{1}{U} = \frac{1}{h_h} + \frac{L}{k} + \frac{1}{h_c}$$

(E6.3A)

Substituting the data into the above equation we obtain the overall heat transfer coefficient as

$$\frac{1}{U} = \frac{1}{75} + \frac{0.001}{405} + \frac{1}{175} \tag{E6.3B}$$

$$U = 52.49 \frac{W}{m^2 \cdot °C}$$

Note: Due to high thermal conductivity and small thickness, the term $\frac{L}{k}$ is negligible and may as well be ignored.

The rate of heat transfer to the cold stream is determined as

$$\dot{q}_c = \dot{m}_c C_{p,c}(T_{c,o} - T_{c,i}) \tag{E6.3C}$$

Using the given data,

$$\dot{q}_c = 0.9 \times 4200 \times (15 - 5) \tag{E6.3D}$$

$$\dot{q}_c = 37,800 \text{ W}$$

The exit temperature of the hot stream can now be calculated as

$$\dot{q}_c = 37,800 \text{ W} = \dot{q}_c = \dot{m}_h C_{p,h}(T_{h,i} - T_{h,o}) \tag{E6.3E}$$

Substituting the given data into the above equation we obtain the outlet temperature of the hot steam

$$37,800 \text{ W} = 0.3 \times 2900 \times (120 - T_{h,o}) \tag{E6.3F}$$

$$T_{h,o} = 76.5°C$$

The temperatures at the inlet and the outlet are depicted in Fig. 6.21.

Using the temperature gradients at the inlet (105°C) and the outlet (71.5°C), we can now determine the area needed using the LMTD approach

$$\dot{q} = 37,800 = UA \frac{\{\Delta T_L - \Delta T_S\}}{\ln\left[\frac{\Delta T_L}{\Delta T_S}\right]} \tag{E6.3G}$$

Substituting the given data into the above equation we obtain the area needed for cooling the oil

$$37,800 = 52.5 \times A \times \frac{\{105 - 71.5\}}{\ln\left[\frac{105}{71.5}\right]} \tag{E6.3H}$$

$$A \approx 8.26 \text{ m}^2$$

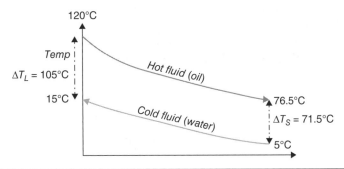

FIGURE 6.21 The temperature profiles for hot and cold streams for the counter-flow plate heat exchanger.

6.8 Approach 2—The Effectiveness-Number of Transfer Units (ε-NTU) Method

In 1950, a team of engineers from Stanford University, W. M. Kays and A. L. London, developed an elegant method to design and analyze heat exchangers. The process involves the formulations of several nondimensional ratios that describe the heat exchanger performance and combining them with the mathematical equations developed in the LMTD approach.

Once again, considering that wide applications of a CFHX in industry, we will consider a tube-in-tube CFHX in which hot stream flows in the inner tube while the cold fluid flows in the annulus in the opposite direction (Fig. 6.22).

First, we define the effectiveness of the heat exchanger in a dimensional ratio as

$$\varepsilon = \frac{\dot{q}}{\dot{q}_{max}} \tag{6.44}$$

where $\dot{q}$ is the actual rate of heat transfer in the heat exchanger system and can be mathematically expressed as below:

$$\dot{q} = \dot{m}_h C_{p,h}\left(T_{h,i} - T_{h,o}\right) = C_h\left(T_{h,i} - T_{h,o}\right) \tag{6.45}$$

or

$$\dot{q} = \dot{m}_c C_{p,c}\left(T_{c,o} - T_{c,i}\right) = C_c\left(T_{c,o} - T_{c,i}\right) \tag{6.46}$$

Because, we have previously defined the heat capacity rates for the hot (C_h) and cold (C_c) streams as $C_h = \dot{m}_h C_{p,h}$ and $C_c = \dot{m}_c C_{p,c}$.

The theoretical maximum heat transfer rate $\dot{q}_{max}$ is a function of the maximum permissible temperature gradient ΔT_{max} and the minimum heat capacity C_{min}. It follows that the maximum permissible temperature gradient occurs at the beginning of the heat transfer process ($t = 0$) and is the difference between the temperatures of the incoming hot and cold streams. Similarly, the smaller of the heat capacity rates of hot ($\dot{m}_h C_{p,h} = C_h$) and cold ($\dot{m}_c C_{p,c} = C_c$) streams will provide the theoretical maximum heat transfer rate. Therefore,

$$\dot{q}_{max} = C_{min}\left(T_{h,i} - T_{c,i}\right) \tag{6.47}$$

where C_{min} is the smaller value of C_h and C_c. Depending on the system configuration, it is possible that either C_c is smaller than C_h ($C_c < C_h$) or C_h is smaller than C_c ($C_h < C_c$) or C_c is the same as C_h ($C_c = C_h$). In this discussion, we will simply call smaller heat capacity rate as C_S and the larger heat capacity rate as C_L.

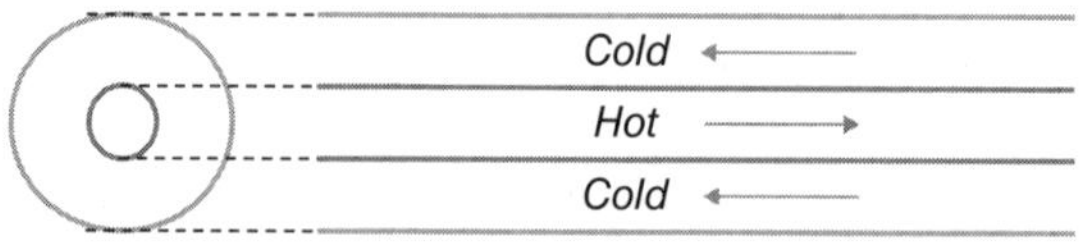

FIGURE 6.22 Schematic for a double-pipe counter-flow exchanger carrying hot fluid in the inner pipe.

Just for the discussion purposes, if we assume that $C_h < C_c$, then $C_h = C_S$, and $C_c = C_L$, then effectiveness (Eq. 6.44) becomes

$$\varepsilon = \frac{\dot{q}}{\dot{q}_{max}} = \frac{\dot{m}_h C_{p,h}(T_{h,i} - T_{h,o})}{C_h(T_{h,i} - T_{c,i})} = \frac{(T_{h,i} - T_{h,o})}{(T_{h,i} - T_{c,i})} \qquad (6.48)$$

In addition, C_h (or C_{small} or simply C_S) and C_c (or C_{large} or simply C_L) can be transformed into another dimensionless ratio called capacity ratio (C) that is expressed as

$$C = \frac{C_S}{C_L} \qquad (6.49)$$

The next step is to revisit the math relations employed in the LMTD method and algebraically convert the equations into nondimensional terms.

Recalling that

$$\dot{q}_h = \dot{q}_c = \dot{q} \qquad (6.50)$$

Because we are assuming that $C_h = \dot{m}_h C_{p,h}$ is the smaller heat capacity (C_S), the actual heat transfer is controlled by the smaller heat capacity.

Therefore, combining this idea with the LMTD, we obtain

$$\dot{q}_h = \dot{q} = C_h(T_{h,i} - T_{h,o}) = C_S(T_{h,i} - T_{h,o}) = UA\frac{\{\Delta T_1 - \Delta T_2\}}{\ln\left[\dfrac{\Delta T_1}{\Delta T_2}\right]} \qquad (6.51)$$

$$(T_{h,i} - T_{h,o}) = \frac{UA}{C_S}\frac{\{\Delta T_1 - \Delta T_2\}}{\ln\left[\dfrac{\Delta T_1}{\Delta T_2}\right]} \qquad (6.52)$$

At this stage, we can see that $\dfrac{UA}{C_S}$ is a dimensionless quantity and is called the number of transfer units or simply NTU, which is a direct function of the area of the heat exchanger.

$$NTU = \frac{UA}{C_S} \qquad (6.53)$$

$$(T_{h,i} - T_{h,o}) = NTU\frac{\{\Delta T_1 - \Delta T_2\}}{\ln\left[\dfrac{\Delta T_1}{\Delta T_2}\right]} \qquad (6.54)$$

The next step is to express ΔT_1 and ΔT_2 in terms of $T_{h,i}$ and $T_{h,o}$. To accomplish this, let us again consider Eq. (6.44)

$$\varepsilon = \frac{\dot{q}}{\dot{q}_{max}} = \frac{\dot{m}_h C_{p,h}(T_{h,i} - T_{h,o})}{C_h(T_{h,i} - T_{c,i})} = \frac{(T_{h,i} - T_{h,o})}{(T_{h,i} - T_{c,i})} \qquad (6.55)$$

The above equation may be rewritten in terms of $T_{c,i}$ as below

$$\varepsilon = \frac{(T_{h,i} - T_{h,o})}{(T_{h,i} - T_{c,i})} \tag{6.56}$$

$$\varepsilon(T_{h,i} - T_{c,i}) = (T_{h,i} - T_{h,o}) \tag{6.57}$$

$$T_{c,i} = T_{h,i} + \frac{T_{h,o}}{\varepsilon} - \frac{T_{h,i}}{\varepsilon} \tag{6.58}$$

Similarly using the heat balance equation $\dot{q}_h = \dot{q}_c$ we obtain

$$\dot{m}_h C_{p,h}(T_{h,i} - T_{h,o}) = \dot{m}_c C_{p,c}(T_{c,o} - T_{c,i}) \tag{6.59}$$

Or

$$C_h(T_{h,i} - T_{h,o}) = C_c(T_{c,o} - T_{c,i}) \tag{6.60}$$

Because $C_h = C_S$ and $C_c = C_L$, we can rearrange the above equation as

$$\frac{C_S}{C_L}(T_{h,i} - T_{h,o}) = (T_{c,o} - T_{c,i}) \tag{6.61}$$

Combining with Eq. (6.49), we obtain the following equation.

$$C(T_{h,i} - T_{h,o}) = (T_{c,o} - T_{c,i}) \tag{6.62}$$

Rearranging the above equation and substituting the expression for $T_{c,i}$ from Eq. (6.58), we can express in terms of $T_{c,o}$ as

$$T_{c,o} = CT_{h,i} - CT_{h,o} + T_{h,i} + \frac{T_{h,o}}{\varepsilon} - \frac{T_{h,i}}{\varepsilon} \tag{6.63}$$

From Fig. 6.16 (counter-flow heat exchanger), we note that

$$(T_{h,i} - T_{c,o}) = \Delta T_1 \tag{6.64}$$

$$(T_{h,o} - T_{c,i}) = \Delta T_2 \tag{6.65}$$

Therefore, after substituting the expressions for $T_{c,o}$ and $T_{c,i}$ in Eqs. (6.64) and (6.65), respectively,

$$\Delta T_1 = (T_{h,i} - T_{c,o}) = T_{h,i} - \left(CT_{h,i} - CT_{h,o} + T_{h,i} + \frac{T_{h,o}}{\varepsilon} - \frac{T_{h,i}}{\varepsilon}\right) \tag{6.66}$$

$$\Delta T_1 = T_{h,i} - CT_{h,i} + CT_{h,o} - T_{h,i} - \frac{T_{h,o}}{\varepsilon} + \frac{T_{h,i}}{\varepsilon} \tag{6.67}$$

$$\Delta T_1 = (T_{h,i} - T_{h,o})\left(\frac{1}{\varepsilon} - C\right) \tag{6.68}$$

And

$$\Delta T_2 = T_{h,o} - T_{c,i} = T_{h,o} - \left(T_{h,i} + \frac{T_{h,o}}{\varepsilon} - \frac{T_{h,i}}{\varepsilon} \right) \tag{6.69}$$

$$\Delta T_2 = T_{h,o} - T_{c,i} = T_{h,o} - T_{h,i} - \frac{T_{h,o}}{\varepsilon} + \frac{T_{h,i}}{\varepsilon} \tag{6.70}$$

$$\Delta T_2 = (T_{h,i} - T_{h,o}) \left(\frac{1}{\varepsilon} - 1 \right) \tag{6.71}$$

Now, plugging the expressions for ΔT_1 and ΔT_2 in Eq. (6.54)

$$(T_{h,i} - T_{h,o}) = \mathrm{NTU} \frac{\{\Delta T_1 - \Delta T_2\}}{\ln\left[\frac{\Delta T_1}{\Delta T_2}\right]} \tag{6.72}$$

$$(T_{h,i} - T_{h,o}) = \mathrm{NTU} \frac{\left\{ \left[(T_{h,i} - T_{h,o})\left(\frac{1}{\varepsilon} - C\right) \right] - \left[(T_{h,i} - T_{h,o})\left(\frac{1}{\varepsilon} - 1\right) \right] \right\}}{\ln\left[\frac{(T_{h,i} - T_{h,o})\left(\frac{1}{\varepsilon} - C\right)}{(T_{h,i} - T_{h,o})\left(\frac{1}{\varepsilon} - 1\right)} \right]} \tag{6.73}$$

$$(T_{h,i} - T_{h,o}) = \mathrm{NTU} \frac{(T_{h,i} - T_{h,o})\left\{ \frac{1}{\varepsilon} - C - \frac{1}{\varepsilon} + 1 \right\}}{\ln\left[\frac{(T_{h,i} - T_{h,o})\left(\frac{1 - \varepsilon C}{\varepsilon}\right)}{(T_{h,i} - T_{h,o})\left(\frac{1 - \varepsilon}{\varepsilon}\right)} \right]} \tag{6.74}$$

After cancelling out and simplification, the following equation is obtained

$$\ln\left[\frac{1 - \varepsilon C}{1 - \varepsilon} \right] = \mathrm{NTU}(1 - C) \tag{6.75}$$

Or

$$\frac{1 - \varepsilon C}{1 - \varepsilon} = e^{\mathrm{NTU}(1-C)} \tag{6.76}$$

Rearranging

$$(1 - \varepsilon C)e^{\mathrm{NTU}(C-1)} = 1 - \varepsilon \tag{6.77}$$

$$e^{\mathrm{NTU}(C-1)} - \varepsilon C e^{\mathrm{NTU}(C-1)} = 1 - \varepsilon \tag{6.78}$$

$$1 - e^{\mathrm{NTU}(C-1)} = \varepsilon(1 - C e^{\mathrm{NTU}(C-1)}) \tag{6.79}$$

$$\varepsilon = \frac{1 - e^{\mathrm{NTU}(C-1)}}{(1 - C e^{\mathrm{NTU}(C-1)})} \tag{6.80}$$

6.9   The ε-NTU Method for a PFHX

Similar to a CFHX, a PFHX (Fig. 6.23) can also be analyzed using the same approach as described by Kays and London.

$$T_{c,i} = T_{h,i} + \frac{T_{h,o}}{\varepsilon} - \frac{T_{h,i}}{\varepsilon} \text{ and } T_{c,o} = CT_{h,i} - CT_{h,o} + T_{h,i} + \frac{T_{h,o}}{\varepsilon} - \frac{T_{h,i}}{\varepsilon} \tag{6.81}$$

From Fig. 6.17 (parallel-flow heat exchanger),

$$(T_{h,i} - T_{c,i}) = \Delta T_1 \tag{6.82}$$

$$(T_{h,o} - T_{c,o}) = \Delta T_2 \tag{6.83}$$

Substituting the expressions for $T_{c,o}$ and $T_{c,i}$ from Eq. (6.81) in Eqs. (6.82) and (6.83), respectively, we obtain

$$\Delta T_1 = \frac{1}{\varepsilon}(T_{h,i} - T_{h,o}) \tag{6.84}$$

and

$$\Delta T_2 = (T_{h,i} - T_{h,o})\left(\frac{1}{\varepsilon} - C - 1\right) \tag{6.85}$$

Now, plugging the expressions for ΔT_1 and ΔT_2 in Eq. (6.54)

$$(T_{h,i} - T_{h,o}) = \text{NTU}\frac{\{\Delta T_1 - \Delta T_2\}}{\ln\left[\dfrac{\Delta T_1}{\Delta T_2}\right]} \tag{6.86}$$

$$(T_{h,i} - T_{h,o}) = \text{NTU}\frac{\left\{\dfrac{1}{\varepsilon}(T_{h,i} - T_{h,o}) - \left(\dfrac{1}{\varepsilon} - C - 1\right)(T_{h,i} - T_{h,o})\right\}}{\ln\left[\dfrac{\dfrac{1}{\varepsilon}(T_{h,i} - T_{h,o})}{\left(\dfrac{1}{\varepsilon} - C - 1\right)(T_{h,i} - T_{h,o})}\right]} \tag{6.87}$$

$$1 = \text{NTU}\frac{(C+1)}{\ln\left[\dfrac{\dfrac{1}{\varepsilon}}{\left(\dfrac{1}{\varepsilon} - C - 1\right)}\right]} \tag{6.88}$$

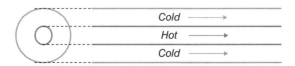

Figure 6.23   Schematic for a double-pipe parallel-flow heat exchanger carrying hot fluid in the inner pipe.

$$1 = \text{NTU} \frac{(C+1)}{\ln\left[\dfrac{1}{1-\varepsilon C-\varepsilon}\right]} \tag{6.89}$$

$$\ln\left[\frac{1}{1-\varepsilon(C+1)}\right] = \text{NTU}(C+1) \tag{6.90}$$

$$\frac{1}{1-\varepsilon(C+1)} = e^{\text{NTU}(C+1)} \tag{6.91}$$

$$\frac{1}{e^{\text{NTU}(C+1)}} = 1-\varepsilon(C+1) \tag{6.92}$$

$$1-e^{-\text{NTU}(C+1)} = \varepsilon(C+1) \tag{6.93}$$

$$\varepsilon = \frac{1-e^{-\text{NTU}(C+1)}}{(C+1)} \tag{6.94}$$

6.10   Special Cases

Case 1 ($C = 1$): In some situations, the heat capacities of the hot and cold streams are same. This usually occurs when the same fluid is employed as the hot and cold streams at the same mass flow rate. In such cases (neglecting the minor effects of temperature on specific heat), $C_c = C_h$ and $C = \dfrac{C_{\text{small}}}{C_{\text{large}}} = 1$. The corresponding equations for the effectiveness and NTU for the parallel- and counter-flow heat exchangers can be derived by applying L'Hospital rule.

Case 2 ($C = 0$): The other extreme case is when the hot and cold steams have dissimilar mass flow rates and specific heats (gas-liquid heat exchanger systems).

Additionally, the relationship between the effectiveness and NTU for a given C value is plotted in Fig. 6.24 for parallel- and counter-flow heat exchangers. For other types of heat exchangers, students are encouraged to refer to the literature.

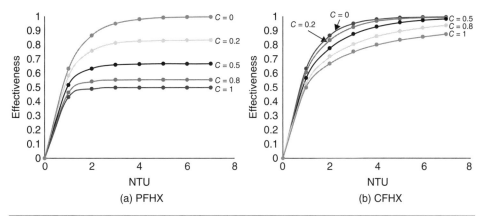

Figure 6.24   Plots between effectiveness, ε and NTU for a given heat capacity ratio, C for parallel-flow (a) and counter-flow (b) heat exchangers.

Example 6.4 (Bioprocessing Engineering—Design of a heat exchanger for the cooling of biodiesel):   Biodiesel (specific heat = 2800 J/kg·°C) at 65°C is to be cooled to 35°C using cold water available at 20°C in a thin-walled, double-pipe counter-flow heat exchanger (Fig. 6.25). The biodiesel enters the inner pipe of 10-cm diameter at 2 kg/s while the water (specific heat = 4200 J/kg·°C) enters the annulus at 1 kg/s. If the overall heat transfer coefficient of the system is 500 W/m²·°C, design a suitable heat exchanger.

Step 1: Schematic
The problem is described in Fig. 6.25.

Step 2: Given data
Inlet temperature of the hot stream, $T_{h,i} = 65°C$; Outlet temperature of the hot stream, $T_{h,o} = 35°C$; Specific heat of the hot stream (biodiesel), $C_{p,h} = 2800$ J/kg·°C; Mass flow rate of the hot stream, $\dot{m}_h = 2 \dfrac{kg}{s}$; Inlet temperature of the cold stream, $T_{c,i} = 20°C$; Specific heat of the cold stream (water), $C_{p,c} = 4200$ J/kg·°C; Mass flow rate of the cold stream, $\dot{m}_c = 1 \dfrac{kg}{s}$; The overall heat transfer coefficient for the inner tube, $U_i = 500 \dfrac{W}{m^2 \cdot °C}$; and The diameter of the inner tube, $D_i = 10$ cm

To be determined: The design (area and length) of the heat exchanger.

Assumptions: Steady-state flow, no loss of heat to the external surroundings, and the effects of temperature on the specific heats of biodiesel and water are negligible, that is, $C_{p,h}$ and $C_{p,c} \neq f$ (temperature).

Approach: Using the ε-NTU method, we will determine the value of NTU and subsequently, design the area of the heat exchanger.

Step 3: Calculations

The heat capacity rate for the cold stream, $C_c = \dot{m}_c C_{p,c} = 1 \times 4200 = 4200 \dfrac{W}{°C} = C_S$

The heat capacity rate for the hot stream, $C_h = \dot{m}_h C_{p,h} = 2 \times 2800 = 5600 \dfrac{W}{°C} = C_L$

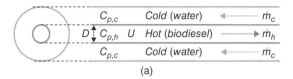

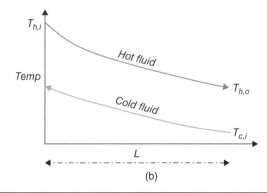

(b)

FIGURE 6.25   The schematic for the counter-flow heat exchanger for Example 6.4 (a) and the temperature profiles for hot and cold streams (b).

Therefore, the value of C becomes,

$$C = \frac{C_S}{C_L} = \frac{4200}{5600} = 0.75$$

The rate of heat transfer from the hot stream is calculated as

$$\dot{q}_h = \dot{m}_h C_{p,h}(T_{h,i} - T_{h,o}) \qquad \text{(E6.4A)}$$

Substituting the data into the above equation, we obtain

$$\dot{q}_h = 2 \times 2800 \times (65 - 35) = 168{,}000 \text{ W}$$

The maximum possible heat transfer rate is determined as

$$\dot{q}_{max} = C_S(T_{h,i} - T_{c,i}) = 4200 \times (65 - 20) = 189{,}000 \text{ W}$$

Now, the effectiveness factor is determined as

$$\varepsilon = \frac{\dot{q}}{\dot{q}_{max}} = \frac{168{,}000}{189{,}000} = 0.889$$

Using the values of $\varepsilon = 0.889$ and $C = 0.75$, we can visually determine the value of NTU as close to 5 from Fig. 6.24.
    Now,

$$\text{NTU} = \frac{UA}{C_S} \qquad \text{(E6.4B)}$$

Substituting the given data into the above equation, we determine the design parameters (area and the length of the HX) as

$$5 = \frac{500 \times A_i}{4200} \qquad \text{(E6.4C)}$$

resulting in

$$A_i = 42 \text{ m}^2$$

$$L \approx 133.\,76 \text{ m (because the area, } A = \pi D_i L)$$

We can also determine the outlet temperature of the water as

$$\dot{q} = \dot{m}_h C_{p,h}(T_{h,i} - T_{h,o}) = 168{,}000 = \dot{m}_c C_{p,c}(T_{c,o} - T_{c,i}) \qquad \text{(E6.4D)}$$

Substituting the data,

$$T_{c,o} = \left(\frac{168{,}000}{1 \times 4200}\right) + 20 \qquad \text{(E6.4E)}$$

The outlet temperature of the water, $T_{c,o} = 60°C$

Example 6.5 (Environmental Engineering: Heating of well water): In a shrimp hatchery, cold water is sourced from a well at 11°C and heated using a hot water available at 80°C (Fig. 6.26). To accomplish this heating, a thin-walled, double-pipe stainless steel parallel-flow heat exchanger equipped with an area of 25 m² is used. The cold-water flows in the inner tube at 0.4 kg/s while the hot water flows in the annulus the same direction at 1.2 kg/s. Assuming a constant specific heat of water as 4200 J/kg·°C and an overall heat transfer coefficient for the system as 125 W/m²·°C, determine the outlet temperatures of the hot and cold streams and the corresponding heat transfer rate. Additionally, what is the impact of switching the flow from parallel to counter flow?

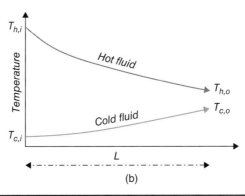

Figure 6.26 The schematic for the parallel-flow heat exchanger for Example 6.5 (a) and the temperature profiles for hot and cold streams (b).

Step 1: Schematic
The problem is described in Fig. 6.26.

Step 2: Given data
Inlet temperature of the hot stream (water), $T_{h,i} = 80°C$; Specific heat of the hot stream (water), $C_{p,h} = 4200$ J/kg·°C; Mass flow rate of the hot stream, $\dot{m}_h = 1.2\ \frac{kg}{s}$; Inlet temperature of the cold stream, $T_{c,i} = 11°C$; Specific heats of the cold and hot streams (water), $C_{p,c} = C_{p,h} = 4200$ J/kg·°C; Mass flow rate of the cold stream, $\dot{m}_c = 0.4\ \frac{kg}{s}$; The overall heat transfer coefficient, $U = 125\ \frac{W}{m^2·°C}$; and The area, $A = 25$ m²

To be determined: Outlet temperatures of the cold and hot streams ($T_{c,o}$ and $T_{h,o}$), heat transfer rate ($\dot{q}$), and the impact of switching from parallel to counter flow.

Assumptions: Steady-state flow, no loss of heat to the external surroundings, and the effects of temperature on the specific heats of water are negligible, that is, $C_{p,h}$ and $C_{p,c} \neq f$(temperature).

Approach: The exit temperatures of the hot and cold fluids are not provided and therefore we cannot use the LMTD approach to solve this problem. In such cases, the ε-NTU method will be appropriate for exit temperatures of the heat exchanger and the corresponding heat transfer rate.

Step 3: Calculations

The heat capacity rate for the cold stream, $C_c = \dot{m}_c C_{p,c} = 0.4 \times 4200 = 1680\ \frac{W}{°C} = C_S$

The heat capacity rate for the hot stream, $C_h = \dot{m}_h C_{p,h} = 1.2 \times 4200 = 5040\ \frac{W}{°C} = C_L$

Therefore, the value of C becomes,

$$C = \frac{C_S}{C_L} = \frac{1680}{5040} = 0.333$$

Now, we will determine the NTU for the heat exchanger as

$$\text{NTU} = \frac{UA}{C_S} \qquad \text{(E6.5A)}$$

Configuration	NTU	Effectiveness (ε)	Eq. #
Double-pipe (parallel flow)	$$\text{NTU} = \frac{-\ln(1 - \varepsilon(C + 1))}{(C + 1)}$$ $$\text{NTU} = -\frac{\ln(1 - 2\varepsilon)}{2} \text{ when } C = 1$$	$$\varepsilon = \frac{1 - e^{-\text{NTU}(C+1)}}{(C + 1)}$$ $$\varepsilon = \frac{1 - e^{(-2\text{NTU})}}{2} \text{ when } C = 1$$	(6.95)
Double-pipe (counter flow)	$$\text{NTU} = \frac{\ln\left[\dfrac{1 - \varepsilon C}{1 - \varepsilon}\right]}{(1 - C)}$$ $$\text{NTU} = \frac{\varepsilon}{1 - \varepsilon} \text{ when } C = 1$$ Using L'Hospital Rule	$$\varepsilon = \frac{1 - e^{\text{NTU}(C-1)}}{(1 - Ce^{\text{NTU}(C-1)})}$$ $$\varepsilon = \frac{\text{NTU}}{1 + \text{NTU}} \text{ when } C = 1$$ Using L'Hospital Rule	(6.96)
All configurations when $C = 0$	$\text{NTU} = -\ln(1 - \varepsilon)$	$\varepsilon = 1 - \exp(-\text{NTU})$	(6.97)

TABLE 6.3 Summary of Equations to Determine the HX Dimensional Parameters (Adapted from Cengel and Ghajar)

Substituting the relevant values in the above equation, we obtain

$$\text{NTU} = \frac{125 \times 25}{1680} = 1.86$$

We can either use Eq. (6.95) from Table 6.3 or use the plot in Fig. 6.24 to determine the effectiveness factor. In this case, we will use the graphical approach which yields the effectiveness factor as

$$\varepsilon \approx 0.69$$

The next step is to determine the heat transfer rate.

$$\varepsilon = 0.69 = \frac{\dot{q}}{\dot{q}_{\max}} = \frac{\dot{q}}{C_{\min}(T_{h,i} - T_{c,i})} = \frac{\dot{q}}{1680 \times (80 - 11)} = 0.69 \tag{E6.5B}$$

$$\dot{q} = \dot{q}_h = \dot{q}_c = 79{,}984.8 \text{ W}$$

The exit temperatures of the hot and cold streams are determined as

$$\dot{q} = \dot{m}_h C_{p,h}(T_{h,i} - T_{h,o}) = 79{,}984.8 = \dot{m}_c C_{p,c}(T_{c,o} - T_{c,i}) \tag{E6.5C}$$

$$T_{c,o} = 58.6°C \quad \text{and} \quad T_{h,o} = 64.1°C$$

Impact of switching the heat exchanger to counter-flow mode
From plot B in Fig. 6.24, we can read the effectiveness of the heat exchanger as

$$\varepsilon \approx 0.786$$

and the corresponding heat transfer rate becomes

$$\dot{q} = \dot{q}_h = \dot{q}_c = 91{,}171 \text{ W}$$

The heat exchanger will be **13.9%** more effective if the flow patterns are switched. The exit temperatures of the hot and cold streams are determined as described previously as

$$T_{c,o} \approx 65.2°C \quad \text{and} \quad T_{h,o} \approx 61.9°C$$

Example 6.6 (Food Processing—Comparison between LMTD and ε-NTU methods): Pasteurization of skim milk occurs in an 89-m long thin-walled, stainless steel double-tube CFHX using steam. The skim milk at 50°C (specific heat = 3960 J/Kg·°C) enters the 7.5-cm diameter inner tube at 0.2 kg/s and exits at 90°C while the steam at 110°C (~143 kPa) flows through the annulus at 0.5 kg/s in the opposite direction. Using LMTD and ε-NTU methods, determine the overall heat transfer coefficient for this HX system.

Step 1: Schematic
The problem is depicted in Fig. 6.27.

Step 2: Given data
Length of the HX, $L = 89$ m; Diameter of the inner tube, $D_i = 7.5$ cm; Mass flow rate of the cold stream, $\dot{m}_c = 0.2 \, \dfrac{\text{kg}}{\text{s}}$; Specific heat of the cold stream, $C_{p,c} = 3960$ J/kg·°C; Inlet temperature of the cold stream, $T_{c,i} = 50$°C; Outlet temperature of the cold stream, $T_{c,o} = 90$°C; Inlet temperature of the hot stream, $T_{h,i} = 110$°C; Outlet temperature of the hot stream, $T_{h,o} = 110$°C; and Mass flow rate of the hot stream, $\dot{m}_h = 0.5 \, \dfrac{\text{kg}}{\text{s}}$

To be determined: The overall heat transfer coefficient of the HX system, U.

Assumptions: Steady-state flow, no loss of heat to the external surroundings, and the effect of temperature on the specific heat of milk is negligible, and steam temperature remains constant after condensation (only phase change occurs).

Approach: Using the available data for the skim milk, we will first employ the LMTD to determine the overall heat transfer coefficient and subsequently compare the results with those of ε-NTU methods.

Step 3: Calculations

LMTD approach

$$\text{Rate of heat transfer to the cold stream, } \dot{q}_c = \dot{m}_c C_{p,c} (T_{c,o} - T_{c,i}) \qquad \text{(E6.6A)}$$

From the given data, we obtain

$$\dot{q}_c = 0.2 \times 3960 \times (90 - 50) = 31{,}680 \text{ W}$$

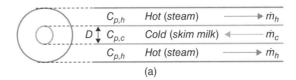

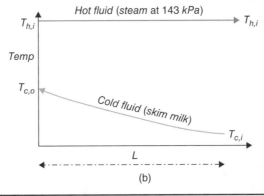

FIGURE 6.27 The schematic for the counter-flow heat exchanger for Example 6.6 for pasteurizing milk (a) and the temperature profiles for hot and cold streams (b).

Therefore, assuming the rate of heat transfer is the same as the rate of the heat transferred to the cold stream, the overall heat transfer coefficient for the inner tube can be determined using the HX design equation as below:

$$\dot{q} = \dot{q}_c = UA\frac{\{\Delta T_L - \Delta T_S\}}{\ln\left[\dfrac{\Delta T_L}{\Delta T_S}\right]} \tag{E6.6B}$$

$$\dot{q} = \dot{q}_c = 31{,}680 = U_i(\pi D_i L)\frac{\{\Delta T_L - \Delta T_S\}}{\ln\left[\dfrac{\Delta T_L}{\Delta T_S}\right]} \tag{E6.6C}$$

We can calculate T_L and T_S using the inlet and outlet temperatures as

$$T_L = 110°C - 50°C = 60°C \text{ and}$$

$$T_S = 110°C - 90°C = 20°C$$

Substituting the data into the above equation, we obtain the value of overall heat transfer coefficient for the inner tube as

$$31{,}680 = U_i \times 3.14 \times 0.075 \times 89 \times \frac{\{60 - 20\}}{\ln\left[\dfrac{60}{20}\right]} \tag{E6.6D}$$

$$U_i = 41.5\ \frac{W}{m^2 \cdot °C}$$

ε-NTU approach

The heat capacity rate for the cold stream, $C_c = \dot{m}_c C_{p,c} = 0.2 \times 3960 = 792\ \dfrac{W}{°C}$

The heat capacity rate for the hot stream, $C_h = \dot{m}_h C_{p,h} = 0.5 \times \infty = \infty$ (phase change)
   Therefore, the value of C becomes,

$$C = \frac{C_S}{C_L} = \frac{792}{\infty} = 0$$

Rate of heat transfer to the cold stream, $\dot{q}_c = \dot{m}_c C_{p,c}(T_{c,o} - T_{c,i})$ (E6.6E)

From the given data, we obtain

$$\dot{q}_c = \dot{q} = 0.2 \times 3960 \times (90 - 50) = 31{,}680\ W$$

The maximum heat transfer rate, $\dot{q}_{max}$ becomes

$$\dot{q}_{max} = C_s \Delta T_{max} = C_s(T_{h,i} - T_{c,i}) = 792 \times (110 - 50) = 47{,}520\ W$$

Next, the value of effectiveness, ε can be determined as

$$\varepsilon = \frac{\dot{q}}{\dot{q}_{max}} = \frac{31{,}680}{47{,}520} \approx 0.667$$

Next, the value of NTU can be determined either via the equation or graphically. In this case, we will use the equation as described below:

$$NTU = \frac{\ln\left[\dfrac{1 - \varepsilon C}{1 - \varepsilon}\right]}{(1 - C)} \tag{E6.6F}$$

Substituting the values of C and ε into the above equation,

$$\text{NTU} = \frac{\ln\left[\dfrac{1 - 0.67 \times 0}{1 - 0.67}\right]}{(1 - 0)} \tag{E6.6G}$$

$$\text{NTU} = 1.1$$

Note: Because $C = 0$, we can directly use Eq. (6.97) from Table 6.3

$$\text{NTU} = -\ln(1 - \varepsilon) = -\ln(1 - 0.677) = 1.098$$

Now, we can determine the heat transfer coefficient for the inner rube as

$$\text{NTU} = 1.098 = \frac{U_i A}{C_{small}} \tag{E6.6H}$$

$$U_i = \frac{1.098 \times 792}{3.14 \times 0.075 \times 89} \tag{E6.6I}$$

$$U_i = 41.5 \, \frac{W}{m^2 \, ^\circ C}$$

which is the same as obtained via the LMTD approach.

If needed, we can also determine the mass flow rate of the steam as

$$\dot{q} = \dot{q}_c = \dot{q}_h = 31{,}680 = h_{fg} \dot{m}_h \tag{E6.6J}$$

where $h_{fg} = 2230$ kJ/kg (from steam tables)

This results in

$$\dot{m}_h = \frac{31{,}680}{2230} = 14.2 \times 10^{-3} \text{ kg/s}$$

Practice Problems for the FE Exam

1. Which type of heat exchangers are the most effective?

   a. Cross-flow

   b. Counter-flow

   c. Parallel-flow

   d. Phase-change

2. A thin-walled highly conductive double-tube counter-flow heat exchanger has an inside and outside heat transfer coefficient of 60 W/m².°C and 400 W/m².°C, respectively. The overall heat transfer coefficient for this heat exchanger is closest to

   a. 52 W/m².°C.

   b. 150 W/m².°C.

   c. 400 W/m².°C.

   d. 60 W/m².°C.

3. The hot-side and cold-side heat transfer coefficients for a 2-mm thick stainless ($k = 15$ W/m·°C) plate heat exchanger with an area of 0.55 m² were given as 75 W/m²·°C and 90 W/m²·°C, respectively. The overall heat transfer coefficient for this heat exchanger is about

   a. 61 W/m²·°C.

   b. 41 W/m²·°C.

   c. 71 W/m²·°C.

   d. 75 W/m²·°C.

4. A thin-walled, counter-flow, double-tube heat exchanger made of copper is used to cool hot bio-oil a wood pyrolysis system. The bio-oil (specific heat = 2500 J/kg·°C) at 100°C enters the inner tube of 3-cm diameter at 1 kg/s and exits the heat exchanger at 55°C while the cold water (specific heat = 4180 J/kg·°C) at 5°C flows in the annulus at 0.5 kg/s. The temperature of the water exiting the heat exchanger is about

   a. 38°C.

   b. 48°C.

   c. 53°C.

   d. 58°C.

5. For the data given in Problem 4, the maximum heat transfer rate the heat transfer can provide is about

   a. 125 kW.

   b. 100 kW.

   c. 150 kW.

   d. 199 kW.

6. For the data given in Problem 4, the representative temperature gradient for the heat exchanger is

   a. 40°C.

   b. 48°C.

   c. 53°C.

   d. 51°C.

7. If the heat exchanger described in Problem 4 is operated in parallel-flow mode, the representative temperature gradient now becomes

   a. 40°C.

   b. 34°C.

   c. 58°C.

   d. None of the above.

8. A tube-in-tube parallel flow heat exchanger is used for cooling the hot air from a gasifier via cold water. Previous experience with this system suggested an overall heat transfer coefficient based on inner tube (5-cm diameter) surface area of 350 W/m²·°C. Based on the measured temperatures at the inlet and the outlet, a log-mean temperature difference of 38°C was calculated. If this heat exchanger is expected to transfer 250 kW, the length of the heat exchanger is closest to

  a. 29 m.

  b. 50 m.

  c. 120 m.

  d. None of the above.

9. In a given heat exchanger, water (specific heat = 4180 J/kg·°C) is used for both cooling and heating at 120 kg/h. The capacity ratio for this heat exchanger is

  a. 1.

  b. 0.5.

  c. 0.

  d. 0.25.

10. The effectiveness for any heat exchanger for a given area maximizes when the capacity ratio approaches

  a. 1.

  b. 0.5.

  c. 0.

  d. 0.25.

Practice Problems for the PE Exam

1. In a fish hatchery, a counter-flow tube-in-tube heat exchanger is used to heat water from 15°C to 22°C using hot air at 120°C from a waste combustion operation. The water flows at 1 kg/s through the inner pipe made of copper ($k = 400$ W/m·°C) and is assumed to be thin-walled with a heat transfer coefficient of 120 W/m²·°C. The hot air flows in the opposite direction in the annulus at 1 kg/s with a heat transfer coefficient of 45 W/m²·°C. Calculate the overall heat transfer coefficient of the heat exchanger.

2. A stainless steel ($k = 15$ W/m·°C) plate heat exchanger is used to cool oil from 90°C to 60°C using cold water at 5°C. If the heat transfer coefficients associated with the hot side and cold side of the 1-mm thick plates are 250 W/m²·°C and 50 W/m²·°C, respectively, determine the overall heat transfer coefficient for the heat exchanger. Also, how will the overall heat transfer coefficient change if the effects of conduction through the plate are ignored?

3. A food processing operation decides to employ a thin-walled double-tube parallel flow heat exchanger to heat the intake water from 15°C to 25°C using hot water. The hot water enters the inner tube of a 5-cm diameter made of copper at 95°C at 2 kg/s

and exits the tube at 70°C. Determine the heat transfer rate and sizing of the heat exchanger needed if the overall heat transfer coefficient based on the inner tube was estimated to be 1000 W/m²·°C. Disregard the effects of temperature on the specific heat of the water.

4. Repeat Problem 3 now for a counter-flow configuration. Compare your results.

5. A 1-mm thick aluminum (K = 250 W/m·°C) plate exchanger is used to cool bio-oil (specific heat = 1500 J/kg·°C) from 150°C to 90°C using cold water available at 2°C. The mass flow rate of bio-oil is 0.2 kg/s while the cold water flows at 1 kg/s. If the heat transfer coefficients associated with the hot and cold plates are 250 and 110 W/m²·°C, respectively, determine the area of the heat exchanger.

6. A double-pipe parallel-flow heat exchanger is used to cool oil (specific heat = 2800 J/kg ·°C) from 80°C to 40°C using cold water available at 10°C. The oil flows in the inner pipe of 8-cm diameter at 1.5 kg/s while the water (specific heat = 4200 J/kg·°C) enters the annulus at 1.9 kg/s. If the heat transfer coefficient based on the inner pipe area is 750 W/m²·°C, design a suitable heat exchanger using the NTU approach.

7. Repeat Problem 6 using the LMTD approach and compare your results.

8. Soup (specific heat = 3000 J/kg·°C) is to be heated from 27°C to 75°C in a 100-long stainless-steel double-tube CFHX using steam. Steam at 105°C flows through the outer tube at 0.8 kg/s while the soup enters the 10-cm diameter inner tube at 0.3 kg/s. Determine the overall heat transfer coefficient for the inner tube of this HX system using both LMTD and ε-NTU methods.

9. Under what conditions will the heat capacity ratio approach 0 and 1? Discuss the effects of $c = 0$ and $c = 1$ on heat exchangers' performance.

CHAPTER 7

Elements of
Thermal Radiation

Chapter Objectives

The topics presented in this chapter will allow students to

- Determine the rate of radiation from black and real bodies.
- Apply the concepts of radiosity and view factors to solve radiation problems.
- Calculate the rate of heat transfer between two surfaces under various conditions.

7.1   Motivation

Thermal radiation or radiation is the third mode of heat transfer that occurs because of the temperature of the body. Radiation occurs all around us at all times. As engineers, our main objective is to determine the amount and the rate of radiation emitted by a given body at a given temperature either in a chosen direction or in all directions. In addition, we are also expected to analyze the radiation exchanged between two bodies at different temperatures and oriented differently. The ideas presented in this chapter will deal with radiation heat transfer applied to various practical situations.

7.2   Understanding Thermal Radiation

Consider a thermos containing hot coffee (Fig. 7.1). If we allow enough time to elapse, we will find the coffee will eventually be cooled down to the room temperature.

However, the thermos is made of insulating materials with vacuum between the walls. Based on our previous knowledge of heat transfer, we know that heat transfer from the coffee to the outside surroundings could not have occurred via conduction or convection because either of these processes would have required a medium (i.e., stagnant air for conduction and moving air for convection) (Fig. 7.1a). This is where radiation heat transfer played a major role in transferring the heat out of the coffee (Fig. 7.1b). The coffee, due to high temperature, emits energy via heat waves which propagate (without movement of the mass) the energy to the surroundings. Therefore, it is appropriate to say that the emission that occurs due to a body's temperature is called thermal radiation. In fact, every object around us (including us and our pets) constantly emits

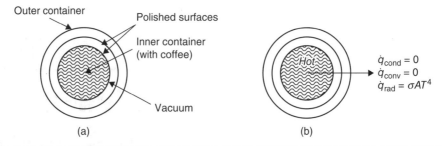

Figure 7.1 A cross-section of a typical thermos flask. A layer of vacuum separates the outer and inner containers.

and receives thermal radiation even though the bodies are at the same temperature. We do not feel this thermal radiation due to thermal equilibrium (Fig. 7.2).

Radiation occurs in all forms of matter in all directions. Theoretically, radiation is a volumetric phenomenon, and this is valid for gases. However, in the case of opaque solids and liquids, the radiation emitted by the molecules located within the bulk of the body will be intercepted and absorbed by the surrounding molecules. Thus, only the surface molecules will be able to transmit the energy to the surroundings. Therefore, for all practical purposes, we can consider the radiation from solids and liquids as surface phenomenon.

Thermal radiation is an extremely important type of heat transfer. Unlike conduction and convection, thermal radiation can occur in vacuum. We experience thermal radiation from the sun which travels through vacuum. When compared to conduction and convection, radiation heat transfer is the fastest mode of heat transfer. Because of the electromagnetic nature of the heat energy waves, the speed of thermal radiation is the same as the speed of the light. Another important difference is the way radiation propagates through thermal gradients. The thermal radiation from the sun reaches the earth despite the fact that the outer space is significantly colder than the earth's surface.

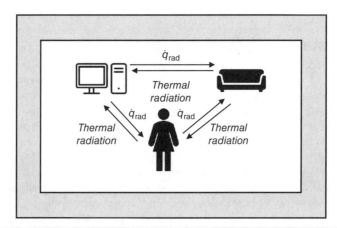

Figure 7.2 Thermal radiation continuously occurs between all bodies.

Radiation	Radio	Microwave	Infrared	Visible	Ultraviolet	X-Ray	Gamma
Wavelength (µm)	>10^5	1000–10^5	0.7–100	0.4–0.7	0.01–0.4	0.01–10^{-5}	<10^{-5}
Frequency (GHz)	<3	3–300	300–4×10^5	4–7.5×10^5	7.5×10^5–3×10^7	3×10^7–3×10^{10}	>3×10^{10}

TABLE 7.1   Summary of the Approximate Wavelengths and Frequencies of the Electromagnetic Spectrum (nasa.gov)

7.3   Thermal Radiation and the Electromagnetic Spectrum

Thermal radiation is type of the electromagnetic wave propagation. Therefore, like any other electromagnetic waves, thermal radiation also possesses wavelength (λ) and frequency (n) and the speed of thermal radiation is given by the speed of light (c) as

$$c = n\lambda \qquad (7.1)$$

To understand where thermal radiation fits in, we will briefly review the basics of the electromagnetic spectrum (Table 7.1).

The radio waves (low energy) are associated with wavelengths of greater than 10^5 µm, while the gamma radiations (very high energy) have extremely short wavelengths, typically, less than 10^{-5} µm. The visible light, which our eyes are adapted to, is a part of the electromagnetic spectrum that falls between 0.4 and 0.7 µm. The emissions with wavelengths between 0.7 and 100 µm are called infrared radiation which all the bodies always emit. Therefore, from a heat transfer standpoint, *thermal radiation* is that part of the electromagnetic spectrum whose wavelengths range between 0.1 and 100 µm.

7.4   The Concept of Blackbody Thermal Radiation

From physics, we are familiar that all bodies above absolute zero will emit radiation over several wavelengths. The radiation emitted will depend on the properties of the material and its temperature. As engineers, we need some standard reference to quantify and compare the radiation emitted by any given body and compare the radiation emissions between two bodies. Therefore, we will choose an imaginary body called a blackbody (or an idealized body) capable of 100% emission for all wavelengths uniformly in all directions (Fig. 7.3).

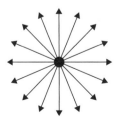

FIGURE 7.3   A blackbody is an ideal body that emits the radiation in all directions uniformly.

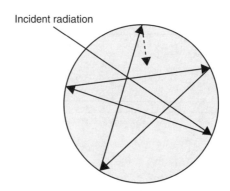

Incident radiation

FIGURE 7.4 Conceptual illustration of a blackbody in which the radiation is entirely absorbed by the surface of the enclosure.

Additionally, a blackbody can absorb 100% of the radiation it receives. Although blackbodies are conceptual, a few bodies such as the sun, carbon black, and large enclosure with a pin hole may be considered as blackbodies for practical purposes. As shown in Fig. 7.4, the radiation after entering the enclosure through the pin hole will be intercepted by a surface wherein a portion of the radiation is absorbed and the remaining is reflected. This absorption–reflection step continues until all the radiation is eventually absorbed by the surface.

For a blackbody at a given temperature, T, the maximum radiation was found to be proportional to T^4 and the area, A of the blackbody.

$$\dot{q}_{\text{max,rad}} \propto A \tag{7.2}$$

$$\dot{q}_{\text{max,rad}} \propto T^4 \tag{7.3}$$

Combining the above two equations and replacing the proportionality sign with a constant, σ called Stefan–Boltzmann constant, we obtain the famous Stefan–Boltzmann law of radiation heat transfer

$$\dot{q}_{\text{max}} = \dot{q}_{\text{max,rad}} = \sigma A T^4 \tag{7.4}$$

where
$$\sigma = 5.67 \times 10^{-8} \, \frac{\text{W}}{\text{m}^2 \cdot \text{K}^4} \tag{7.5}$$

Note: *Because this chapter is all about radiation, we will drop the subscript from $\dot{q}_{rad}$ and simply use $\dot{q}_{max}$ or $\dot{q}$ (as appropriate) to represent radiation heat transfer throughout this chapter.*

In conduction and convection problems, temperature can be expressed either in degree Celsius or Kelvin. However, it is very important to note that in the context of radiation, all calculations related to temperatures are always performed in Kelvin.

Example 7.1 (General Engineering–Total emission from the sun):   Assuming the sun as a blackbody, its surface temperature as 5507°C, and a total surface area of 6 trillion km², estimate the total amount of energy radiated from the sun about all wavelengths (Fig. 7.5).

Step 1: Schematic
The problem is depicted in Fig. 7.5.

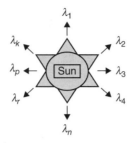

FIGURE 7.5 Schematic of the emission from the sun about all the wavelengths (λ_1, λ_2, ... λ_k).

Step 2: Given data

Temperature of the sun's surface, $T = 5507°C$; Total surface area, $A = 6 \times 10^{12}$ km^2

To be determined: Total emission from the sun about all wavelengths.

Assumptions: The sun is a blackbody and therefore a perfect emitter, $\varepsilon = 1$.

Analysis: Because of the blackbody assumption, we can use original form of the Planck's equation to determine the maximum amount of radiation energy.

Step 3: Calculations

The maximum radiation energy (emission) from the sun is determined as

$$\dot{q}_{max} = \sigma A T^4 = 5.67 \times 10^{-8} \times 6 \times 10^{12} \times (10^3)^2 \times (5,507 + 273)^4$$

$$\dot{q}_{max} = 3.79 \times 10^{26} \text{ W}$$

It should be noted that this energy is for all possible wavelengths.

7.5 Radiation from a Real Body

A blackbody is an ideal body that completely absorbs and emits all the radiation. However, the real bodies that we work with in our daily lives do not absorb 100% of all the radiation nor do them emit 100% of all the radiation. Therefore, as we defined in Chap. 1, we will use a nondimensional parameter called emissivity, ε to compare a real body with a blackbody. Hence,

$$\varepsilon = \frac{\dot{q}}{\dot{q}_{max}} \tag{7.6}$$

The emissivity of a real body's surface depends on the temperature. However, the effect of temperature on the emissivity is not very significant at temperatures less than about 300°C. Because most of the biological and agricultural engineering applications are associated with temperatures less than 300°C, we assume a constant emissivity for initial calculations. Likewise, the wavelength also influences the emissivity of a surface of a real body. Some surfaces, however, are not affected by the wavelengths and the value of the emissivity remains uniform about all wavelengths. Such surfaces are called gray surfaces (Table 7.2). Similarly, when the emissivity remains uniform in all directions, such surfaces are called diffuse surfaces.

Most importantly, the topography (smoothness, extent of polish, chemical treatment) of the surface will greatly affect the emissivity and therefore correct values are to be used in design calculations. The emissivities of commonly used materials are summarized in Table 7.3.

Surface	Emissivity
Blackbody	$\varepsilon = 1$, $\varepsilon \neq f(\lambda)$, and $\varepsilon \neq f(\text{direction})$
Gray body	$\varepsilon < 1$ and $\varepsilon \neq f(\lambda)$
Real body	$\varepsilon < 1$ and $\varepsilon = f(\lambda)$
Diffuse	$\varepsilon < 1$ and $\varepsilon \neq f(\text{direction})$
Diffuse and gray	$\varepsilon < 1$, $\varepsilon \neq f(\lambda)$, and $\varepsilon \neq f(\text{direction})$

TABLE 7.2 Summary Characteristics of Various Surfaces Emitting Radiation (Cengel & Ghajar, Heat and Mass Transfer: Fundamentals and Applications, Fifth Edition. McGraw Hill, 2016. Reprinted with permission.)

Material	Emissivity
Construction materials	0.9–0.95
Water	0.96
Human and animal skin	0.95–0.97
Metals (oxidized)	0.25–0.75
Metals (unpolished)	0.1–0.4
Metals (polished)	0.05–0.18
Carbon	0.8–0.95

TABLE 7.3 Emissivity of Common Materials Used in Biological and Agricultural Engineering at Room Temperature (adapted from Cengel & Ghajar, Heat and Mass Transfer: Fundamentals and Applications, Fifth Edition. McGraw Hill, 2016. Reprinted with permission.)

Example 7.2 (Environmental Engineering—Heat lost by pigs via radiation): A finisher hog barn holds 500 pigs of average weight of 120 kg. If the average temperature and the area of the pig are 38°C and 2.1 m², respectively, determine the total energy lost by the pigs assuming an emissivity of 0.96 (Fig. 7.6).

Step 1: Schematic
The problem is depicted in Fig. 7.6.

Step 2: Given data
Number of pigs = 500; Average weight, $W = 120$ kg; Average temperature of skin surface, $T = 38$°C; Average surface area, $A = 2.1$ m²; Emissivity of the pig skin, $\varepsilon = 0.96$

To be determined: Total emission from all the pigs for all wavelengths.

Assumptions: All pigs are uniform, and the emissivity is the same throughout the body.

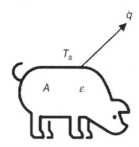

FIGURE 7.6 Schematic of a sample pig emitting radiation out of its body about all the wavelengths (note that there are 500 pigs).

Analysis: Because the pigs are considered real bodies, with an emissivity is 0.96, we can determine the total emission by multiplying the emissivity by the total blackbody emission.

Step 3: Calculations
The total emission by all the pigs is given by

$$\dot{q}_{Total} = \text{No. of pigs} \times \varepsilon\sigma AT^4 \tag{E7.2A}$$

$$\dot{q}_{Total} = 500 \times 0.96 \times 5.67 \times 10^{-8} \times 2.1 \times (273 + 38)^4$$

Total emission, $\dot{q}_{Total} = 534.67$ kW

7.6 Spectral Blackbody Emissive Power

We have thus far considered radiation from black and real bodies for all wavelengths at a given temperature. However, engineers also need to determine the radiation for a specific wavelength. For example, we want to know the emission of a headlight of an automobile at a specific wavelength at a given temperature to determine its suitability. Fortunately for us, Prof. Max Planck, the famous German theoretical physicist, derived an expression to determine the radiation (or emissive power) for a given wavelength for a given temperature. Plank's law showed that the spectral blackbody emissive power is

$$E_{b,\lambda} = \frac{2\pi hc^2 \lambda^{-5}}{\left\{\exp\left(\dfrac{hc}{k_B \lambda T}\right) - 1\right\}} \tag{7.7}$$

where $E_{b,\lambda}$ is the spectral blackbody emissive power ($\frac{W}{m^2 \cdot \mu m}$), h is the Planck's constant ($J \cdot s$), c is the speed of light in vacuum ($\frac{m}{s}$), λ is the wavelength (μm), k_B is the Boltzmann's constant (ratio of universal gas constant to the Avogadro number) ($\frac{J}{K}$), and T is the absolute temperature (K). The values of the constants are summarized in Table 7.4.
 Substituting the values of the constants in Eq. (7.7), we obtain

$$E_{b,\lambda} = \frac{3.72 \times 10^8 \lambda^{-5}}{\left\{\exp\left(\dfrac{14,300}{\lambda T}\right) - 1\right\}} \frac{W}{m^2 \cdot \mu m} \tag{7.8}$$

Constant	Value
Boltzmann's constant, $k_B = \dfrac{\text{Gas constant}}{\text{Avogadro number}}$	$\dfrac{8.314}{6.022 \times 10^{23}} = 1.38 \times 10^{-23} \dfrac{J}{K}$
Planck's constant, h	6.62×10^{-34} J·s
Speed of light (vacuum), c	$2.99 \times 10^8 \dfrac{m}{s}$

TABLE 7.4 Values of the Constants Used in Planck's Equation to Determine the Spectral Blackbody Emissivity

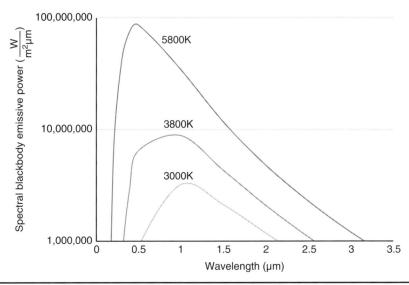

FIGURE 7.7   The spectral blackbody emissive power at various temperatures.

The spectral blackbody emissive power as a function of wavelengths for various temperatures is shown in Fig. 7.7.

Alternatively, using the relation, $c = n\lambda$ and differentiating wavelength with respect to frequency, Planck's law described above may also be expressed in a frequency (n) form as follows:

$$E_{b,n} = \frac{2\pi h n^3}{c^2 \left\{ \exp\left(\dfrac{hn}{k_B T} \right) - 1 \right\}}$$ (7.9)

Example 7.3 (Environmental Engineering):   A carbon black coated metal plate is heated to a temperature of 200°C. Determine its emissive power at a wavelength of 10 microns (Fig. 7.8).

Step 1: Schematic
The problem is depicted in Fig. 7.8.

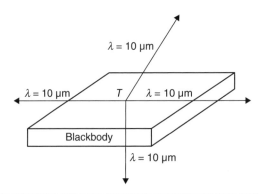

FIGURE 7.8   Emission from a hot plate at 200°C (note that the plate is emitting radiation in all directions).

Step 2: Given data
Temperature of the plate, $T = 200°C$; Wavelength, $\lambda = 10\ \mu m$

To be determined: Emissive power from the plate at 200°C for a wavelength of 10 μm.

Assumptions: The plate is a blackbody and a diffusive emitter.

Analysis: Because the plate is a blackbody, we can determine the emissive power at 200°C for 10 μm via Plank's equation.

Step 3: Calculations
The blackbody spectral emissive power is given by

$$E_{b,\lambda} = \frac{3.72 \times 10^8 \lambda^{-5}}{\left\{\exp\left(\dfrac{14,300}{\lambda T}\right) - 1\right\}} \tag{E7.3A}$$

Substituting the relevant values from the given data will result in

$$E_{b,10\mu m} = \frac{3.72 \times 10^8 \times (10)^{-5}}{\left\{\exp\left(\dfrac{14,300}{10 \times (273 + 200)}\right) - 1\right\}}$$

$$E_{b,10\mu m} = 190.2\ \frac{W}{m^2\,\mu m}$$

7.7 Total Blackbody Emissive Power

To obtain the total blackbody emissive power for all wavelengths, either of Eq. (7.8) or (7.9) is integrated from $\lambda = 0$ and $\lambda = \infty$ (for wavelength) or from $n = 0$ and $n = \infty$ (for frequency) as follows:

$$E_b = \int_0^\lambda \frac{2\pi h c^2 \lambda^{-5}}{\left\{\exp\left(\dfrac{hc}{k_B \lambda T}\right) - 1\right\}}\, d\lambda \tag{7.10}$$

Evaluating the integral, using a gamma function, and simplifying we obtain

$$E_b = \frac{2}{15}\frac{k_B^4 \pi^5}{c^2 h^3} T^4 \tag{7.11}$$

Substituting the values of the constants summarized in Table 7.4 in the above equation will result in

$$E_b = 5.67 \times 10^{-8}\ T^4 = \sigma T^4 \tag{7.12}$$

$$E_b = \sigma T^4 \tag{7.13}$$

where E_b is the total emissive power $(\frac{W}{m^2})$ and $\sigma = 5.67 \times 10^{-8}\ \frac{W}{m^2 \cdot K^4}$ is the Stefan–Boltzmann constant that we used in the previous section. The above equation is identical to Eq. (7.7) that we discussed in blackbody radiation (on a normalized area basis).

Example 7.4 (Environmental Engineering):   Determine the total emissive power of the carbon black coated metal plate described in Example 7.3 and whose area was measured to be 0.48 m².

Step 1: Schematic
The same schematic provided in Example 7.3 is applicable. However, the plate is considered to emit all possible wavelengths (0 μm–∞).

Step 2: Given data
Temperature of the plate, $T = 200°C$; Area, A= 0.48 m²; Wavelength, $\lambda = 0-\infty$

To be determined: Total emissive power from the plate at 200°C for all wavelengths.

Assumptions: The plate is a blackbody and a diffusive emitter.

Analysis: Because the plate is a blackbody, we can determine the total emissive power at 200°C from the integration of Planck's equation.

Step 3: Calculations
The blackbody spectral emissive power per unit area is given by

$$E_b = \sigma T^4 \tag{E7.4A}$$

Substituting the values of the Stefan–Boltzmann constant and the temperature we obtain from the given data and Table 7.4 will result in

$$E_b = 5.67 \times 10^{-8} \times (273 + 200)^4$$

$$E_b = 2838.1 \ \frac{W}{m^2}$$

The total emissive power is calculated as

$$E = 2838.1 \ \frac{W}{m^2} \times 0.48 \ m^2 = 1362.3 \ W$$

7.8   Wien's Law

In Plank's equation plot, for a given temperature, each curve reaches a maximum emission (which corresponds a wavelength called λ_{max}). We can therefore determine, for a given temperature, the wavelength at which the spectral blackbody emissive power reaches its maximum by setting

$$\frac{dE_{b\lambda}}{d\lambda} = 0 \tag{7.14}$$

$$\frac{d\left(\dfrac{2\pi hc^2 \lambda^5}{\left\{ \exp\left(\dfrac{hc}{k_B \lambda T} \right) - 1 \right\}} \right)}{d\lambda} = 0 \tag{7.15}$$

Equation (7.15) after differentiation and simplification will result in a remarkable equation called Wien's law (Eq. (7.16)), named after a famous German physicist. Wien's law represents the locus of the straight line connecting the wavelengths at which the blackbody emissive power reaches its maximum for each temperature.

$$T\lambda_{max} = 2898 \ \lambda m \cdot K \tag{7.16}$$

Wien's law is very helpful in estimating the temperatures of objects that are inaccessible. For example, the radiation from astronomical bodies' (e.g., sun) for various wavelengths can be collected and plotted as a function of wavelengths. Subsequently, after determining the wavelength at which the curve reaches its peak (λ_{max}), Wien's law can be used to determine the temperature. In Fig. 7.7, we see that the wavelength at which the solar emission peaks is approximately 0.5 μm. Therefore, the surface of the sun's temperature is about $T \times 0.5 = 2898$ μm·K or $T = 5796$ K.

Example 7.5 (Use of Wien's law): Determine the temperature to which a typical home heater is to be heated to obtain peak infrared radiation at 7 μm.

Step 1: Given data

The target wavelength of the heater, $\lambda = 7$ μm

To be determined: Temperature of the heater to emit infrared radiation at 7 μm.

Assumptions: The heater maintains a constant temperature.

Analysis: We can use Wien's law to determine the temperature the heater needs to reach to maximize the infrared radiation.

Step 2: Calculations

Wien's law is written as

$$T\lambda_{max} = 2898 \text{ μm·K} \tag{E7.5A}$$

Substituting the value of λ_{max} as 10 μm in the above equation, we obtain

$$T \times 7 = 2898 \text{ μm·K} \tag{E7.5B}$$

$$T = 414 \text{ K} = 141°C$$

The heater must reach a temperature of 141°C to emit a peak radiation of 7 μm.

7.9 Blackbody Radiation Fraction Function

Sometimes, it will be necessary to determine the net radiation between two chosen wavelengths. For instance, in Example 7.1, we determined the total radiation from the sun at 5507°C which included all the wavelengths using the Stefan–Boltzmann law. However, if we are only interested to know the radiation in the microwave region ($\lambda = 0.001 - 0.1$ m), we would have to integrate the Planck's curve between $\lambda_a = 0.001$ m and $\lambda_b = 0.1$ m as follows:

$$\int_{\lambda_a = 0.001 \text{ m}}^{\lambda_a = 0.1 \text{ m}} E_{b,\lambda} \, d\lambda \tag{7.17}$$

which is not so trivial each time. This is where the concept of nondimensional *blackbody radiation fraction function* (commonly called blackbody radiation function or BRF) will be very helpful. We can use Planck's law and Stefan–Boltzmann's law to determine the fraction of the energy emitted between two given wavelengths. For example, the net radiation between two wavelengths 0 and λ_a (Fig. 7.9) is given by

$$\int_{0}^{\lambda_a} E_{b,\lambda} \, d\lambda \tag{7.18}$$

which is the shaded area under the Planck's curve as shown in Fig. 7.9a. Now, if we divide the radiation by the total radiation from the body for all wavelengths (Stefan–Boltzmann law), we obtain

$$\frac{\int_{0}^{\lambda_a} E_{b,\lambda} \, d\lambda}{\sigma T^4} = \frac{\text{Area of the shaded portion}}{\text{Total area under the curve}} \tag{7.19}$$

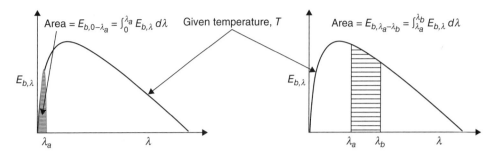

FIGURE 7.9 Schematic of the radiation between two wavelengths: (a) radiation between 0 and λ_a and (b) radiation between λ_a and λ_b.

This fraction is called *BRF* and denoted by $F_{0-\lambda_a}$.

Therefore,

$$F_{0-\lambda_a} = \frac{\text{Area of the shaded portion}}{\text{Total area under the curve}} = \frac{\int_0^{\lambda_a} E_{b,\lambda}\, d\lambda}{\sigma T^4} \tag{7.20}$$

Similarly, as shown in Fig. 7.9b, to determine the net radiation between two wavelengths λ_a and λ_b, we can write the BRF as

$$F_{\lambda_a-\lambda_b} = \frac{\text{Area of the shaded portion}}{\text{Total area under the curve}} = \frac{\int_0^{\lambda_b} E_{b,\lambda}\, d\lambda - \int_0^{\lambda_a} E_{b,\lambda}\, d\lambda}{\lambda T^4} \tag{7.21}$$

The values of the BRF have been evaluated and presented in Table 7.5.

Using these fractions, one can directly determine what percentage of radiation occurs between the given wavelengths. The following example illustrates the application of the BRF.

λT, $\mu m \cdot K$	f_λ	λT, $\mu m \cdot K$	f_λ
200	0.000000	6200	0.754140
400	0.000000	6400	0.769234
600	0.000000	6600	0.783199
800	0.000016	6800	0.796129
1000	0.000321	7000	0.808109
1200	0.002134	7200	0.819217
1400	0.007790	7400	0.829527
1600	0.019718	7600	0.839102
1800	0.039341	7800	0.848005
2000	0.066728	8000	0.856288
2200	0.100888	8500	0.874608
2400	0.140256	9000	0.890029

TABLE 7.5 The Values of Blackbody Radiation Functions for a Given λT (Cengel & Ghajar, Heat and Mass Transfer: Fundamentals and Applications, Fifth Edition. McGraw Hill, 2016. Reprinted with permission.)

λT, μm·K	f_λ	λT, μm·K	f_λ
2600	0.183120	9500	0.903085
2800	0.227897	10,000	0.914199
3000	0.273232	10,500	0.923710
3200	0.318102	11,000	0.931890
3400	0.361735	11,500	0.939959
3600	0.403607	12,000	0.945098
3800	0.443382	13,000	0.955139
4000	0.480877	14,000	0.962898
4200	0.516014	15,000	0.969981
4400	0.548796	16,000	0.973814
4600	0.579280	18,000	0.980860
4800	0.607559	20,000	0.985602
5000	0.633747	25,000	0.992215
5200	0.658970	30,000	0.995340
5400	0.680360	40,000	0.997967
5600	0.701046	50,000	0.998953
5800	0.720158	75,000	0.999713
6000	0.737818	100,000	0.999905

Table 7.5 The Values of Blackbody Radiation Functions for a Given λT (Cengel & Ghajar, Heat and Mass Transfer: Fundamentals and Applications, Fifth Edition. McGraw Hill, 2016. Reprinted with permission.) (*Continued*)

Example 7.6 (BRF): Biochar is being carbonized at a temperature of 600°C in presence of nitrogen. Determine the proportion of radiation emitted by the hot pellet of biochar within the entire infrared region (0.7–100 μm), assuming that the biochar acts as a blackbody (Fig. 7.10).

Step 1: Schematic
The problem is depicted in Fig. 7.10.

Step 2: Given data
Temperature of the biochar pellet, $T = 600°C = 873K$; First wavelength $\lambda_a = 0.7$ μm; Second wavelength $\lambda_b = 100$ μm

To be determined: The fraction of the radiation emitted by the biochar pellet between 0.7 and 100 μm.

Assumptions: The biochar serves as a blackbody and therefore is a complete emitter of energy.

Approach: We will use BRF to determine the fractional radiation at $\lambda_a = 0.7$ μm and $\lambda_b = 100$ μm, and compute the difference between the fraction of the radiation between 100 and 0.7 μm.

Step 3: Calculations
Fraction of radiation for 0.7 μm

$$\lambda_a T = 0.7 \times 873 = 611.1 \text{ μm·K}$$

From the BRF chart, the value of f_{λ_a} for 611.1 μm·K is ≈ 0.0.
    Fraction of radiation for 100 μm

$$\lambda_a T = 100 \times 873 = 87,300 \text{ μm·K}$$

From the BRF chart, the value of f_{λ_b} for 87,300 μm·K is ≈ 0.9999

$$f_{\lambda_a - \lambda_b} = 0.9999 = 99.99\%$$

Therefore, the biochar is emitting almost all its energy as infrared radiation.

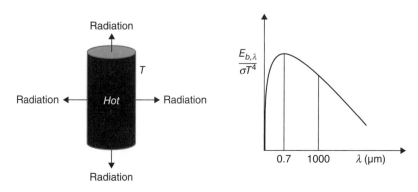

Figure 7.10 Schematic of biochar described in Example 7.6 whose radiation in the infrared range is to be determined.

7.10 Energy Balance in Radiation

Whenever a body is subjected to radiation, a portion of the incident radiation (also called irradiation) will be reflected away from the body (Fig. 7.11). Next, a portion of the irradiation is absorbed while the balance may be transmitted through the body.

If we denote incident radiation as G, reflected portion as G_R, absorbed portion as G_A, and transmitted portion as G_T, we can write the energy balance as

$$G = G_R + G_A + G_T \tag{7.22}$$

or

$$\frac{G_R}{G} + \frac{G_A}{G} + \frac{G_T}{G} = 1 \tag{7.23}$$

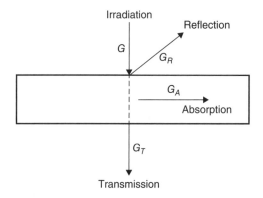

Figure 7.11 Radiation on a body is either absorbed or transmitted, or reflected or a combination of absorption, transmission, and reflection.

Now we define three nondimensional fractions called reflectivity, ρ, absorptivity, α, and transmitivity, τ such that

$$\rho = \frac{\text{Amount reflected}}{\text{Total irradiation}} = \frac{G_R}{G} \tag{7.24}$$

$$\alpha = \frac{\text{Amount absorbed}}{\text{Total irradiation}} = \frac{G_A}{G} \tag{7.25}$$

$$\tau = \frac{\text{Amount transmitted}}{\text{Total irradiation}} = \frac{G_T}{G} \tag{7.26}$$

where

$$0 \leq \rho, \alpha, \tau \leq 1 \tag{7.27}$$

Substituting the above relations in the above equation will yield

$$\rho + \alpha + \tau = 1 \tag{7.28}$$

The above equation may be simplified further depending on the nature of the body (medium) exposed to the radiation. While working with air and gases, the effect of reflectivity is negligible (due to atomic arrangement), and therefore, $\rho = 0$. Similarly most of the materials we work with including water are considered opaque, resulting in $\tau = 0$. If we are working with the blackbodies, there will be no reflectivity and transmissivity making the analysis straight forward as $\rho = \tau = 0$.

7.11   Radiation Intensity

Previously, we learned that a blackbody emits radiation uniformly in all directions. However, for real surfaces the emitted radiation is not uniform in all directions. So, how will we be able to quantify the amount of radiation in any given direction? For example, imagine a cold winter night and you are in front of a heat source. Your level of comfort will directly depend on how much of the radiation from the heat source is emitted in your direction. This is where the concept of radiation intensity will be very helpful.

Because the body emits radiation in all directions, the influence of radiation is best depicted as a hemisphere for a two-dimensional body and a sphere for three-dimensional body. Therefore, we will employ spherical coordinate system (r, θ, ϕ) for our analysis. Consider a small area, dA_1 that emits radiation in all directions in a hemisphere (Fig. 7.12a). The projected area vector of the surface dA_1 along the direction of view is $dA_1 \times \cos\theta$.

The differential solid angle, $d\Omega$ formed by the area, dA_2 over dA_1 on the hemisphere (Fig. 7.12b) is simply the ratio of the area formed by dA_2 to the square of the radius of the hemisphere, r and is given by

$$d\Omega = \frac{dA_2}{r^2} = \frac{(r \times \sin\theta \times d\varnothing) \times (r \times d\theta)}{r^2} = d\theta \times \sin\theta \times d\phi \tag{7.29}$$

Now, we introduce the radiation intensity, I_e as the energy flux emitted in the polar (θ) and azimuthal (ϕ) directions per unit differential solid angle. Mathematically,

$$I_e = \frac{d\dot{q}}{(dA_1 \times \cos\theta) \times d\Omega} \tag{7.30}$$

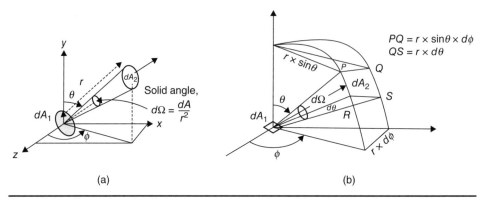

Figure 7.12 Illustration of radiation intensity (a) and the solid angle subtended by a differential area on the hemisphere (b).

We can rearrange the radiation intensity in the form of differential emission flux ($\frac{W}{m^2}$)

$$dE = \frac{d\dot{q}}{dA_1} = I_e \times \cos\theta \times d\Omega = I_e \times \cos\theta \times d\theta \times \sin\theta \times d\varnothing \qquad (7.31)$$

The emissive power, E can be determined by integrating the above expression for $\theta = 0 \rightarrow \frac{\pi}{2}$ and $\varnothing = 0 \rightarrow 2\pi$.

$$E = \int_{\theta=0}^{\frac{\pi}{2}} \int_{\varnothing=0}^{2\pi} I_e \times \cos\theta \times d\theta \times \sin\theta \times d\varnothing \qquad (7.32)$$

If we assume that the surface is diffuse, the value of I_e becomes constant and the above equation after integration will simplify as

$$E = \pi I_e \qquad (7.33)$$

Further, if the surface behaves like a blackbody, the above equation becomes

$$E = \sigma T^4 = \pi I_b \qquad (7.34)$$

and

$$I_b = \frac{\sigma T^4}{\pi} \qquad (7.35)$$

Conversely, if a diffuse surface is receiving radiation, I_i, the irradiation intensity, G is expressed as

$$G = \pi I_i \qquad (7.36)$$

Example 7.7 (General Engineering): A small disc of 1 cm diameter is radiating heat at 750°C to a square plate of area 5 cm² located 1 m away at an angle of 45° to the normal (Fig. 7.13). What will be the amount of radiation received by the square plate after 3 minutes if it subtends an angle of 45° with its normal?

Step 1: Schematic
The problem is depicted in Fig. 7.13.

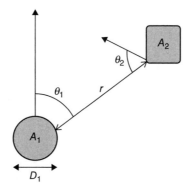

FIGURE 7.13 Schematic of heat transfer from the disc to the plate described in Example 7.7.

Step 2: Given data
Diameter of the emitting (first) surface (disc), $D_1 = 1$ cm; Temperature of the disc, $T_1 = 750°C$; Area of the receiving surface (square plate), $A_1 = 5$ cm^2; The distance between the plates, $r = 1$ m; Angle between the disc and plate, $\theta_1 = 45°$; the angle of the plate with its normal, $\theta_2 = 45°$

To be determined: The amount of radiation received by the square plate after 3 minutes.

Assumptions: The disc acts as a blackbody and the surface radiation properties remain the same.

Approach: We will first determine the solid angle subtended by the square plate when viewed from the disc and subsequently determine the amount of radiation the square plate receives from the disc. In addition, because both the areas are small, we can directly use the definition of solid angle for this problem.

Step 3: Calculations
The solid angle subtended by the plate viewed from disc is given by

$$d\Omega = \frac{dA_2 \cos(\theta_2)}{r^2}$$ (E7.7A)

Substituting the data into the above equation,

$$d\Omega = \frac{5 \times 10^{-4} \times \cos(45)}{1^2} = 2.62 \times 10^{-4} \text{ steradian}$$

Because the hot disc (area 1) is assumed to be a blackbody, the intensity of radiation, I_1 emitted by the disc through the solid angle is calculated as

$$E_b = \sigma T_1^4 = \pi I_1$$ (E7.7B)

$$I_1 = \frac{\sigma T_1^4}{\pi}$$ (E7.7C)

or

$$I_1 = \frac{5.67 \times 10^{-8} \times (750 + 273)^4}{3.14} = 19{,}776.8 \ \frac{W}{m^2 sr}$$

Now, the rate of radiation received by the plate from the disc is determined as

$$\dot{q} = I_1 \times A_1 \cos\theta_1 \times d\Omega = 19{,}776.8 \times 3.14 \times \frac{0.01^2}{4} \times \cos(45) \times 2.62 \times 10^{-4} = 2.14 \times 10^{-4} \text{ W}$$

The amount of radiation received by the plate in 3 minutes is calculated as

$$2.14 \times 10^{-4} \times 3 \times 60 = 0.038 \text{ J}$$

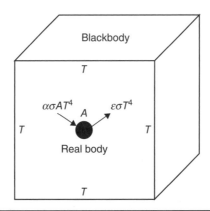

Figure 7.14 Schematic describing Kirchhoff's law that states that emissivity and absorptivity are identical for an enclosed real body.

7.12   Kirchhoff's Law of Radiation

The famous German physicist, Gustav Kirchhoff proposed that under certain conditions, the emissivity and absorptivity of a real body are equal. To explain this, we will consider a real body at temperature, T with an area, A, emissivity, ε, and absorptivity, α placed in an enclosure (source) at temperature, T large enough to be considered as a blackbody (Fig. 7.14).

If the total radiation flux emitted by the blackbody is σT^4, the radiation adsorbed by the real body is

$$\dot{q}_{\text{Absorbed}} = I_A = \alpha\sigma AT^4 \tag{7.37}$$

Additionally, we know that the real body also emits radiation (because it is at a temperature, T) whose value is given by

$$\dot{q}_{\text{emission}} = \varepsilon\sigma AT^4 \tag{7.38}$$

After thermal equilibrium condition is established, $\dot{q}_{\text{Absorbed}} = \dot{q}_{\text{emission}_A}$

i.e.,

$$\alpha\sigma AT^4 = \varepsilon\sigma AT^4 \tag{7.39}$$

or

$$\alpha = \varepsilon \tag{7.40}$$

The above equation is very helpful in analyzing the radiation interaction between a blackbody and a real body and usually works for situations when the temperatures of the source and the recipient are not very different.

7.13   Radiosity

The radiation emitted by a surface is a combination of its own emission and the reflected portion of the radiation that the surface is subjected to. For example, take a gray and opaque body of unit surface at a temperature, T with a reflectivity, ρ and an emissivity, ε that is exposed to an incident radiation, G (Fig. 7.15). Now, the surface reflects some of the incident radiation which is given by ρG. In addition, the surface also emits radiation (due to the temperature of the surface) that is equivalent to εE_b, where E_b is the maximum possible blackbody radiation.

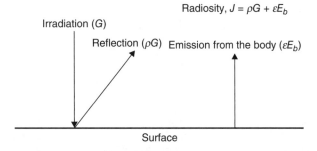

FIGURE 7.15 The concept of radiosity combines the emission from a body with its reflected radiation.

Therefore, radiosity, J, is the total radiation emitted from a surface per unit area and is mathematically

$$J = \rho G + \varepsilon E_b \tag{7.41}$$

Because the surface is gray and opaque ($\tau = 0$), the following relations apply

$$\rho + \alpha = 1 \text{ and } \varepsilon = \alpha \tag{7.42}$$

Substituting in the above equation and simplifying will yield the expression for radiosity

$$J = G - \varepsilon(G - E_b) \tag{7.43}$$

For a required radiosity, we can determine the incident radiation as

$$G = \frac{J - \varepsilon E_b}{1 - \varepsilon} \tag{7.44}$$

If the surface acts as a blackbody ($\varepsilon = 1$ and $\rho = 0$), the radiosity will reduce to

$$J = E_b = \sigma T^4 \tag{7.45}$$

We can also use the concept of radiosity to determine the net radiation from any gray surface by calculating the difference between the radiosity and the irradiation for the entire area of the surface.

$$\dot{q} = \text{Radiosity} - \text{Irradiation} \tag{7.46}$$

$$\dot{q} = JA - GA = A(J - G) \tag{7.47}$$

Using the expression for irradiation from Eq. (7.44) will result in

$$\dot{q} = A\left[J - \left(\frac{J - \varepsilon E_b}{1 - \varepsilon} \right) \right] \tag{7.48}$$

Which can further be simplified as

$$\dot{q} = \frac{E_b - J}{\left(\dfrac{1 - \varepsilon}{A\varepsilon} \right)} \tag{7.49}$$

Recalling from the thermal resistance networks from Chap. 3, the radiation heat transfer can now be expressed in the familiar form

$$\dot{q} = \frac{E_b - J}{R} = \frac{\text{Driving force}}{\text{Resistance}} \qquad (7.50)$$

where $R = \dfrac{1-\varepsilon}{A\varepsilon}$ is the resistance offered by the surface.

Example 7.8 (Total radiation transfer from poultry):   A poultry farm uses a radiator to heat the birds of the farm in the winter months. Measurements indicated that the birds were irradiated with a heat flux of 100 W/m². Determine the radiosity of the farm birds assuming a surface temperature of 35°C and an average emissivity of 0.94.

Step 1: Given data

Irradiation, $G = 100 \ \dfrac{\text{W}}{\text{m}^2}$; Surface temperature of the bird, $T = 35°C$; Average emissivity, $\varepsilon = 0.94$

To be determined: The amount of radiosity from the farm birds.

Assumptions: The surface radiation properties remain the same.

Approach: We will determine the value of radiosity by solving the expression for radiosity as outlined in Section 7.13.

Step 3: Calculations
The equation for radiosity is expressed as

$$J = G - \varepsilon(G - E_b) \qquad (E7.8A)$$

or

$$J = G - \varepsilon(G - \sigma T^4) \qquad (E7.8B)$$

Substituting the data into the above equation, we obtain

$$J = 100 - 0.94 \times [100 - (5.67 \times 10^{-8} \times (273 + 35)^4)]$$

$$J = 456.8 \ \frac{\text{W}}{\text{m}^2}$$

7.14   Radiation between Surfaces: General Analysis

To analyze and determine the radiation heat exchange between two surfaces, we require information on the emission properties, intensity of radiation, and how each body is oriented with respect to the other. We previously discussed the emission properties such as transmittivity, absorptivity, transmittivity, and radiosity and intensity of radiation. In this section, we will focus on the orientation aspects pertaining to radiation.

To elucidate the importance of the orientation in radiation, we will use a simple example. Imagine two identical flat plates (diffuse surfaces), plate 1 and plate 2 at different temperatures, T_1 and T_2 facing each other (Fig. 7.16). In the first configuration (Fig. 7.16a), the orientation of both plates is normal to each other.

Therefore, in the first case, all the radiation (100%) emitted by plate 1 is intercepted by plate 2 and vice versa. In the second configuration (Fig. 7.16b), however, the orientation of the plates is different from the first case. Therefore, only a small portion of radiation from plate 1 will be intercepted by plate 2 while a larger portion of radiation from plate 2 is received by plate 1. Therefore, the net heat transfer between the plates is different for case 1 and case 2, even though the plate radiation properties and their temperature gradients are identical. Thus, the orientation between the two surfaces becomes very important.

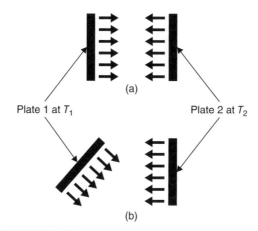

Plate 1 at T_1 Plate 2 at T_2

FIGURE 7.16 Two identical flat plates directly facing each other (a) and facing at an angle (b) exhibit different radiation exchange properties.

7.15 The Concept of View Factors

To quantify the effects of orientation between two surfaces m and n, we will define another dimensional parameter called view factor (F) that quantifies the fraction of radiation emitted by surface m and intercepted by surface n. Because there are two surfaces, a view factor is associated with two subscripts—one for each surface. Mathematically,

$$F_{m,n} = \frac{\text{Radiation from } m \text{ received by } n}{\text{Radiation emitted by } m} = \frac{\dot{q}_{m,n}}{\dot{q}_m} \tag{7.51}$$

where $F_{m,n}$ is the view factor and should be interpreted as the fraction of radiation from m that is received by n, $\dot{q}_{m,n}$ is the radiation emitted by m that is received by n, and $\dot{q}_m$ is the radiation from m. Similarly, we can write the second view factor $F_{n,m}$ as the fraction of radiation emitted by n and is received by m, and is given by

$$F_{n,m} = \frac{\text{Radiation from } n \text{ received by } m}{\text{Radiation emitted by } n} = \frac{\dot{q}_{n,m}}{\dot{q}_n} \tag{7.52}$$

To continue our discussion on the view factors further, let us choose two diffuse surfaces A_1 and A_2 separated by a distance, r. We will also choose infinitesimally small areas dA_1 and dA_2 with solid angles $d\Omega_{2,1}$ and $d\Omega_{1,2}$ and the angles with the normal are α_1 and α_2 (Fig. 7.17).

If the radiation intensity of dA_1 is I_1, using the concepts from Section 7.11, we can write the rate of heat transfer from dA_1 to dA_2 as

$$\dot{q}_{dA_1, dA_2} = I_1 \cos\alpha_1 \, dA_1 \, d\Omega_{21} \tag{7.53}$$

We know that

$$d\Omega_{21} = \frac{dA_2 \cos\alpha_2}{r^2} \tag{7.54}$$

$$\dot{q}_{dA_1, dA_2} = \frac{I_1 \cos\alpha_1 \, dA_1 \, dA_2 \cos\alpha_2}{r^2} \tag{7.55}$$

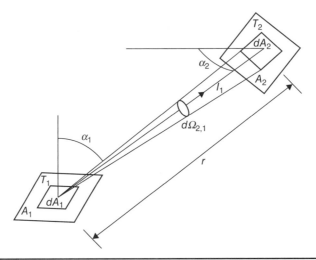

Figure 7.17 Schematic of two differently oriented surfaces at different temperatures exchanging radiation.

We also know that

$$\dot{q}_{dA_1} = \pi dA_1 I_1 \tag{7.56}$$

Therefore, we can write the differential view factor $dF_{1,2}$ as

$$dF_{1,2} = \frac{\dot{q}_{dA_1, dA_2}}{\dot{q}_{dA_1}} = \frac{\dfrac{I_1 \cos\alpha_1 dA_1 dA_2 \cos\alpha_2}{r^2}}{\pi dA_1 I_1} = \frac{dA_2 \cos\alpha_1 \cos\alpha_2}{\pi r^2} \tag{7.57}$$

$$F_{1,2} = \frac{1}{A_1} \int \int \frac{\cos\alpha_1 \cos\alpha_2 dA_1 dA_2}{\pi r^2} \tag{7.58}$$

or

$$F_{1,2} A_1 = \int \int \frac{\cos\alpha_1 \cos\alpha_1 dA_1 dA_2}{\pi r^2} \tag{7.59}$$

Similarly, the view factor for fraction of radiation from A_2 that is intercepted by A_1 can be calculated as

$$F_{2,1} A_2 = \int \int \frac{\cos\alpha_1 \cos\alpha_1 dA_1 dA_2}{\pi r^2} \tag{7.60}$$

We can see that the right-hand sides of Eqs. (7.59) and (7.60) are identical. Therefore, we obtain the famous reciprocity law as follows:

$$F_{1,2} A_1 = F_{2,1} A_2 \tag{7.61}$$

The view factors are used to quantify the fraction of radiation and always range between 0 and 1. For convex and plane surfaces (Fig. 7.18a and b), the value of view factor is always zero because the surface radiation emitted will never be intercepted. However, for a concave surface (Fig. 7.18c), the view factor can be greater than zero.

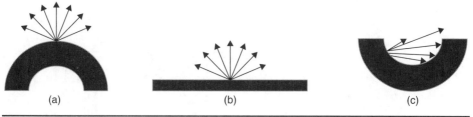

FIGURE 7.18   Illustration of view factors for convex (a), flat (b), and concave (c) surfaces.

For closed enclosures with n surfaces, based on the energy balance, the sum of all view factors for a given surface must be 1. This is called the summation rule. Mathematically,

$$\dot{q}_1 = \dot{q}_{1,1} + \dot{q}_{1,2} + \dot{q}_{1,3} + \cdots + \dot{q}_{1,n} \tag{7.62}$$

$$1 = \frac{\dot{q}_{1,1}}{\dot{q}_1} + \frac{\dot{q}_{1,2}}{\dot{q}_1} + \frac{\dot{q}_{1,3}}{\dot{q}_1} + \cdots + \frac{\dot{q}_{1,n}}{\dot{q}_1} \tag{7.63}$$

$$1 = F_{1,1} + F_{1,2} + F_{1,3} + \cdots + F_{1,n} \tag{7.64}$$

For an enclosure with n surfaces, the summation rule can be written in general terms as

$$\sum_{j=1}^{n} F_{i,j} = 1 \tag{7.65}$$

For example, for a room with six surfaces, we can write the view factors for all surfaces as

$$1 = F_{1,1} + F_{1,2} + F_{1,3} + F_{1,4} + F_{1,5} + F_{1,6} \tag{7.66}$$

$$1 = F_{2,1} + F_{2,2} + F_{2,3} + F_{2,4} + F_{2,5} + F_{2,6} \tag{7.67}$$

$$1 = F_{3,1} + F_{3,2} + F_{3,3} + F_{3,4} + F_{3,5} + F_{3,6} \tag{7.68}$$

$$1 = F_{4,1} + F_{4,2} + F_{4,3} + F_{4,4} + F_{4,5} + F_{4,6} \tag{7.69}$$

$$1 = F_{5,1} + F_{5,2} + F_{5,3} + F_{5,4} + F_{5,5} + F_{5,6} \tag{7.70}$$

$$1 = F_{6,1} + F_{6,2} + F_{6,3} + F_{6,4} + F_{6,5} + F_{6,6} \tag{7.71}$$

Therefore, we can see that the number of view factors is the square of the number of surfaces and might look intimidating at first to analyze these types of problems. However, by combining the principles of reciprocity and summation we can solve for the unknown view factors. The following example will illustrate the procedure.

Example 7.9 (General Engineering—All concentrations):   A 25-m piping system consists of two concentric cylinders. The outer cylinder is 2 m in diameter and the inner cylinder is 1.5 m. Determine the view factors for all the surfaces.

Step 1: Schematic
The problem is depicted in Fig. 7.19.

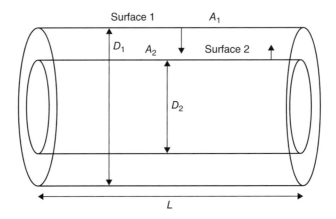

FIGURE 7.19   Schematic of concentric cylinders radiating heat to each other as described in Example 7.9.

Step 2: Given data

Length of the concentric pipe system, $L = 25$ m; Diameter of the outer pipe, $D_1 = 2$ m; Diameter of the inner pipe, $D_2 = 1.5$ m

To be determined: View factors.

Assumptions: The radiation properties of all the surfaces remain the same.

Approach: There are two surfaces and therefore the number of view factors will be $2^2 = 4$. We will employ the concepts of reciprocity and summation to determine all the four view factors.

Step 3: Calculations

Using the summation rule, the view factors for the outer surface (i.e., surface 1)

$$F_{1,1} + F_{1,2} = 1 \qquad \text{(E7.9A)}$$

Similarly, for inner surface (surface 2)

$$F_{2,1} + F_{2,2} = 1 \qquad \text{(E7.9B)}$$

But we know that $F_{2,2} = 0$ (because surface 2 does not receive any of its own radiation) and therefore, $F_{2,1} = 1$.

From the reciprocity rule,

$$F_{1,2} A_1 = F_{2,1} A_2 \qquad \text{(E7.9C)}$$

or

$$F_{1,2} = \frac{F_{2,1} A_2}{A_1} = \frac{1 \times \pi \times D_2 \times L}{\pi \times D_1 \times L} = \frac{D_2}{D_1} = \frac{1.5}{2} = 0.75$$

From the summation rule,

$$F_{1,1} + 0.75 = 1 \qquad \text{(E7.9D)}$$

$$F_{1,1} = 0.25$$

The four view factors are: $F_{1,1} = 0.25$, $F_{1,2} = 0.75$, $F_{2,1} = 1$, $F_{2,2} = 0$

7.16 Heat Transfer between Two Surfaces

Now we will apply the concepts of view factors to determine the radiation exchange between surfaces. Because of simplicity, we will first analyze blackbodies. Imagine two blackbody surfaces 1 and 2 with areas A_1 and A_2 maintained at constant temperatures T_1 and T_2 with view factors, $F_{1,2}$ and $F_{2,1}$ separated by a distance (Fig. 7.20). Using the view factors, we can write the net radiation from surface 1 to surface 2 as

$$\dot{q}_{1,2} = \text{Radiation from 1 intercepted by 2} - \text{Radiation from 2 intercepted by 1} \quad (7.72)$$

$$\dot{q}_{1,2} = F_{1,2}A_1E_{b1} - F_{2,1}A_2E_{b2} \quad (7.73)$$

Using the reciprocity law, the above equation will be simplified as

$$\dot{q}_{1,2} = A_1F_{1,2}\sigma\left(T_1^4 - T_2^4\right) \quad (7.74)$$

For diffuse gray surfaces 1 and 2 (Fig. 7.21), with view factors, $F_{1,2}$ and $F_{2,1}$ and radiosities, J_1 and J_2 respectively, the rate of radiation exchange can be written as

$$\dot{q}_{1,2} = \text{Radiation from 1 intercepted by 2} - \text{Radiation from 2 intercepted by 1} \quad (7.75)$$

$$\dot{q}_{1,2} = F_{1,2}A_1J_1 - F_{2,1}A_2J_2 \quad (7.76)$$

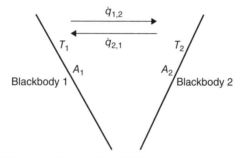

FIGURE 7.20 Schematic of two flat blackbodies facing each other at different temperatures.

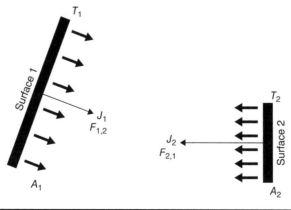

FIGURE 7.21 Schematic of two diffuse and gray surfaces with different view factors and radiosities facing each other at different temperatures.

Again, due to the reciprocity law, the equation will become

$$\dot{q}_{1,2} = F_{1,2}A_1(J_1 - J_2)$$ (7.77)

Or in terms of thermal resistance network

$$\dot{q}_{1,2} = \frac{(J_1 - J_2)}{\left(\dfrac{1}{F_{1,2}A_1}\right)}$$ (7.78)

$$\dot{q}_{1,2} = \frac{(J_1 - J_2)}{R_{1,2}} = \frac{\text{Driving force}}{\text{Resistance}}$$ (7.79)

In summary, we can write the resistances as follows:

Resistance offered by the body's surface $= \dfrac{1 - \varepsilon}{A\varepsilon}$

Resistance offered by the spacing between the bodies $= \dfrac{1}{F_{1,2}A_1}$

Now, we can also combine the ideas from view factors and thermal resistances to analyze the heat transfer rates between any two surface enclosures 1 and 2 (Fig. 7.22).

Assuming the T_1, A_1, ε_1, and E_{b1} and T_2, A_2, ε_2, and E_{b2} as temperature, area, emissivity, and blackbody (maximum) emission associated with surfaces 1 and 2, we can easily see the two surfaces are separated by three resistances: (i) the surface resistance offered by 1, (ii) the resistance offered by the spacing between the two surfaces, and (iii) the surface resistance offered by 2. Therefore, the net radiation exchange can be written as

$$\dot{q}_{1,2} = \frac{\text{Driving force}}{\text{Total resistance}} = \frac{E_{b1} - E_{b2}}{R_{\text{surface 1}} + R_{\text{space}} + R_{\text{surface 2}}} = \frac{\sigma\left[T_1^4 - T_2^4\right]}{\dfrac{1 - \varepsilon_1}{A_1\varepsilon_1} + \dfrac{1}{F_{1,2}A_1} + \dfrac{1 - \varepsilon_2}{A_2\varepsilon_2}}$$ (7.80)

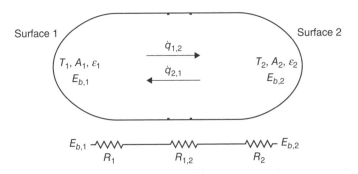

FIGURE 7.22   Net radiation between two surface enclosures at different temperatures can be determined using the concept of thermal resistances in series.

Example 7.10 (Environmental Engineering—Radiation during composting):   The rectangular floor (10-m wide and 40-m long) of a waste treatment facility with a semi-circular ceiling is composting manure that was spread flat on the floor (Fig. 7.23). Recent measurements have indicated that the average ceiling and the floor (manure) temperatures are 35°C and 15°C, respectively. Determine the radiation heat transfer rate if the emissivity of the ceiling and manure are 0.82 and 0.94, respectively.

Step 1: Schematic

The problem is depicted in Fig. 7.23.

Step 2: Given data

Surface 1 is the ceiling and surface 2 is the floor, Length of the floor, $L = 40$ m, Width of the floor = diameter of the ceiling, $W = 10$ m; Temperature of the ceiling, $T_1 = 35°C$; Temperature of the manure surface (floor), $T_2 = 15°C$; Emissivity of the ceiling, $\varepsilon_1 = 0.82$; emissivity of the manure, $\varepsilon_2 = 0.94$

To be determined: Rate of radiation from the ceiling to the floor.

Assumptions: The temperatures are uniform, both surfaces are gray and diffuse emitters, and convection effects are negligible.

Analysis: Because both the surfaces are assumed to be gray and diffuse emitters, can directly use the relation that we developed in the preceding section.

Step 3: Calculations

First, we need to determine the view factors for the surfaces. From the summation rule, we can write the view factors as

$$F_{1,1} + F_{1,2} = 1 \tag{E7.10A}$$

$$F_{2,1} + F_{2,2} = 1 \tag{E7.10B}$$

Because, the floor is a flat surface,

$$F_{2,2} = 0 \text{ and therefore } F_{2,1} = 1$$

Using reciprocity, $A_1 F_{1,2} = A_2 F_{2,1}$ resulting in

$$F_{1,2} = \frac{A_2}{A_1} F_{2,1} = \frac{L \times W}{\left(\dfrac{\pi \times W \times L}{2}\right)} \times F_{2,1} = \frac{2 \times F_{2,1}}{\pi} \tag{E7.10C}$$

$$F_{1,2} = \frac{2}{\pi} \text{ and therefore, } F_{1,1} = 1 - \frac{2}{\pi}$$

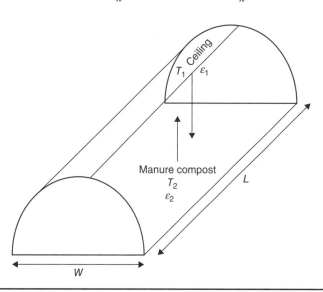

FIGURE 7.23   Schematic of the compositing room with semi-circular ceiling described in Example 7.10.

Now, the rate of radiation heat transfer from the ceiling to the floor is given by

$$\dot{q}_{1,2} = \frac{\sigma\left[T_1^4 - T_2^4\right]}{\dfrac{1-\varepsilon_1}{A_1\varepsilon_1} + \dfrac{1}{F_{1,2}A_1} + \dfrac{1-\varepsilon_2}{A_2\varepsilon_2}} \qquad\qquad \text{(E7.10D)}$$

$$\dot{q}_{1,2} = \frac{5.67 \times 10^{-8} \times [(273+35)^4 - (273+15)^4]}{\left(\dfrac{1-0.82}{\dfrac{3.14 \times 10 \times 40}{2}} \times 0.82\right) + \left(\dfrac{1}{\dfrac{2}{3.14} \times \dfrac{3.14 \times 10 \times 40}{2}}\right) + \left(\dfrac{1-0.94}{40 \times 10 \times 0.94}\right)}$$

The heat transferred via radiation from the ceiling to the floor is 39,936.6 W.

Practice Problems for the FE Exam

1. Which of the following is not true about radiation heat transfer?

   a. Radiation needs a medium for heat transfer.

   b. All objects above absolute zero emit radiation.

   c. Radiation is always the fastest form of heat transfer.

   d. Larger objects emit more radiation than smaller objects of the same material at a given temperature.

2. The total amount of heat emitted by a spherical blackbody of 10-cm diameter at 0°C will be about

   a. 50 W.

   b. 10 W.

   c. 80 W.

   d. None of the above.

3. A pig's skin temperature was measured to be 40°C. Assuming the skin to be blackbody, the emission will be maximum at

   a. 52.5 μm.

   b. 25.5 μm.

   c. 9.25 μm.

   d. 100 μm.

4. The spectral blackbody emissivity from a hot biochar pellet assumed to be a blackbody at 400°C about a wavelength of 100 μm is closest to

   a. $90 \dfrac{W}{m^2 \cdot \mu m}$.

   b. $0.15 \dfrac{W}{m^2 \cdot \mu m}$.

   c. $0.20 \dfrac{W}{m^2 \cdot \mu m}$.

   d. None of the above.

5. Which of the following statements is true about radiation heat transfer?

    a. For blackbodies, the value of reflectivity, $\alpha = 0$.

    b. Metals will have the value of transmissivity, $\tau = 0$.

    c. Gases have the maximum values of reflectance, ρ.

    d. Emissivity of a real body is always a constant.

6. The temperature a blackbody must be heated to obtain a radiosity $1000 \dfrac{W}{m^2}$ is

    a. 92.3°C.

    b. 71.7°C.

    c. 102.5°C.

    d. 62.3°C.

7. A metal object receives an incident radiation of 250 W/m² of which 28% is reflected. The amount of radiation absorbed by the body is

    a. $250 \dfrac{W}{m^2 \mu m}$.

    b. $100 \dfrac{W}{m^2 \mu m}$.

    c. $125 \dfrac{W}{m^2 \mu m}$.

    d. $180 \dfrac{W}{m^2 \mu m}$.

8. A pig at 40°C is placed in a room whose walls are maintained at 25°C. If the emissivity of the pig skin is given as 0.92, the net radiation heat exchange flux between the pig and the room is about

    a. $0.11 \dfrac{W}{m^2}$.

    b. $100 \dfrac{W}{m^2}$.

    c. $90 \dfrac{W}{m^2}$.

    d. None of the above.

9. Two diffuse surfaces of areas 5 m² and 3.4 m² are separated by a distance and face each other at an angle. If the view factor of the first surface with respect to the second surface ($F_{1,2}$) is 0.7, the value of the view factor $F_{2,1}$ will be approximately

    a. 0.9.

    b. 1.0.

    c. 0.5.

    d. None of the above.

10. The temperature a blackbody must be heated to obtain a radiosity 1000 $\frac{W}{m^2}$ is

   a. 92.3°C.

   b. 71.7°C.

   c. 102.5°C.

   d. 62.3°C.

Practice Problems for the PE Exam

1. A hog barn maintained at 22°C houses 700 pigs with each pig weighing about 165 kg each and an average body surface area of 2 m². If the average skin temperature of the pigs was measured to be 42°C, calculate the total heat transfer rate assuming a heat transfer coefficient of 50 W/m²·°C.

2. Determine the rate of heat loss from a hot stainless steel (emissivity = 0.1) cylindrical gasifier whose diameter and length were measured to be 1 m and 2 m, respectively, and maintained at a temperature of 1100°C.

3. Earth's annual energy consumption is around 6×10^{20} Joules. Compare this value with the energy radiated by the sun in the infra-red range.

4. A metal plate is heated to a temperature of 700°C in a furnace. Determine the emissive power at 20 μm.

5. A cylindrical stainless steel (emissivity = 0.1) pyrolysis reactor (30-cm diameter and 60-cm long) is heated to a temperature of 1100°C. What fraction of the radiation falls in the infra-red region?

6. The surface of a space heater reaches 900°C. What fraction of the radiation does infrared account for?

7. In a turkey unit, heaters are used to provide warmth to the birds. If an irradiation of 200 W/m² is provided, what will the radiosity of the farm birds be if the surface temperature of the turkeys was found to be 38°C with an emissivity of 0.94?

8. A poultry farm uses a radiator to heat the birds of the farm in the winter months. Measurements indicated that the birds were irradiated with a heat flux of 100 W/m². Determine the radiosity of the farm birds assuming a surface temperature of 35°C and an average emissivity of 0.94.

9. A 70-m long industrial copper tubular piping system consists of two concentric pipes. If the outer pipe is 70 cm in diameter and the inner cylinder is 50 cm, calculate the view factors for all the surfaces.

10. A composting facility has a rectangular floor of dimensions 60 m × 20 m with a semi-circular ceiling. If the temperatures of the surface of the compost and the ceiling were determined as 20°C and 38°C, what will be the radiation heat transfer assuming emissivities for the compost and the ceiling as 0.9 and 0.8, respectively?

CHAPTER 8

Fundamentals of Fluid Flow

Chapter Objectives

This chapter will prepare students to

- Identify the flow regimes given the physical properties of the flow systems.
- Apply continuity equation to determine the flow rates and velocities in conduits.
- Use energy equation to solve numerical problems related to fluid flow.

8.1 Motivation

As engineers, we are expected to analyze, design, and select HVAC equipment that carry refrigerants, heat exchangers, and piping systems, conduits, and pumps carrying different types of fluids. Engineering design of any of the aforementioned equipment requires a thorough understanding of principles of fluid motion. Therefore, this chapter provides a quick overview of the relevant ideas of fluid flow with examples.

8.2 Viscosity

Viscosity (also called dynamic viscosity) is a very important property of a fluid that offers a resistance to flow. A fluid may be visualized as arrangement of layers of fluid molecules. When an external force is applied, the fluid layers start to move, during which a frictional resistance is developed between the adjacent layers that acts against the applied force. This frictional resistance that opposes the flow is called viscosity. It will be easier to understand the concept of viscosity by a simple everyday example. Imagine three beakers, each containing water, honey, and cooking oil at room temperature (say 25°C). If we try to stir the contents of the beakers with a spoon (here we will imagine that we are stirring at the same speed to maintain uniformity), we will realize that it will be easiest to stir the water, followed by oil and honey, which will be hardest to stir. This is because, the resistance offered by honey is the strongest (oil resistance is medium) while the water offers the least resistance. Therefore, we can conclude that honey is the most viscous relative to oil and water. If we look up the viscosity values, we will find that honey has a viscosity of 8.5 Pa·s, while soybean oil and water have viscosities 0.05 Pa·s and 8.9×10^{-4} Pa·s, respectively.

Viscosity is caused by the combination of intermolecular cohesive forces and the intermolecular momentum transfer. In liquids, the molecules are packed more closely (than gases) and, therefore, it is the cohesive forces that are responsible for the viscosity. In gases, however, the molecules are not as densely packed as liquids. But the molecules are constantly involved in random collisions (intermolecular momentum transfer), and this imparts the viscosity to the gases.

The mathematical definition of viscosity is given by Newton's law of viscosity, which states that the shear stress is linearly related to shear rate and is expressed in its differential form as

$$\tau = \mu \frac{dv}{dy} \tag{8.1}$$

where τ is the shear stress (Pa), μ is the viscosity or dynamic viscosity (Pa·s), and $\frac{dv}{dy}$ is the shear rate (1/s). The derivation of the above equation can be found in any fluid mechanics textbook. Viscosity can be viewed as the amount of shear stress required to be applied to obtain one unit of shear rate. In engineering literature, another form of viscosity called kinematic viscosity (v) $\left(\dfrac{m^2}{s}\right)$ is also used, which is defined as

$$\text{Kinematic viscosity, } v = \frac{\text{Dynamic viscosity}}{\text{Density}} \tag{8.2}$$

or

$$v = \frac{\mu}{\rho} \tag{8.3}$$

It is very important to note that viscosity strongly depends on the temperature. Liquids, in which the molecules are closely packed (relative to gases), are dominated by the intermolecular cohesive forces (forces of attraction). At higher temperatures, the molecules in the liquid are equipped with higher energies that can overcome the cohesive forces. Therefore, at higher temperatures, due to the decrease in the molecular cohesion, the viscosity of a liquid is reduced. We can easily relate to this phenomenon. When cooking liquid foods such as soups and broths, we can stir the liquid easily when on the stove. However, after cooling, the viscosity increases resulting in thickening of the food. For the gases, however, the viscosity is due to the momentum transfer between the gas molecules. When temperature is increased, the kinetic energy of the molecules is increased resulting in the increased collisions and velocities thereby increasing the viscosity of the given gas.

Example 8.1 (Engineering—All concentrations): In a viscosity experiment, two parallel plates at 30°C are separated by soybean oil $\left(\rho = 900 \ \dfrac{kg}{m^3}\right)$. The distance between the plates is 5 mm and a force of 5 N is applied to record a velocity of 5 m/s. Using this information, determine the kinematic viscosity if the plate dimensions are 40 cm × 20 cm (Fig. 8.1).

Step 1: Schematic
The problem is depicted in Fig. 8.1.

Step 2: Given data
Temperature, $T = 30°C$; Applied force, $F = 5$ N; Velocity, $v = 5 \ \dfrac{m}{s}$; Distance between the plates, $z = 5$ mm; Area of the plates, $A = 0.4 \times 0.2 = 0.08 \ m^2$; Density, $\rho = 900 \ \dfrac{kg}{m^3}$

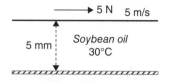

FIGURE 8.1 Schematic of the viscosity experiment described in Example 8.1.

To be determined: Kinematic viscosity of soybean oil used in the experiment.

Assumptions: The velocity gradient is linear and temperature remains constant at 30°C.

Approach: We will use the Newton's law of viscosity and determine the dynamic viscosity. Subsequently, we will convert the dynamic viscosity into kinematic viscosity.

Step 3: Calculations
The Newton's law of viscosity is given by

$$\tau = \mu \frac{dv}{dz} \tag{E8.1A}$$

$$\tau = \frac{F}{A} = \mu \times \frac{dv}{dz} \tag{E8.1B}$$

$$\frac{5}{0.08} = \mu \times \frac{5}{\left(\dfrac{5}{1000}\right)} \tag{E8.1C}$$

$$\mu = 0.0625 \text{ N} \frac{\text{s}}{\text{m}^2} = 0.0625 \text{ Pa·s}$$

The kinematic viscosity can be calculated as

$$v = \frac{\mu}{\rho} = \frac{0.0625}{900} = 6.95 \times 10^{-5} \frac{\text{m}^2}{\text{s}}$$

8.3 Pressure (*P*)

Pressure is the force that the fluid molecules exert per unit area in the normal direction under equilibrium conditions. Mathematically,

$$dP = \frac{dF}{dA} \tag{8.4}$$

The standard unit of pressure is $\frac{\text{N}}{\text{m}^2}$, also called Pascal although atm, bar, psi, and Torr are commonly used. We are commonly exposed to 1 atm pressure. However, in engineering calculations, three types of pressures are employed: absolute pressure (P_a), gage pressure (P_g), and vacuum pressure (P_v). For easy understanding let us imagine complete vacuum where the pressure is absolute zero pressure. Absolute pressure is the pressure which is measured with reference to absolute zero pressure. The gage pressure, on the other hand, measures pressure with reference to the atmospheric pressure. When a pressure sensor measures zero gage pressure, it refers to 1 atm. Finally, when the pressure is less than 1 atm, it is called vacuum pressure. However, in engineering calculations, it is customary to use absolute pressure. Table 8.1 helps us with the pressure conversions.

Pressure Unit	atm	kPa	psi	bar	Torr
atm	1	101.32	14.69	1.01	760
kPa	0.0099	1	0.145	0.01	7.5
psi	0.068	6.89	1	0.069	51.71
bar	0.99	100	14.5	1	750.06
Torr	0.0013	0.133	0.019	0.0013	1
	$P_a = P_{atm} + P_g$			$P_v = P_{atm} - P_a$	

TABLE 8.1 Conversion Table from One Pressure Unit to Other. To Obtain a Pressure in a Different Unit, Multiply with the Number in the Intersection Cell

Example 8.2 (Engineering—All concentrations): The regulator of a gas tank shows the pressure as 8000 kPa at room temperature (27°C). What is the absolute pressure in bar, atm, Torr, and psi?

Step 1: Given data
Temperature, $T = 27°C$; Pressure, $P = 8000$ kPa

To be determined: Absolute pressure in bar, atm, Torr, and psi.

Assumptions: Temperature remains constant at 27°C.

Approach: We will use Table 8.1 to convert the data into various units of pressure.

Step 2: Calculations
We know that

$$P_a = P_{atm} + P_g \tag{E8.2A}$$

Therefore, the absolute pressure is

$$P_a = 101.325 + 8000 = 8101.325 \text{ kPa}$$

We can now easily convert the absolute pressure into other units as follows:

$$P_a(\text{bar}) = 8101.325 \times 0.01 = 81.01 \text{ bar}$$

$$P_a(\text{atm}) = 8101.325 \times 0.0099 = 80.2 \text{ atm}$$

$$P_a(\text{Torr}) = 8101.325 \times 7.5 = 60,912 \text{ Torr}$$

$$P_a(\text{psi}) = 8101.325 \times 0.145 = 1175.32 \text{ psi}$$

8.4 Flow Velocity (v)

From fluid mechanics, we can recall that when a fluid moves through a pipe or a channel, the velocity of the fluid is different at different locations. For example, if we closely observe the flow of water in a stream, the edges experience a slow velocity than the water near the centerline of the stream. Similarly, in a pipe, the velocity of the fluid will be smaller (in fact, the fluid comes to a complete stop at the walls due to the no-slip situation) near the pipe walls relative to the center of the pipe. It is impractical to measure the fluid velocities at various locations along a given cross section. Therefore, for all practical purposes, we routinely use an average velocity, also called velocity (v), to denote the rate (and the direction) at which the fluid is moving across a reference cross section of the given conduit. The standard unit of fluid velocity is $\frac{m}{s}$.

8.5 Volumetric Flow Rate (Q)

The volumetric flow rate, or simply flow rate (in certain geographical locations called discharge), is the quantity (volume) of a fluid flowing per unit time across a reference cross section. Mathematically,

$$Q = A \times v \tag{8.5}$$

The standard unit of flow rate is $\dfrac{m^3}{s}$.

8.6 Mass Flow Rate ($\dot{m}$)

In certain applications, we are interested in knowing how much mass (number of molecules) has crossed a reference point per unit time. Therefore, mass flow rate ($\dot{m}$) is defined as the mass of the fluid flowing per unit time across a reference cross section.

$$\dot{m} = \rho \times Q \tag{8.6}$$

The standard unit of mass flow rate is $\dfrac{kg}{s}$.

Example 8.3 (Environmental Engineering—Air flow rate to a poultry unit): A set of exhaust fans move $25 \dfrac{m^3}{min}$ of air on a summer day with an average temperature of 35°C at a pressure of 1 atm. Determine the mass flow rate of air (Fig. 8.2).

Step 1: Schematic
The problem is described in Fig. 8.2.

Step 2: Given data
Flow rate, $Q = 25 \dfrac{m^3}{min}$; Temperature, $T = 35$°C; Pressure, $P = 1$ atm

To be determined: Mass flow rate of air.

Assumptions: (1) Temperature remains constant and (2) air acts as an ideal gas.

Step 3: Calculations
We will take the molecular weight of air as 28.96 g/mol and use the ideal gas law to determine the mass of air the fan is moving per minute at 35°C.
The ideal gas law is written as

$$PV = nRT \tag{E8.3A}$$

where P = pressure (atm), V = volume of the gas (air) (m³), R = gas constant (0.082 L atm mol^{-1} K^{-1}), and T = temperature (K).
Substituting the values after appropriate unit conversions on 1-minute basis

$$n = \frac{PV}{RT} = \frac{1 \times 25 \times 10^3}{0.082 \times (273 + 35)} = 989.86 \text{ moles of air/min}$$

Exhaust fans

FIGURE 8.2 Schematic of the poultry unit with exhaust fans described in Example 8.3.

Mass flow rate of air, $\dot{m} = 989.86 \ \dfrac{\text{mol}}{\text{min}} \times 28.96 \ \dfrac{\text{g}}{\text{mol}} = 28{,}666.45 \ \dfrac{\text{g}}{\text{min}} = \dfrac{28{,}666.45}{1000 \times 60} \ \dfrac{\text{kg}}{\text{s}}$

The fans together move about $0.48 \ \dfrac{\text{kg}}{\text{s}}$ of air.

8.7   Fluid Flow Regimes

When a fluid stream is in motion, the molecules within the fluid can behave in two ways: orderly or chaotic. To illustrate these ideas, imagine a march of an army platoon. The soldiers always move in a perfect path. We realize that although all the soldiers are moving continuously, their relative positions remain the same. In the context of a fluid flow, when the molecules of the fluid are moving in an *orderly manner* (with no transverse mixing), such flow is called *laminar flow* (Fig. 8.3). In other words, in laminar flow, we can visualize the fluid mass as thin layers of fluid, superimposed on one another, with *each layer (lamina) moving parallel* to its adjacent layer of fluid (Fig. 8.4). Because the flow is orderly, we can use mathematics to predict the fluid behavioral parameters such as pressure drop between two points, flow and velocity.

Now, let us consider a flow of fluid in which the molecules are travelling in a complete *chaotic manner*. In other words, the molecules cross each other and collide with other molecules, resulting in *random mixing* of the fluid. Such random and chaotic motion is called *turbulent flow* (Fig. 8.5).

To understand turbulent flow, imagine a soccer game in which spectators are running into the stadium after the home team wins the match. The fans run randomly in all directions at different speeds to congratulate their favorite players (Fig. 8.6). Because turbulent flow is chaotic, its analyses are complicated and therefore is based on empirical approaches.

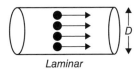

Laminar

FIGURE 8.3   Schematic of fluid in a laminar flow with all the molecules undergoing an orderly motion.

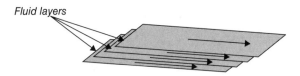

Fluid layers

FIGURE 8.4   Schematic of fluid in a laminar flow in which adjacent layers of fluid move in an orderly path.

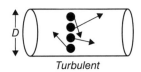

Turbulent

FIGURE 8.5   Schematic of fluid in a turbulent flow with all the molecules undergoing random motion.

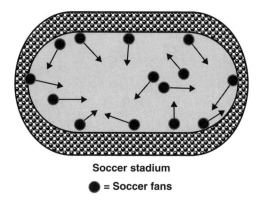

Soccer stadium

● = Soccer fans

FIGURE 8.6 Random movement of soccer fans in a stadium is analogous to a turbulent flow regime.

The concept of laminar and turbulent flow was proposed by the famous Irish engineer, Osborne Reynolds, who, through extensive experimentation, provided evidence of these regimes. Reynolds studied the flow patterns using ingeniously simple experiments (also called Reynolds experiments) in which he injected a dye in a flow of water through a pipe. Reynolds observed that during low velocities the dye and the water were flowing separately (unmixed) and appeared as two though two different flow streams were flowing in parallel–an indication of laminar flow. However, when the velocity was increased, the dye stream broke apart and the water and the dye appear to get mixed and eventually at higher velocities, they were intimately mixed (turbulent flow). Reynolds continued his experiments by varying all the flow and physical parameters and determined that the type of flow regime is dependent on the physical properties of the fluid (flow velocity, density, and viscosity) and the conduit (diameter, coarseness of the conduit). Based on his experiments, Reynolds determined that in any fluid flow, the inertial forces (that facilitate the random molecular motion during the fluid flow) and the viscous forces (that resist the random molecular motion during the fluid flow) are constantly competing against each other. Further, Reynolds proposed a mathematical dimensionless parameter, called *Reynolds number,* relating the inertial forces to the viscous forces as described below:

$$N_{Re} = \frac{\rho V D}{\mu} \tag{8.7}$$

where $\rho \left(\frac{\text{kg}}{\text{m}^3} \right)$ is the density of the fluid, $V \left(\frac{\text{m}}{\text{s}} \right)$ is the mean velocity of the fluid, D (m) is the allowable travel distance of the molecules or characteristic length of the conduit, and μ (Pa·s) is the viscosity of the fluid.

In Reynolds' experiments, the conduit employed was a tube and therefore the characteristic length of the conduit turned out to be the diameter. However, in reality, the characteristic length will depend on the configuration of the flow system. For example, for the flow around a rectangle plate, the characteristic length is the length of the plate (L) from the leading edge parallel to the flow. Similarly, if we are interested to determine the Reynolds number for flow around a vertical cylinder, the length serves as the characteristic length, while Reynolds number for the same cylinder placed horizontally in the direction of the flow will include the diameter as its characteristic length. Therefore, in general terms, the Reynolds number is written as

$$N_{Re} = \frac{\rho V L_c}{\mu} \tag{8.8}$$

where $\rho \left(\dfrac{\text{kg}}{\text{m}^3} \right)$ is the density of the fluid, $V \left(\dfrac{\text{m}}{\text{s}} \right)$ is the mean velocity of the fluid, $\mu \, (\text{Pa} \cdot \text{s})$ is the viscosity of the fluid, and L_c (m) is the characteristic length.

The numerator in Eq. (8.8) corresponds to the inertial forces and the denominator represents the viscous forces. When kinematic viscosity $v \left(\dfrac{\text{m}^2}{\text{s}} \right)$ is used, the Reynolds number takes the familiar form

$$N_{Re} = \frac{V L_c}{v} \tag{8.9}$$

where $v = \dfrac{\mu}{\rho}$.

Sometimes, in lieu of the velocity, we will have information on flow rate of the fluid (Q). The Reynolds number for a fluid in a pipe of diameter (D) can be rewritten as

$$N_{Re} = \frac{\rho \dfrac{Q}{\left(\pi D^2 / 4 \right)} D}{\mu} = \frac{4 Q \rho}{\pi D \mu} \tag{8.10}$$

In some industrial applications, however, mass flow rates ($\dot{m}$) are commonly used. Recalling that $\dot{m} = Q\rho$, we can express the Reynolds number as

$$N_{Re} = \frac{4 \dot{m}}{\pi D \mu} \tag{8.11}$$

Often, engineers encounter situations in which the fluids are flown in noncircular conduits. For example, in HVAC applications, square- or rectangular-shaped ducts are employed. In such cases, the hydraulic diameter is used in lieu of the characteristic length as follows:

$$N_{Re} = \frac{\rho V D_H}{\mu} \tag{8.12}$$

where $D_H = 4 \dfrac{A}{P_W}$, A = area of cross section, and P_W = wetted perimeter of the conduit.

It was determined that for Reynolds number less than about 2100 (commonly called critical Reynolds number), the flow with a pipe will be laminar. At Reynolds numbers <2100, the viscous forces are dominant and can control and minimize the random motion of the molecules in the fluid and maintain an orderly flow regime. However, when the Reynolds number exceeds about 4000, the applied inertial forces on the molecules can easily overcome the viscous resistance and the flow becomes completely mixed and therefore turbulent. However, when the Reynolds numbers are between 2100 and 4000, the flow is said to be fluctuating between laminar and turbulent and hence called transitional flow. As engineers, it is very important to note that we should always avoid designs in which flow falls in the transition regime. Transition regime involves flows randomly changing between laminar and turbulent and using

Configuration	Laminar	Turbulent
Flow around a sphere	<10	>200,000
Flow past cylinder	<400,000	>400,000
Flow within cylinder	<2100	>4000
Flow in a packed bed cylinder	<10	>2000
Stirred cylinder	<10	>10,000
Open channel flow	500	>2000
Flow around a flat plate	<300,000	>500,000

Note: While calculating the Reynolds number, one must exercise caution to choose an appropriate characteristic length.

TABLE 8.2   Range of Reynolds Numbers for Various Geometries*

a Reynolds number between 2100 and 4000 results in significantly erratic and inconsistent flow delivery performances.

One must also realize that the range of Reynolds numbers (laminar <2100, and turbulent >4000), we just discussed, is applicable only for fluids flowing through pipes. For other common configurations, the general range of Reynolds numbers is provided in Table 8.2.

Example 8.4 (Environmental Engineering): A fan duct is to be installed in a poultry barn stocked with 50,000 birds. Based on previous experience, each bird requires 4 L/min of fresh air for optimum growth. If the linear air velocity of the fan is limited to 10 m/s, determine the duct size (Fig. 8.7).

Step 1: Schematic
The problem is illustrated in Fig. 8.7.

Step 2: Given data
Limiting linear velocity, $v = 10 \left(\dfrac{m}{s} \right)$; Required air flow per bird, $Q_b = 4 \dfrac{L}{min}$

To be determined: Diameter of the duct to supply air.

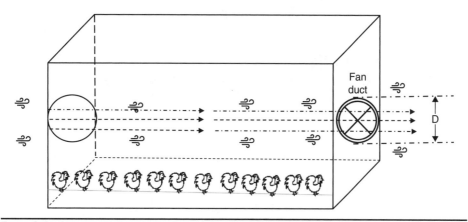

FIGURE 8.7   Schematic of a poultry barn with an exhaust duct for Example 8.4.

Step 3: Calculations

This is a simple case of continuity equation.

Total air flow needed, $Q_T = Q_b \times$ Number of birds $= 4 \times 50,000 = 200,000 \ \dfrac{L}{min}$

We know that $Q_T = Av = \dfrac{\pi D^2}{4} v$

Therefore, we can substitute the values of Q_T and v in the above equation with appropriate unit conversions and solve for the diameter D.

$$\frac{200,000}{1000 \times 60} = \frac{3.14 \times D^2}{4} \times 10 \qquad \text{(E8.4A)}$$

The required size for the duct is 0.65 m.

Example 8.5 (Agricultural Engineering): Determine the limiting diameter for a pipe to transport biodiesel $\left(\rho = 0.9 \ \dfrac{kg}{m^3}, \text{ Kinematic viscosity, } v = 6 \times 10^{-6} \ \dfrac{m^2}{s} \right)$ at a flow rate of 100 L/min of under laminar flow conditions.

Step 1: Given data

Density, $\rho = 900 \ \dfrac{kg}{m^3}$; Kinematic viscosity, $v = 6 \times 10^{-6} \ \dfrac{m^2}{s}$; Discharge, $Q = 100 \ \dfrac{L}{min}$

Step 2: Calculations

The equation for Reynolds number for a circular pipe* is $N_{Re} = N_{Re} = \dfrac{VD}{v}$. The limiting value (also called critical value) of the N_{Re} for the flow to remain laminar is 2100

$$2100 = \frac{VD}{v} = \frac{\left(\dfrac{Q}{\dfrac{\pi D^2}{4}} \right) D}{v} = \frac{4Q}{\pi D v} \qquad \text{(E8.5A)}$$

Substituting the values after appropriate unit conversions will yield

$$2100 = \frac{4 \times \left(\dfrac{100}{1000 \times 60} \right)}{3.14 \times D \times 6 \times 10^{-6}} \qquad \text{(E8.5B)}$$

$$D = 0.168 \text{ m} = 16.8 \text{ cm}$$

Note: The characteristic length for a circular pipe is always the diameter.

8.8 The Continuity Equation

We are familiar with the idea that mass can neither be created nor destroyed and will remain the same within a system. We can apply the same concept and arrive at very useful mathematical relations in fluid transport. Consider a pipe in any arbitrary variable cross section in which an incompressible fluid of density ρ flows under steady-state conditions (Fig. 8.8). Now, while the fluid is moving, we will select a small fluid element of thickness Δx_1 and an area of A_1 and follow its movement. After a time interval Δt, the fluid element will have moved to a new location with a thickness Δx_2 and having an area A_2. From the law of conservation of mass, we know that the mass of the fluid will remain the same at both the locations 1 and 2. Therefore,

$$m_1 = m_2 \qquad \text{(8.13)}$$

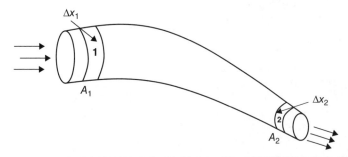

FIGURE 8.8 The mass flowing through a conduit always remains constant.

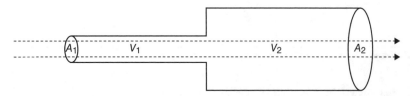

FIGURE 8.9 An increase in the cross-sectional area will decrease the velocity of the fluid and vice versa.

Writing mass in terms of ρV, we obtain

$$A_1 \Delta x_1 = A_2 x_2 \tag{8.14}$$

or

$$\frac{A_1 \Delta x_1}{\Delta t} = \frac{A_2 x_2}{\Delta t}$$

$$A_1 v_1 = A_2 v_2 \tag{8.15}$$

$$Q_1 = Q_2 \tag{8.16}$$

Equations (8.15) and (8.16) are called continuity equations. From Eq. (8.15), it can easily be deduced that a piping system with a larger diameter along its flow (Fig. 8.9) will experience a reduction in velocity in order to maintain the same flow rate.

Example 8.6 (Agricultural Engineering—Flow through branched duct): Warm air is flowing through a 0.6-m diameter circular main duct at an average linear velocity of 200 m/min. The main duct is now branched into two ducts (Fig. 8.10). The first branch has a diameter of 0.2 m while the second branch has a diameter of 0.3 m and carries 75% of the warm air from the main duct. Determine the flow velocity in the branched ducts.

Step 1: Schematic
The problem is depicted in Fig. 8.10.

Step 2: Given data
Velocity in the main duct, $v = 200 \frac{m}{min}$; Diameter of the main duct, $D = 0.6$ m; Diameter of the first branch, $D_1 = 0.2$ m; Diameter of the second branch, $D_2 = 0.3$ m; Flow in the first branch, $Q_1 = 0.25 \times Q$; Flow in the second branch, $Q_2 = 0.75 \times Q$

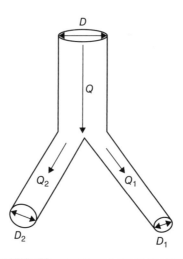

Figure 8.10   Schematic of branched duct carrying air.

To be determined: Velocities in branched ducts.

Step 3: Calculations

This is a simple application of continuity equation. We can calculate the flow in the main duct as

$$Q = vA = \frac{200}{60} \times \frac{3.14 \times 0.6^2}{4} = 0.942 \ \frac{m^3}{s}$$

Assuming no losses and leaks, $Q = Q_1 + Q_2$ and we are also given that $Q_1 = 0.25Q$ and $Q_2 = 0.75Q$. Therefore, $Q_1 = 0.25Q = v_1 A_1$

or

$$0.25 \times 0.942 = v_1 \times \frac{\pi \times 0.2^2}{4} \tag{E8.6A}$$

$$v_1 = 7.5 \ \frac{m}{s}$$

Similarly, $Q_2 = 0.75Q = v_2 A_2$

$$0.75 \times 0.942 = v_2 \times \frac{\pi \times 0.3^2}{4} \tag{E8.6B}$$

$$v_2 = 10 \ \frac{m}{s}$$

8.9   The Energy Equation

Assuming that the fluid flow is steady, inviscid, incompressible, frictionless, and irrotational, let us consider a small fluid element of length dl and an area of A along the streamline (Fig. 8.11) subjected to a pressure P and $\left(P + \frac{\partial P}{\partial l} dl\right)$ on either direction. Performing a force balance ($F = m \times a$) on the element, we will obtain

$$\underbrace{(P \times A) - \left[\left(P + \frac{\partial P}{\partial l} dl\right) \times A\right] - (m \times g \times \cos\alpha)}_{\text{Net force in l-direction}} = m \times a \tag{8.17}$$

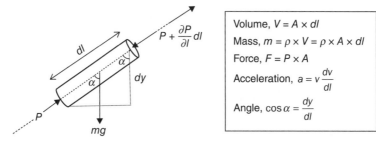

Figure 8.11 Force balance on the element of the fluid along its streamline.

After substituting the relevant expressions in the above equation and simplifying,

$$-\frac{\partial P}{\partial l}\,dl\times A-\rho\times A\times dl\times g\times\frac{dy}{dl}=\rho\times A\times dl\times v\frac{dv}{dl} \tag{8.18}$$

Dividing the equation by $dl\times A\times\rho$ and simplifying

$$\frac{\partial P}{\rho}+(g\times dy)+(v\times dv)=0 \tag{8.19}$$

Equation (8.19) is called Euler equation of motion along the streamline. After integrating Eq. (8.19), we obtain the well-known Bernoulli's equation as follows:

$$\frac{P}{\rho}+gy+\frac{v^2}{2}=\text{Constant} \tag{8.20}$$

where C is a constant.

This is the Bernoulli's equation in the **standard** form that quantifies the energy on a mass basis $\left(\dfrac{\text{N}\cdot\text{m}}{\text{kg}}\right)$.

After multiplying the equation by ρ, Bernoulli's equation in the **pressure** form that quantifies the energy on a volume mass $\left(\dfrac{\text{N}\cdot\text{m}}{\text{m}^3}\right)$ is obtained (below).

$$P+\gamma y+\frac{\rho v^2}{2}=\text{Constant} \tag{8.21}$$

Further, dividing Eq. (8.18) by g will result in the most used Bernoulli's equation in the head form that quantifies the energy on a weight basis $\left(\dfrac{\text{N}\cdot\text{m}}{\text{N}}\right)$ (below) (Fig. 8.12).

$$\frac{P}{\gamma}+y+\frac{v^2}{2g}=\text{Constant} \tag{8.22}$$

Because the specific weight of the fluid, $\gamma=\rho g$.

If we consider two locations (say 1 and 2) along the streamline, we can expand the Bernoulli's equation as

$$\frac{P_1}{\gamma}+y_1+\frac{v_1^2}{2g}=\frac{P_2}{\gamma}+y_2+\frac{v_2^2}{2g} \tag{8.23}$$

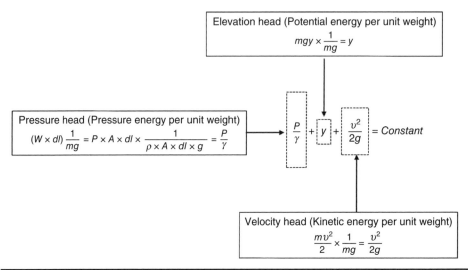

Figure 8.12 Bernoulli's equation in the head form.

Conventionally, Bernoulli's equation is expressed as

$$\frac{P_1}{\gamma} + h_1 + \frac{v_1^2}{2g} = \frac{P_2}{\gamma} + h_2 + \frac{v_2^2}{2g}$$

where P_1, v_1, h_1 and P_2, v_2, h_2 represent the pressure, velocity, and height of the fluid at locations 1 and 2, while γ is the specific weight of the fluid. However, when using Bernoulli's equation, one must note that the flow must be steady, inviscid, frictionless, and irrotational.

Example 8.7 (Aquacultural Engineering—Flow through a tank): A storage pond of 1 hectare filled with seawater is used as a reservoir for exchanging water from shrimp ponds (Fig. 8.13). The water level in the reservoir was measured to be 1.5 m from the outlet at the bottom that consisted of a 0.5-m diameter concrete pipe. Determine the discharge at the outlet.

Step 1: Schematic
The problem is described in Fig. 8.13.

Step 2: Given data
Height, $h_1 = 1.5$ m; Area, $A = 1$ hectare $= 10,000$ m²; Diameter, $D = 0.5$ m

To be determined: Flow rate from the outlet pipe.

Step 3: Calculations
We will apply the Bernoulli's equation at locations 1 and 2 and solve for the velocity at point 2. We will also choose point 2 as the datum (i.e., $h_2 = 0$).
   The Bernoulli's equation is written as

$$\frac{P_1}{\gamma} + \frac{v_1^2}{2g} + h_1 = \frac{P_2}{\gamma} + \frac{v_2^2}{2g} + h_2 \qquad \text{(E8.7A)}$$

We also know that $P_1 = 1$ atm $= P_2$ and $h_1 = 1.5$ m.

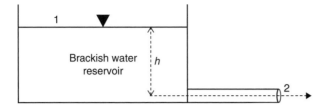

Figure 8.13   Schematic of brackish water supply to shrimp ponds.

In addition, the velocity of the brackish water at the surface is negligible relative to the velocity of the water at the outlet point. Therefore, $v_1 \approx 0$.

Therefore, after cancelling out, Eq. (E8.7A) is simplified as

$$h_1 = \frac{v_2^2}{2g} \tag{E8.7B}$$

$$v_2 = \sqrt{2gh_1} \tag{E8.7C}$$

The discharge is given by
$$Q = A_2 v_2 = \frac{\pi D^2}{4} \times \sqrt{2gh_1} \tag{E8.7D}$$

Substituting the values for D and h_1, we obtain the discharge as

$$Q = \frac{3.14 \times 0.5^2}{4} \times \sqrt{2 \times 9.81 \times 1.5} = 1.06 \; \frac{\text{m}^3}{\text{s}}$$

Example 8.8 (Environmental Engineering–Flow through a tank):   A smooth PVC pipeline in a wastewater treatment facility is 60-m long with an upward slope of 1 in 30 m (Fig. 8.14). The pipe with an inlet diameter of 60 cm tapers down to 45 cm at the outlet and carries a wastewater at 1000 LPM. Determine the pressure drop for this pipeline if the energy losses are negligible.

Step 1: Schematic
The problem is illustrated in Fig. 8.14.

Step 2: Given data
We will denote inlet as location 1 and outlet as location 2.

Upward slope, $m = 1/30$; Length, $L = 60$ m; Inlet diameter, $D_{\text{inlet}} = D_1 = 0.6$ m; Outlet diameter, $D_{\text{outlet}} = D_2 = 0.45$ m; Discharge, $Q = 1000 \; \dfrac{\text{L}}{\text{min}}$

To be determined: Pressure differential.

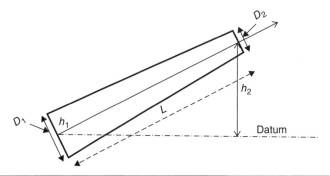

FIGURE 8.14 Schematic of pipeline carrying the wastewater.

Step 3: Calculations
We will apply the Bernoulli's equation at inlet (location 1) and outlet (location 2) and solve for the pressure drop.

$$\frac{P_1}{\gamma} + \frac{v_1^2}{2g} + h_1 = \frac{P_2}{\gamma} + \frac{v_2^2}{2g} + h_2 \tag{E8.8A}$$

$$\left(\frac{P_1 - P_2}{\gamma}\right) = \left(\frac{v_2^2 - v_1^2}{2g}\right) + (h_2 - h_1) \tag{E8.8B}$$

From continuity equation, we know that $Q_1 = Q_2 = Q$. We also know that

$$v_1 = \frac{Q}{A_1} = \frac{Q}{\left(\pi D_1^2 \Big/ 4\right)} = \frac{4Q}{\pi D_1^2} \tag{E8.8C}$$

$$v_2 = \frac{Q}{A_2} = \frac{Q}{\left(\pi D_2^2 \Big/ 4\right)} = \frac{4Q}{\pi D_2^2} \tag{E8.8D}$$

Substituting expressions for v_1 and v_2 into Eq. (E8.8B), we obtain

$$\left(\frac{P_1 - P_2}{\gamma}\right) = \left(\frac{\left(\frac{4Q}{\pi D_2^2}\right)^2 - \left(\frac{4Q}{\pi D_1^2}\right)^2}{2g}\right) + (h_2 - h_1) \tag{E8.8E}$$

$$\frac{P_1 - P_2}{\rho g} = \frac{16Q^2}{2g\pi^2}\left[\frac{1}{D_2^4} - \frac{1}{D_1^4}\right] + (h_2 - h_1) \tag{E8.8F}$$

We are given the values of Q, D_1, and D_2. It is also given that the slope is 1 m in 30 m. Therefore, $h_1 = 0$ m (datum) and $h_2 = 2$. Now, substituting the values in Eq. (E8.8F)

$$\left(\frac{P_1 - P_2}{1000 \times 9.81}\right) = \frac{16 \times \left(\dfrac{1000}{1000 \times 60}\right)^2}{2 \times 9.81 \times 3.14^2}\left[\frac{1}{0.45^4} - \frac{1}{0.6^4}\right] + (2 - 0) \tag{E8.8G}$$

The pressure drop, $P_1 - P_2 = \Delta P \approx 19{,}620.5$ Pa.

Example 8.9 (Agricultural Engineering—Energy generation):   A stream of water is falling from a height of 40 m. Flow measurements indicated that 30 m³/s water flows through crest of dimensions 5 m × 2 m. How much hydropower can be produced from the falling water if an overall efficiency of 78% is expected?

Step 1: Given data

Flow rate, $Q = 30 \dfrac{m^3}{s}$; Height of the fall, $y = 40$ m, Crest area, $A = 5$ m × 2 m = 10 m²; Equipment efficiency, $\eta = 78\%$; Density, $\rho = 1000 \dfrac{kg}{m^3}$

To be determined: Hydropower potential.

Step 2: Calculations

Assumptions: The water behaves as an inviscid, frictional losses, and operates in a steady-state mode. The Bernoulli's equation with respect to energy is expressed as

$$\text{Total available energy, } E = \frac{P}{\rho} + \frac{v^2}{2} + gy \tag{E8.9A}$$

$$E = \frac{P}{\rho} + \frac{\left(\dfrac{Q}{A}\right)^2}{2} + gy \tag{E8.9B}$$

$$E = \frac{101,325}{1000} + \frac{\left(\dfrac{30}{10}\right)^2}{2} + (40 \times 9.81) = 498.2 \ \frac{J}{kg} \tag{E8.9C}$$

$$\text{Hydropower potential, } P = E \times \dot{m} = E \times (\rho \times Q) = 498.2 \times 1000 \times 30 = 14.94 \times 10^6 \ \frac{J}{s} = 14.94 \text{ MW}$$

Given that an overall efficiency of the hydropower system is 78%, the actual hydropower that can be harvested is $P_a = P \times \eta = 14.94 \times 0.78 \approx 11.66$ MW.

Practice Problems for the FE Exam

1. An experiment was performed to determine the viscosity of synthetic oil (density = 0.92 g/cc) at 25°C by placing biodiesel between two parallel plates (area = 0.06 m²) separated by 5 mm. Subsequently when a force of 3 N was applied, the upper plate's velocity was recorded as 5 m/s.

   a. 0.5 Pa·s.

   b. 0.005 Pa·s.

   c. 0.05 Pa·s.

   d. None of the above

2. The kinematic viscosity of the synthetic oil tested in Problem 1 is closest to

   a. 0.0005 m²/s.

   b. 0.005 Pa·s.

   c. 0.05 m²/s.

   d. None of the above

3. Turbulent flow in a circular pipe is not characterized by
   a. Chaotic movement of water.
   b. Reynolds numbers greater than 10,000.
   c. Density of the fluid lower than 1.0.
   d. Excellent mixing of the fluid.

4. Laminar flow in a circular pipe is associated with
   a. Flow of oils.
   b. Reynolds numbers less than 10,000.
   c. Density of the fluid lower than 1.0.
   d. Ordered movement of water molecules.

5. The flow in a 2-cm-diameter pipe carrying 50 L/min of soybean oil (density = 900 kg/m^3 and viscosity = 0.08 Pa·s) at 25°C is
   a. Laminar.
   b. Turbulent.
   c. Transition.
   d. Fully turbulent.

6. Warm air (density = 1.27 kg/m^3) at 40°C flows through a square duct of dimensions 0.5 m × 0.5 m at a flow rate of 2800 LPM. The mass flow rate of the air is approximately
   a. 0.9 kg/s.
   b. 0.6 kg/s.
   c. 1 kg/s.
   d. 0.06 kg/s.

7. Air (density = 1.2 kg/m^3 viscosity = 20 × 10^{-6} Pa·s) is flowing in a rectangular duct of dimensions 0.5 m × 0.3 m at 100 L/s. The Reynolds number for this system is
   a. 1500.
   b. 5000.
   c. 15,000.
   d. None of the above

8. A 1-cm-diameter pipe carrying water at 24 L/min abruptly enlarges into a 2.5-cm-diameter pipe. The decrease in the velocity in the 2.5-cm-diameter pipe relative to the 1-cm-diameter pipe is closest to
   a. 40%.
   b. 50%.
   c. 85%.
   d. None of the above

9. In a fluid flowing in a horizontal pipe, a decrease in the velocity head will

    a. Make the flow laminar.

    b. Sometimes increase in the pressure head.

    c. Always increase in the pressure head.

    d. None of the above

10. An open cylindrical tank filled with water to a height of 4 m has a 2-cm-diameter orifice open to the atmosphere. The discharge through the orifice is about

    a. 30 LPM.

    b. 0.3 LPM.

    c. 6 LPM.

    d. 167 LPM.

Practice Problems for the PE Exam

1. Flow measurements in a swine barn indicated that the fans move $12\dfrac{m^3}{min}$ of air on a spring day with an average temperature of 20°C at a pressure of 1 atm. Calculate the mass flow rate of air by the fans if the density of the air is given as 1.2 kg/m³.

2. A 5-cm-diameter pipe carrying water at 100 L/min abruptly contracts into a 2.5-cm-diameter pipe. Compare the velocities in the 5- and 2.5-cm-diameter pipes.

3. It was decided to install a fan duct in a hog barn populated with 1000 finisher pigs. Based on previous experience, each pig requires 5 LPM of fresh air for optimum growth. If the linear air velocity of the fan is limited to 5 m/s, determine the duct size.

CHAPTER 9

Fluid Flow through Pipes

Chapter Objectives

The content presented in this chapter will equip students to

- Calculate the frictional losses in pipes for laminar and turbulent flow systems.
- Determine the pumping power needed to transport a given volume of fluid between two locations.
- Analyze pipes in series and parallel and determine the pumping requirements.

9.1 Motivation

We frequently encounter engineering problems that involve transport of fluids in pipes and ducts. While designing and selecting a piping system, we need answers to questions, such as how much fluid will be transported, what will be the required pressure, and most importantly how much energy will the pumping system consume. Besides, while moving in conduits, fluids always encounter frictional resistances that oppose the flow. As engineers we are expected to understand and quantify these resistances and determine additional power required to overcome these losses so that the target flow rates are always met. Therefore, this chapter will exclusively focus on the movement of fluids in pipes and ducts for laminar and turbulent flow regimes with emphasis on the pressure, discharge, and frictional losses. In addition, practical situations such as pipes connected in series and parallel are also presented.

9.2 Laminar Flow through Pipes

Let us consider a fully developed, incompressible, steady-state, laminar fluid flow having a viscosity, μ, in a pipe of length, l and radius, R. We now choose a small element of the fluid of length, dl, radius, r subjected to pressures P and $P + \frac{\partial P}{\partial l} dl$ on either side and shear force, τ on the top and bottom of the element (Fig. 9.1). The cross-sectional area,

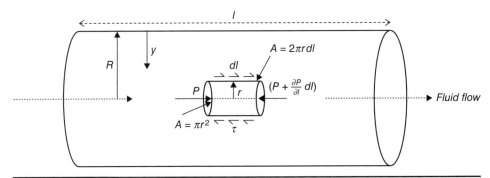

Figure 9.1 Schematic of a small cylindrical element of fluid subjected to a pressure and a shear force.

A_c and the longitudinal area, A_l are πr^2 and $2\pi r dl$, respectively. If we perform a force balance around the element,

$$\boxed{\tau = \mu \frac{dv}{dy} = -\mu \frac{dv}{dr}, \quad \because y = R - r; \; dy = -dr}$$

$$P \times A_c - \left(P + \frac{\partial P}{\partial l} dl\right) \times A_c - \tau \times A_l = 0 \tag{9.1}$$

$$P \times \pi r^2 - \left(P + \frac{\partial P}{\partial l} dl\right) \times \pi r^2 - \left(-\mu \frac{dv}{dr}\right) \times 2\pi r dl = 0 \tag{9.2}$$

After simplifying

$$-\frac{\partial P}{\partial l} \times r + \mu \frac{dv}{dr} \times 2 = 0 \tag{9.3}$$

$$\frac{dP}{dl} = \frac{-2\tau}{r} \tag{9.4}$$

$$\frac{dP}{dl} = \frac{2\mu}{r} \frac{dv}{dr} \tag{9.5}$$

$$\frac{dv}{dr} = r \frac{1}{2\mu} \frac{dP}{dl} \tag{9.6}$$

Integrating the above equation,

$$v = \frac{r^2}{4\mu} \frac{dP}{dl} + C \tag{9.7}$$

When $r = R$, $v = 0$ ($\because$ it is a no-slip condition). Therefore,

$$C = -\frac{R^2}{4\mu} \frac{dP}{dl} \tag{9.8}$$

And the velocity v becomes

$$v = \frac{r^2}{4\mu} \frac{dP}{dl} - \frac{R^2}{4\mu} \frac{dP}{dl} \tag{9.9}$$

or

$$v = \frac{-1}{4\mu} \frac{dP}{dl}(R^2 - r^2) \qquad (9.10)$$

or

$$v = -\left(\frac{dP}{dl}\right)\frac{R^2}{4\mu}\left(1 - \frac{r^2}{R^2}\right) \qquad (9.11)$$

The above equation indicates that the velocity follows a *parabolic distribution* (Fig. 9.2). We are also interested in determining the maximum velocity and its location. Therefore, we set $\frac{dv}{dr} = 0$.

After differentiating Eq. (9.11) with respect to r, we arrive at

$$\frac{dv}{dr} = 0 = -\left(\frac{dP}{dl}\right)\frac{R^2}{4\mu}\left(0 - \frac{2r}{R^2}\right) \qquad (9.12)$$

$$\left(\frac{dP}{dl}\right)\frac{r}{2\mu} = 0 \qquad (9.13)$$

or
$$r = 0 \qquad (9.14)$$

Therefore, the maximum velocity occurs at $r = 0$, that is, at the center of the pipe, where the shear stress is zero (Fig. 9.2), which we will verify later. The value of the maximum velocity, v is

$$v_{\max} = -\left(\frac{dP}{dl}\right)\frac{R^2}{4\mu} \qquad (9.15)$$

Now to estimate the flow rate, Q, we will need to multiply the area with the velocity. To accomplish this, we will consider an infinitesimally small circular ring of thickness dr and multiply the area of the element $(2\pi r dr)$ and the velocity $\left(v = \frac{-1}{4\mu}\frac{dP}{dl}(R^2 - r^2)\right)$.

$$dQ = (2\pi r dr)\times\frac{-1}{4\mu}\frac{dP}{dl}(R^2 - r^2) \qquad (9.16)$$

$$\int_0^Q dQ = \int_0^R (2\pi r\, dr)\times\frac{-1}{4\mu}\frac{dP}{dl}(R^2 - r^2) \qquad (9.17)$$

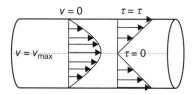

FIGURE 9.2 Velocity and shear stress distribution for a fluid flowing through a pipe.

After evaluating the integral 9.17, we obtain

$$Q = -\frac{2\pi}{4\mu}\frac{dP}{dl}\left[R^2\frac{r^2}{2} - \frac{r^4}{4}\right] \tag{9.18}$$

The limits for Q are from 0 to Q and for 0 to R for r

$$Q = -\frac{2\pi}{4\mu}\frac{dP}{dl}\frac{R^4}{4} = -\frac{\pi R^4}{8\mu}\frac{dP}{dl} \tag{9.19}$$

The average velocity, v_{ave} can easily be determined by dividing the discharge by the area of the pipe.

Using area, $A = \pi R^2$, we determine the average velocity as $\dfrac{Q}{A}$

$$v_{ave} = \frac{-\dfrac{\pi R^4}{8\mu}\dfrac{dP}{dl}}{\pi R^2} \tag{9.20}$$

Simplifying,

$$v_{ave} = -\frac{R^2}{8\mu}\frac{dP}{dl} \tag{9.21}$$

By comparing Eqs. (9.15) and (9.21), we can conclude that

$$\frac{v_{ave}}{v_{max}} = 0.5 \tag{9.22}$$

If we recall our discussion from Chap. 8, we called v_{ave} as flow velocity. Therefore, rearranging Eq. (9.20) and integrating from 0 to L and P_1 to P_2 will result in

$$\int_0^L \frac{8\mu}{R^2}v\,dl = \int_{P_1}^{P_2} -dP \tag{9.23}$$

$$\frac{8\mu v}{R^2}l = P_1 - P_2 \tag{9.24}$$

Equation (9.23) can be rearranged as

$$v = \frac{\Delta P}{8\mu l}R^2 \tag{9.25}$$

or in terms of diameter

$$v = \frac{\Delta P}{32\mu l}D^2 \tag{9.26}$$

The flow velocity, therefore, is directly proportional to the pressure drop across the pipe and the square of the diameter. However, the velocity is inversely proportional to the viscosity and the pipe length. This is expected because the increase in pipe length will increase the frictional losses due to molecular collisions as the fluid moves through the pipe. We can also determine the volumetric flow through a pipe by multiplying the fluid velocity by its surface area. Therefore,

$$Q = \left(\pi\frac{D^2}{4}\right)\left(\frac{\Delta P}{32\mu l}D^2\right) \qquad (9.27)$$

$$Q = \frac{\pi\Delta P D^4}{128\mu l} \qquad (9.28)$$

Equation (9.28) is the famous **Hagen–Poiseuille equation** and can also be rearranged as

$$Q = \frac{\Delta P}{\dfrac{128\mu l}{\pi D^4}}$$

The numerator in the above equation can be viewed as the driving force for the fluid to initiate the movement in the pipe. The denominator, however, is the resistance to the flow that comes from the viscosity, length, and, most importantly, the diameter. It is easily seen that increasing the diameter will have a great influence in increasing the volumetric flow rate. This equation is analogous to fundamental heat transfer laws that we discussed in the previous module.

$$\text{Flow rate}\left(\frac{\text{m}^3}{\text{s}}\right) = Q = \frac{\Delta P}{\left(\dfrac{128\mu l}{\pi D^4}\right)} = \frac{\text{Driving force}}{\text{Resistance}} \qquad (9.29)$$

$$\text{Conduction rate}\left(\frac{\text{J}}{\text{s}}\right) = \frac{\Delta T}{\left(\dfrac{a}{Ak}\right)} = \frac{\text{Driving force}}{\text{Resistance}} \qquad (9.30)$$

$$\text{Convection rate}\left(\frac{\text{J}}{\text{s}}\right) = \frac{\Delta T}{\left(\dfrac{1}{Ah}\right)} = \frac{\text{Driving force}}{\text{Resistance}} \qquad (9.31)$$

$$\text{Radiation rate}\left(\frac{\text{J}}{\text{s}}\right) = \frac{\Delta T}{\left(\dfrac{1}{A\sigma(T_1+T_2)\left(T_1^2+T_2^2\right)}\right)} = \frac{\text{Driving force}}{\text{Resistance}} \qquad (9.32)$$

As the fluid moves along the pipe, it encounters frictional resistance to flow and can result in the decrease in the pressure of the fluid. However, in our analysis of flow through pipes, we assumed that the flow was laminar and fully developed. This means that the frictional resistance to the flow is only due to the viscous forces or the viscosity of the fluid. Therefore, we can rewrite Eq. (9.26) to determine the reduction in the pressure over the length of the pipe.

$$\Delta P = \frac{32\mu v l}{D^2} \qquad (9.33)$$

We can also divide Eq. (9.33) by the specific weight of the fluid ($\rho \times g$) to determine the pressure loss over the length of the pipe:

$$\frac{\Delta P}{\rho g} = \frac{32\mu vl}{\rho g D^2} \tag{9.34}$$

We can see that the left-hand term in the above equation is the head loss in the pipe and is commonly denoted as h_L. Therefore, the Hagen–Poiseuille equation can also be expressed as follows in a head-loss form:

$$\frac{\Delta P}{\rho g} = h_L = \frac{32\mu vl}{\rho g D^2} \tag{9.35}$$

For those situations where the pipes are inclined at an angle θ, modified versions of Eqs. (9.26) and (9.28) are used:

$$v = \frac{(\Delta P - \rho g L \sin\theta)}{32\mu L} D^2 \tag{9.36}$$

$$Q = \frac{\pi(\Delta P - \rho g L \sin\theta) D^4}{128\mu L} \tag{9.37}$$

Example 9.1 (General Engineering–All concentrations): For a steady state, incompressible, fully developed laminar flow in a pipe, determine the location at which the velocity reaches the average value (Fig. 9.3).

Step 1: Schematic
The problem is depicted in Fig. 9.3.

Step 2: Given data
The flow is laminar, steady-state, incompressible, and fully developed.

Step 3: Calculations
We have previously derived the expressions for velocity and maximum velocity as

$$v = -\left(\frac{dP}{dl}\right)\frac{R^2}{4\mu}\left(1 - \frac{r^2}{R^2}\right) \tag{E9.1A}$$

$$v_{max} = -\left(\frac{dP}{dl}\right)\frac{R^2}{4\mu} \tag{E9.1B}$$

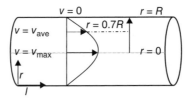

Figure 9.3 Schematic of the pipe with laminar flow described in Example 9.1.

Combining Eqs. (E9.1A) and (E9.1B) will result in

$$v = v_{max}\left(1 - \frac{r^2}{R^2}\right) = \frac{v_{max}}{2} \qquad \text{(E9.1C)}$$

$$\frac{r^2}{R^2} = 0.5 \qquad \text{(E9.1D)}$$

$$r = R\sqrt{0.5}$$

The velocity will reach its average at a distance $r = 0.7R$ from the center.

Example 9.2 (Bioprocessing Engineering—Flow of glycerin in a pipe):   Glycerin with a density 1.26 g/cm³ and a viscosity of 0.95 Pa·s is being pumped in a 25-m long pipe of 10 cm diameter using a pump. If the measured pressure drop was found to be 20,000 Pa, determine the expected discharge through the pipe (Fig. 9.4).

Step 1: Schematic
The problem is described in Fig. 9.4.

Step 2: Given data
Diameter, $D = 10$ cm; Length, $l = 25$ m; Pressure drop, $\Delta P = 20,000$ Pa; Density of the glycerin, $\rho = 1.26 \dfrac{g}{cm^3} = 1260 \dfrac{kg}{m^3}$; Viscosity of glycerin, $\mu = 0.95$ Pa·s

To be determined: Discharge, Q.

Assumptions: The flow is isothermal, laminar, steady state, incompressible, and fully developed.

Step 3: Calculations
We are not given the velocity of the glycerin to determine or verify flow regime. However, from experience we know that the flow of glycerin is usually laminar. Therefore, we will use Hagen–Poiseuille equation to determine the discharge of the glycerin through the pipe. Subsequently we will calculate the Reynolds number to verify the validity of the Hagen–Poiseuille equation. The form of the Hagen–Poiseuille equation for the calculation of discharge through a pipe is given by

$$Q = \frac{\pi \Delta P D^4}{128 \mu L} = \frac{3.14 \times 20,000 \times 0.05^4}{128 \times 0.95 \times 50} = 0.002 \ \frac{m^3}{s} = 123.9 \ \frac{L}{min}$$

Verification of Reynolds number
Now that we calculated the flow rate, we can easily verify whether our equation is valid. The Reynolds number for the flow can be determined by using the Reynolds number expressed in the form of the flow rate, Q (Eq. 8.10, Chapter 8):

$$N_{Re} = \frac{4Q\rho}{\pi D \mu} = \frac{4 \times (0.002) \times 1260}{3.14 \times 0.1 \times 0.95} = 34.9 < 2100 \ \text{(Laminar)}$$

Therefore, the flow is laminar, and our approach is correct.

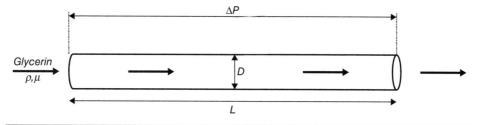

FIGURE 9.4   Schematic for Example 9.2 describing the flow of glycerin through a pipe.

9.3 Shear Stress Distribution

Sometimes, we are also interested to determine how the shear stress varies across the pipe. This can easily be accomplished by rearranging Eq. (9.4) as

$$\frac{dP}{dl} = \frac{-2\tau}{r} \tag{9.38}$$

$$\tau = -\frac{r}{2}\frac{dP}{dl} \tag{9.39a}$$

The above equation indicates that shear stress is a linear function of r. When $r = 0$, the shear stress is zero and maximizes when $r = R$ (at the wall of the pipe) and is expressed as

$$\tau_{max} = -\frac{R}{2}\frac{dP}{dl} \tag{9.39b}$$

The shear stress distribution is shown in Fig. 9.2.

Example 9.3 (Environmental Engineering—Shear stress in a pipe): Biodiesel is being pumped in a horizontal pipeline of 10-m long and 7.5-cm diameter. The pressure sensors attached between the two ends of the pipe read a pressure drop of 30 kPa. Determine the shear stress distribution, the maximum viscous force, and the velocity through the pipe if the viscosity and the density of biodiesel are given as $6 \times 10^{-6}\ \frac{m^2}{s}$ and $890\ \frac{kg}{m^3}$, respectively (Fig. 9.5).

Step 1: Schematic
The problem is described in Fig. 9.5.

Step 2: Given data
Diameter, $D = 7.5$ cm; Length, $l = 10$ m; Pressure drop, $\Delta P = 30,000$ Pa; Density of the biodiesel, $\rho = 0.89\ \frac{g}{cm^3} = 890\ \frac{kg}{m^3}$. Also, the units of the biodiesel provided in the problem suggest that the given viscosity is the kinematic viscosity. Therefore, $v = 6 \times 10^{-6}\ \frac{m^2}{s}$.

To be determined: Shear stress distribution ($\tau = f(r)$), viscous force (F), and velocity (v).

Assumptions: The flow is laminar, steady state, incompressible, and fully developed.

Step 3: Calculations
The shear stress (τ) in a laminar flow regime in a pipe is given by

$$\frac{dP}{dl} = \frac{-2\tau}{r} \tag{E9.3A}$$

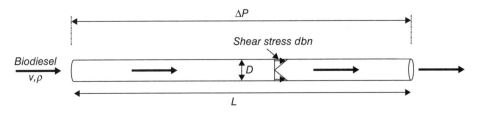

Figure 9.5 Schematic of the pipe carrying biodiesel described in Example 9.3.

The shear stress maximizes at $r = R$ (pipe wall) and minimizes at $r = 0$ (center of the pipe) and varies linearly.

$$\tau_{min} = \left(\frac{-dP}{dl}\right)\frac{r}{2} = \left(\frac{30,000}{10}\right) \times \frac{0}{2} = 0$$

$$\tau_{max} = \left(\frac{-dP}{dl}\right)\frac{r}{2} = \left(\frac{30,000}{10}\right) \times \frac{0.0375}{2} = 56.25 \text{ Pa}$$

The maximum shear (viscous) force, F occurs at the wall of the pipe.

$$F = \tau_{max} \times A = \tau_{max} \times 2 \times \pi \times R \times l = 84.375 \times 2 \times 3.14 \times 0.0375 \times 10 = 132.46 \text{ N}$$

Velocity, v can be calculated by the velocity form of the Hagen–Poiseuille equation as follows:

$$v = \frac{\Delta P}{32\mu L}D^2 = \frac{\Delta P}{32(v \times \rho) \times L}D^2 = \frac{30,000}{32 \times 6 \times 10^{-6} \times 890 \times 10}0.075^2$$

Because $\mu = v \times \rho$,

$$v = 98.8 \frac{\text{m}}{\text{s}}$$

We can also calculate the discharge as $Q = v \times A = 98.8 \times 3.14 \times \frac{0.0375^2}{4} = 0.436 \frac{\text{m}^3}{\text{s}}$.

9.4 Pressure Drop in Pipes (Major Losses)

Whenever a fluid flows through pipes, due to friction, the fluid encounters a resistance that results in decrease in pressure or loss of head (head loss in the pipe) between two locations along the flow length. Darcy and Weisbach performed several experiments and developed an empirical relationship between the pressure drop across the length of the pipe as a function of the flow velocity and the pipe dimensions.

$$\Delta P = f\frac{\rho l v^2}{2D} \tag{9.40}$$

Dividing on both sides of the equation by the specific weight of the fluid (ρg) will result in the famous Darcy–Weisbach equation:

$$h_L = \frac{f l v^2}{2gD} \tag{9.41}$$

where l, v, and D are the pipe length, fluid velocity, and the pipe diameter, respectively. The f is the nondimensional constant called the **friction factor** that depends on the surface properties of the pipe material and the flow characteristics. Darcy–Weisbach equation is applicable for all types of flow as long as the values of the friction factors are determined correctly. In the case of laminar flow through pipes, we can easily determine the friction factor by equating the head loss provided by the Darcy–Weisbach equation to the head loss resulting from the pressure drop form of the Hagen–Poiseuille equation and solving for the friction factor as below:

$$\frac{f l v^2}{2gD} = \frac{32\mu l v}{\rho g D^2} \tag{9.42}$$

or
$$f = \frac{64\mu}{\rho v D} \tag{9.43}$$

Noting that for a pipe, $\dfrac{\rho v D}{\mu} = N_{Re}$, we can rewrite the above equation as

$$f = \frac{64}{N_{Re}} \tag{9.44}$$

It is clear from Eq. (9.44) that the friction factor in laminar flow regime is completely Reynolds number dependent.

It is important to remember that the loss of pressure in the pipe or the head loss in a horizontal pipe of constant diameter can be determined using the following equation:

$$h_L = \frac{\Delta P}{\rho g} = \frac{f l v^2}{2gD} \tag{9.45}$$

The head loss in the pipe, h_L represents a major loss in the energy due to friction and is therefore called major loss and is generally represented by F_{major}. Additionally, the fluid transport systems often consist of several pipes of different diameters, and hence the major losses are additive, and the total loss can be expressed as

$$F_{major,total} = \sum\nolimits_{all\ pipes} \frac{f l v^2}{2gD} \tag{9.46}$$

When all the major losses are included, the energy equation can now be expressed as

$$\frac{P_1}{\gamma} + \frac{v_1^2}{2g} + h_1 - W_P + F_{major,total} = \frac{P_2}{\gamma} + \frac{v_2^2}{2g} + h_2 \tag{9.47}$$

or

$$\frac{P_1}{\gamma} + \frac{v_1^2}{2g} + h_1 + W_P - \sum\nolimits_{all\ pipes} \frac{f l v^2}{2gD} = \frac{P_2}{\gamma} + \frac{v_2^2}{2g} + h_2 \tag{9.48}$$

Example 9.4 (Agricultural Engineering—Head loss in a pipe): Vegetable oil is pumped through a 5-cm-diameter, 25-m-long pipe at a flow rate of 1.5 kg/s. Determine the head loss and pressure drop over the length of the pipe. It is given that the density and viscosity of the vegetable oil are 920 $\dfrac{kg}{m^3}$ and 0.03 Pa·s, respectively (Fig. 9.6).

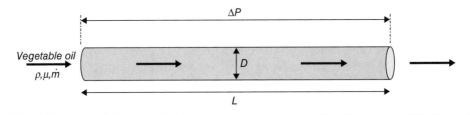

FIGURE 9.6 Schematic for Example 9.4 in which vegetable oil flows through a pipe.

Step 1: Schematic

The problem is illustrated in Fig. 9.6.

Step 2: Given data

Diameter, $D = 5$ cm; Length, $l = 25$ m; Mass flow rate, $\dot{m} = 1.5\ \frac{\text{kg}}{\text{s}}$; Density of the oil, $\rho = 920\ \frac{\text{kg}}{\text{m}^3}$; Viscosity, $\mu = 0.03$ Pa·s

To be determined: Head loss and the pressure drop.

Assumptions: The flow is laminar, steady state, incompressible, and fully developed.

Step 3: Calculations

We will verify the laminar flow condition and apply the appropriate equations to determine the head loss and pressure drop.

Verification of Reynolds number

The problem involves mass flow rate of oil. Therefore, the Reynolds number for the flow can be determined by using the mass flow rate expression, $\dot{m}$ (Eq. (8.11), Chap. 8):

$$N_{Re} = \frac{4\dot{m}}{\pi D \mu} = \frac{4 \times 1.5}{3.14 \times 0.05 \times 0.03} = 1273.88 < 2100 \text{ (Laminar)}$$

We can calculate the head loss, h_L using the Darcy–Weisbach equation as follows:

$$h_L = \frac{f l v^2}{2gD} \tag{E9.4A}$$

However, we need to estimate the value of friction factor, f. In our case, the flow is laminar. Therefore, we can determine the value of f as

$$f = \frac{64}{N_{Re}} = \frac{64}{1273.88} = 0.05$$

The head loss after expression velocity in terms of mass flow rate becomes

$$h_L = \frac{f l v^2}{2gD} = \frac{f l \left(\dfrac{\dot{m}/\rho}{\pi D^2 / 4} \right)^2}{2gD} = \frac{0.05 \times 25 \times \left(\dfrac{\left(1.5/920\right)}{\left(3.14 \times 0.05^2 / 4\right)} \right)^2}{2 \times 9.81 \times 0.05}$$

Because $Q = \dfrac{\dot{m}}{\rho}$ and $v = \dfrac{Q}{A}$

$$h_L = 0.88 \text{ m}$$

The pressure drop can be calculated as

$$\Delta P = h_L \times \rho \times g = 0.88 \times 920 \times 9.81 \approx 7938 \text{ Pa}$$

Note: The same result can also be obtained from Hagen–Poiseuille equation and dividing the pressure drop by ρg.

9.5 Pumping Power

Once we determine the head loss through a pipe, we can determine the pumping power (net power or water horsepower) needed to transport the fluid in the pipe by any of the following equations:

$$P_W = Q \times \Delta P \tag{9.49}$$

Because we know that $\Delta P = \rho g h_L$, the above equation can be expressed as

$$P_W = Q \times \rho \times g \times h_L = Q \times \gamma \times h_L \qquad (9.50)$$
$$\because \gamma = \rho \times g$$

We can also replace $Q \times \rho$ with the mass flow rate, $\dot{m}$ and Eq. (9.50) becomes

$$P_W = \dot{m} \times g \times h_L \qquad (9.51)$$

It is important to note that P_W is the net power required by the fluid. Because, all pumping devices (including new) experience losses due to friction, wear and tear, age, and result in a drop in the efficiency. Usually, the power provided to the pump is higher than the net power requirement. The power provided to the pump is called the brake horsepower (or commonly BHP) that accounts for the loss in its efficiency.

Example 9.5 (Bioprocessing Engineering—Transport of bio-oil): Bio-oil is fuel-like liquid produced via pyrolysis (intense heating in absence of oxygen) of organic materials. An engineer in a bioenergy facility decides to transport bio-oil from the manufacturing facility to the stocking room that are 100 m apart. Given the density and viscosity of bio-oil as 1100 kg/m³ and 80×10^{-3} Pa·s, determine the pumping power needed to transport 500 $\dfrac{\text{L}}{\text{min}}$ bio-oil through an 8-cm diameter smooth frictionless horizontal stainless steel pipe (Fig. 9.7).

Step 1: Schematic
The problem is described in Fig. 9.7.

Step 2: Given data
Diameter, $D = 8$ cm; Length, $l = 100$ m; Flow rate, $Q = 500\dfrac{\text{L}}{\text{min}}$; Density of the bio-oil, $\rho = 1100\dfrac{\text{kg}}{\text{m}^3}$; Viscosity of bio-oil, $\mu = 80 \times 10^{-3}$ Pa·s

To be determined: Pumping power.

Assumptions: The flow is laminar, steady state, incompressible, and fully developed.

Step 3: Calculations
We will verify the laminar flow condition and apply the appropriate equations to determine the pressure drop, ΔP from which the pumping power can be determined.

Verification of Reynolds number
The Reynolds number for the flow can be determined by using the Reynolds number expressed in the form of the flow rate, Q (Eq. (8.10), Chap. 8)

$$N_{\text{Re}} = \frac{4Q\rho}{\pi D \mu} = \frac{4 \times \left(\dfrac{500}{1000 \times 60}\right) \times 1100}{3.14 \times 0.08 \times 80 \times 10^{-3}} = 1825 < 2100 \text{ (Laminar)}$$

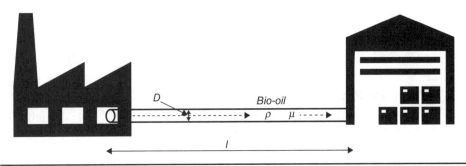

FIGURE 9.7   Schematic for Example 9.5 describing the transport of bio-oil.

Because the flow is laminar, the pressure drop can directly be calculated using the Hagen–Poiseuille equation

$$\Delta P = \frac{128 \mu l Q}{\pi D^4} \tag{E9.5A}$$

$$\Delta P = \frac{128 \times 80 \times 10^{-3} \times 100 \times \left(\frac{500}{1000 \times 60} \right)}{3.14 \times 0.08^4}$$

$$\Delta P = 66{,}348.2 \text{ Pa}$$

The power requirement can now be estimated as

$$P_W = \Delta P \times Q \tag{E9.5B}$$

$$P_W = 66{,}348.2 \times \frac{500}{1000 \times 60}$$

$$P_W = 552.9 \text{ W}$$

It may be noted that the actual power required (shaft power) may be higher than the calculated power due to loss in efficiency.

Example 9.6 (Environmental Engineering—Pumping costs of swine manure slurry):   A hog producer wanted to pump swine manure slurry (viscosity = 0.011 Pa·s) from an underground tank in a hog barn into an anaerobic digestor located at a higher elevation at 30°C. The pumping setup consisted of a pump attached to a 5-cm-diameter, 70-m-long pipe of 70-m long as shown in Fig. 9.8. Determine the monthly pumping costs incurred by the producer for pumping $50 \frac{L}{min}$ of slurry for 4 hours per day, if the density of the slurry is found to be $1050 \frac{kg}{m^3}$ assuming a pump efficiency of 88%. The cost of the electricity may be assumed to be 12 cents/kWh.

Step 1: Schematic
The schematic has already been provided as part of the problem.

Step 2: Given data
Diameter, $D = 5$ cm; Length, $l = 70$ m; Flow rate, $Q = 50 \frac{L}{min}$; Density of the slurry, $\rho = 1050 \frac{kg}{m^3}$; Viscosity of slurry, $\mu = 0.011$ Pa·s; Monthly pumping time, $t = 30 \times 4 = 120$ h/month

To be determined: Pumping power and costs.

Assumptions: The flow is laminar, steady state, incompressible, and fully developed.

Step 3: Calculations
We will verify the laminar flow condition and apply the appropriate equations to determine the pressure drop, ΔP from which the pumping power can be determined.

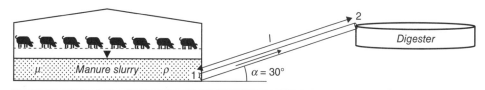

FIGURE 9.8   Schematic describing pumping of swine manure described in Example 9.6.

Verification of Reynolds number

The Reynolds number for the flow can be determined by using the Reynolds number expressed in the form of the flow rate, Q (see Eq. (8.10)).

$$N_{Re} = \frac{4Q\rho}{\pi D\mu} = \frac{4 \times \left(\dfrac{50}{1000 \times 60}\right) \times 1050}{3.14 \times 0.05 \times 0.011} = 2026.6 < 2100 \text{ (Laminar)}$$

If we assign the source and delivery points as 1 and 2, the Bernoulli's equation can now be expressed as

$$\frac{P_1}{\gamma} + \frac{v_1^2}{2g} + y_1 + W_P - F_{major} = \frac{P_2}{\gamma} + \frac{v_2^2}{2g} + y_2 \tag{E9.6A}$$

Additional considerations

(1) Because, both the locations 1 and 2 are at atmospheric pressure ($P_1 = P_2 = P_{atm}$)

(2) Velocity at location 1 is very small relative to velocity at location 2 ($v_1 = 0$)

(3) If we consider y_1 as the datum ($y_1 = 0$), then $y_2 = l \times \sin(\alpha°) = 70 \times \sin(30°) = 35$ m

(4) The velocity of the manure slurry at location 2 can be determined as $v_2 = \dfrac{Q}{A} = \dfrac{Q}{\left(\pi D^2\big/4\right)} = \dfrac{\left(50\big/1000 \times 60\right)}{\left(3.14 \times 0.05^2\big/4\right)} = 0.42 \ \dfrac{\text{m}}{\text{s}}$. Subsequently, from Darcy–Weisbach equation, we can determine the head loss, h_L as major loss,

$$h_L = F_{major} = \frac{flv^2}{2gD} \tag{E9.6B}$$

where $f = \dfrac{64}{N_{Re}}$

$$h_L = \frac{flv^2}{2gD} = \frac{64 \times l \times v^2}{N_{Re} \times 2 \times g \times D} = \frac{64 \times 70 \times 0.42^2}{2026.6 \times 2 \times 9.81 \times 0.05} = 0.4 \text{ m}$$

Therefore, after substituting the relevant values, Eq. (E9.6A) becomes

$$W_P - h_L = \frac{v_2^2}{2g} + y_2 \tag{E9.6C}$$

Because h_L is the same as F_{major}.

The power required by the pump, W_P is

$$W_P = \frac{v_2^2}{2g} + y_2 + h_L = \frac{0.42^2}{2 \times 9.81} + 35 + 0.4 = 35.4 \text{ m}$$

Pumping power, $P_W = W_P \times Q \times \gamma = W_P \times Q \times \rho \times g = 35.4 \times \left(\dfrac{50}{1000 \times 60}\right) \times 1050 \times 9.81$

$$P_W = 303.99 \text{ W}$$

The actual pumping power to be provided will be $\dfrac{303.99}{0.88} = 345.45$ W.

The cost of the pumping will be

$$\text{Cost} = \frac{247.9 \times 120}{1000} \times 0.12 = \$4.97/\text{month}$$

The hog producer is expected to incur \$4.97 per month for transporting the manure slurry from the stirring tank to the digester.

9.6   Turbulent Flow through Pipes

As seen previously, laminar flow generally occurs when highly viscous fluids (e.g., honey and oils) are transported at relatively low velocities. However, most of the fluids we encounter in practice are associated with turbulent flow regimes. For example, water that flows through pipes, rivers, channels, and streams is usually turbulent. Similarly flow of air and other gases through pipes and ducts are also turbulent. Unlike laminar flow, where the flow is orderly, the fluid molecules in turbulent flow regime are subjected to random and highly chaotic movements (Fig. 9.9). Therefore, it has been difficult for researchers to develop equations from theory alone and a combination of semi-empirical approaches has been proposed to analyze turbulent flows. The detailed mathematical treatment of turbulent flow in pipes is beyond the scope of this discussion. However, we will discuss some on the most common empirical approaches used by engineers to deal with turbulent flow regimes.

Despite the chaotic movement of the molecules, from a practical perspective, rapid and random mixing of molecules in turbulent flow will result in greatly increasing the exchange of heat, mass, and momentum in fluids. Therefore, turbulent flows are preferred in heat exchangers, chemical reactors, and food-processing operations. In such situations, we are frequently required to analyze the flows in pipes under turbulent flow conditions. Especially, we are interested in the frictional losses due to the random and rapid motion of the fluid molecules in the flow and how the friction will impact the head losses in pipes.

For turbulent flow regimes, the head loss can be determined using the same Darcy–Weisbach equation (discussed previously in laminar flow)

$$h_L = \frac{flv^2}{2gD} \tag{9.52}$$

However, the friction factor, f in turbulent flow was found to be a function of Reynolds number (N_{Re}), and surface roughness of the pipe (ε) (the average length of the protrusions present on the surface) through which the fluid is flowing (Fig. 9.10).

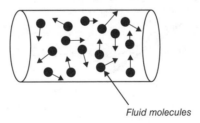

Fluid molecules

FIGURE 9.9   Fluid molecules in a turbulent flow regime move in random directions resulting in a chaotic flow pattern.

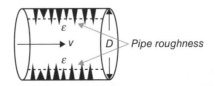

FIGURE 9.10   The roughness (protrusion) of the inner surface of the pipe also significantly influences the friction factor.

As mentioned earlier, due to the random and disorderly nature of the flow, developing exact mathematical formulations from the first principles for friction factors in turbulent flow is rather difficult. Therefore, several scientists and engineers such as Nikuradse, Rouse, and others began to develop correlations from experimental data in which friction factors were measured for various surface and flow conditions. By far the most commonly used correlation by all the engineers was developed by Colebrook (1921) which related the friction factor with Reynolds number and surface relative roughness in turbulent flow regimes.

Colebrook equation

$$\frac{1}{\sqrt{f}} = -2\log\left\{ \frac{\varepsilon/D}{3.7} + \frac{2.51}{N_{Re}\sqrt{f}} \right\}$$ (9.53)

Subsequently, Moody transformed all the historical experimental data and produced a versatile graphical nondimensional plot using the Colebrook equation. The graph developed by Moody is called Moody's diagram (1944) (Fig. 9.11) and is one of the most famous plots used by engineers involved in fluid dynamics. The plot relates the friction factor (nondimensional) with Reynolds number (nondimensional) and relative roughness of the pipe $\left(\dfrac{\varepsilon}{D}\right)$ (nondimensional) and therefore can be used for a wide range of flow rates, pipe diameters, and pipe surface roughness. Most importantly the plot is extremely simple to use and therefore is widely accepted by technical and nontechnical personnel alike. A step-by-step procedure to demonstrate the working of the Moody's diagram is described in the following example.

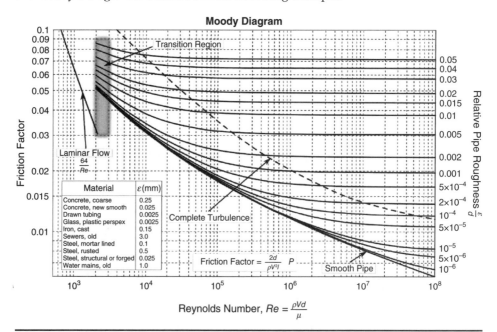

FIGURE 9.11 Moody's diagram for determining the nondimensional friction factor using the relative roughness and Reynolds number. (Cengel & Ghajar, Heat and Mass Transfer: Fundamentals and Applications, Fifth Edition. McGraw Hill, 2016. Reprinted with permission.)

Example 9.7 (General Engineering—All concentrations—Illustration of Moody diagram): Determine the friction factor in a 10-cm diameter, 25-long cast iron pipe whose average roughness is 0.1 mm and carrying water at 20 m/s at 25°C. Use the properties of water at 25°C as $\rho = 997 \, \frac{kg}{m^3}$, $\mu = 9 \times 10^{-4}$ Pa·s. Also, using the friction factor obtained, determine the frictional head loss in the pipe (Fig. 9.12).

Step 1: Schematic
The problem is described in Fig. 9.12.

Step 2: Given data
Diameter, $D = 10$ cm; velocity, $v = 20 \, \frac{m}{s}$; Average roughness, $\varepsilon = 0.1$ mm; Length of the pipe, $l = 25$ m;

Density of water at 25°C, $\rho = 997 \, \frac{kg}{m^3}$; Viscosity of water at 25°C, $\mu = 9 \times 10^{-4}$ Pa·s

To be determined: Friction factor, f.

Step 3: Calculations
Let us first calculate the Reynolds number,

$$N_{Re} = \frac{\rho v D}{\mu} = \frac{997 \times 20 \times 0.1}{9 \times 10^{-4}} = 2.2 \times 10^6$$

Now let us calculate the value of roughness ratio, $\frac{\varepsilon}{D}$

$$\frac{\varepsilon}{D} = \frac{(0.1/1000)}{0.1} = 0.001$$

Using the values $\frac{\varepsilon}{D} = 0.001$ and $N_{Re} = 2.2 \times 10^6$ on the Moody's diagram, determine the value of friction factor, f (from the Y-axis) as follows (Fig. 9.13):

The Friction factor, f on the Y-axis is found to be 0.02.

Now, the head loss in the pipe can be determined using the Darcy–Wiesbach equation.

$$h_L = F_{major} = \frac{f l v^2}{2gD} = \frac{0.02 \times 25 \times 20^2}{2 \times 9.81 \times 0.1}$$

The pipe will experience a friction head loss of 101.9 m.

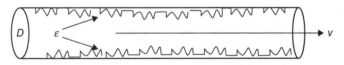

FIGURE 9.12 Schematic of the cast iron pipe in Example 9.7 exhibiting surface roughness.

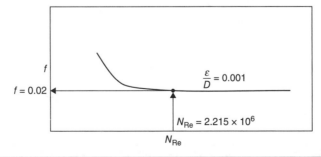

FIGURE 9.13 Illustration of determination of the friction factor using Moody's diagram.

Example 9.8 (Bioprocessing Engineering—Height of water in a disinfectant tank): It is a common practice in seafood processing industry to wash down all the food handling surfaces including floors (with hot water) at the end of each shift. Consider a large tank containing hot water at 80°C. The tank is connected at the bottom to a 40-m long smooth hosepipe of 2-cm diameter. Determine the minimum height of the water to be maintained to allow a flow rate of 15 L/min through the pipe during washing (Fig. 9.14). Use the viscosity and density of water at 80°C as 3.57×10^{-4} Pa·s and $972 \frac{kg}{m^3}$.

Step 1: Schematic
The problem is illustrated in Fig. 9.14.

Step 2: Given data
Diameter, $D = 2$ cm; Length, $l = 40$ m; Flow rate, $Q = 15 \frac{L}{min}$; Density of water at 80°C, $\rho = 972 \frac{kg}{m^3}$; Viscosity of water at 80°C, $\mu = 3.57 \times 10^{-4}$ Pa·s

To be determined: Height of the water level in the tank.

Assumptions: The flow is fully developed, incompressible, and steady. We cannot assume laminar flow because the fluid is water and most of the water flow is turbulent. Nonetheless, we will determine the Reynolds number to verify the flow regime.

Step 3: Calculations
Reynolds number,

$$N_{Re} = \frac{4Q\rho}{\pi D \mu} = \frac{4 \times \left(\frac{15}{1000 \times 60}\right) \times 972}{3.14 \times 0.02 \times 3.57 \times 10^{-4}} = 43,354.9 > 2100 \text{ (Turbulent)}$$

Now applying Bernoulli's equation between points 1 and 2, we obtain

$$\frac{P_1}{\gamma} + \frac{v_1^2}{2g} + y_1 + W_P - F_{major} = \frac{P_2}{\gamma} + \frac{v_2^2}{2g} + y_2 \tag{E9.8A}$$

Additional considerations

(1) Both the locations 1 and 2 are at atmospheric pressure ($P_1 = P_2 = P_{atm}$)

(2) There is no pump between locations 1 and 2 and hence, $W_P = 0$

(3) Velocity at location 1 is very small relative to velocity at location 2 ($v_1 = 0$)

(4) If we consider y_2 as the datum (level zero), then $y_1 =$ to be determined

After substituting the above considerations, Eq. (E9.8A) now becomes

$$y_1 = \frac{v_2^2}{2g} + F_{major} \tag{E9.8B}$$

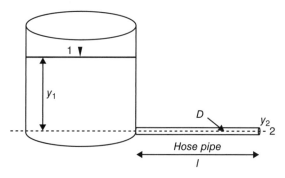

FIGURE 9.14 Schematic of the disinfectant tank connected to a hose at the bottom as described in Example 9.8.

or

$$y_1 = \frac{v_2^2}{2g} + \frac{flv_2^2}{2gD} \tag{E9.8C}$$

$$y_1 = \frac{v_2^2}{2g}\left(1 + \frac{fl}{D}\right) \tag{E9.8D}$$

Because the velocity $v = v_2$.

The velocity of the hot water at location 2 can be determined as

$$v_2 = \frac{Q}{A} = \frac{Q}{\left(\pi D^2/4\right)} = \frac{\left(15/1000 \times 60\right)}{\left(3.14 \times 0.02^2/4\right)} = 0.79 \ \frac{\text{m}}{\text{s}}$$

Subsequently, from Darcy–Weisbach equation, we can determine the head loss, after estimating the friction factor, f from the Moody's diagram. From the Moody's plot, the value of f ($N_{Re} = 43,354.9$ Smooth pipe) for the pipe can be read as ≈ 0.021.

Now, the

$$y_1 = \frac{0.796^2}{2 \times 9.81}\left[1 + \left(\frac{0.021 \times 40}{0.02}\right)\right] = 1.39 \ \text{m}$$

The minimum height of the water in the tank is about 1.39 m.

Considering the implicit nature for the Colebrook equation, several simple-to-use explicit equations have been developed which can provide results with reasonably good accuracy. Some of the equations are summarized in the table below.

Source	Equation	Conditions	Equation #
Swamee and Jain (1976)	$f = 0.25\left\{\log\left[\left(\frac{\varepsilon}{D}\right) + \frac{5.74}{N_{Re}^{0.9}}\right]\right\}^{-2}$	Smooth and rough pipes	(9.54)
Blasius	$f = \frac{0.316}{N_{Re}^{0.25}}$	Smooth pipes only $N_{Re} < 100,000$	(9.55)
Haaland equation (1983)	$\frac{1}{f^{0.5}} = -1.8\log\left(\frac{6.9}{N_{Re}} + \left(\frac{\left(\frac{\varepsilon}{D}\right)}{3.7}\right)^{1.17}\right)$	Smooth and rough pipes	(9.56)
Nikuradse	$f = 0.0032 + \frac{0.221}{N_{Re}^{0.237}}$	Smooth pipes only $10,000 < N_{Re} < 3 \times 10^6$	(9.57)

Example 9.9 (General Engineering–All concentrations):   Using Example 9.8 as basis (Reynolds number = 43,354.9 and smooth pipe), compare the friction factors using other implicit equations.

Step 1: Given data
Reynolds number, $N_{Re} = 43,354.9$; Smooth pipe, $\varepsilon \approx 0$

Step 2: Calculations

As per Swamee and Jain (1976),
$$f = 0.25\left\{\log\left(\frac{\left(\frac{\varepsilon}{D}\right)}{3.7} + \frac{5.74}{N_{Re}^{0.9}}\right)\right\}^{-2} \tag{E9.9A}$$

$$f = 0.25\left\{\log\left(\frac{\left(\frac{\varepsilon}{D}\right)}{3.7} + \frac{5.74}{N_{Re}^{0.9}}\right)\right\}^{-2} = 0.25\times\left\{\log\left(\frac{\left(\frac{0}{0.02}\right)}{3.7} + \frac{5.74}{43,354.9^{0.9}}\right)\right\}^{-2} = 0.214$$

As per Blasius,
$$f = \frac{0.316}{N_{Re}^{0.25}} \tag{E9.9B}$$

$$f = \frac{0.316}{N_{Re}^{0.25}} = \frac{0.316}{43,354.9^{0.25}} = 0.0218$$

As per Haaland,
$$\frac{1}{f^{0.5}} = -1.8\log\left(\frac{6.9}{N_{Re}} + \left(\frac{\left(\frac{\varepsilon}{D}\right)}{3.7}\right)^{1.17}\right) \tag{E9.9C}$$

$$\frac{1}{f^{0.5}} = -1.8\log\left(\frac{6.9}{N_{Re}} + \left(\frac{\left(\frac{\varepsilon}{D}\right)}{3.7}\right)^{1.17}\right) = -1.8\log\left(\frac{6.9}{43,354.9} + \left(\frac{\left(\frac{0}{0.02}\right)}{3.7}\right)^{1.17}\right) \Rightarrow f = 0.213$$

As per Nikuradse,
$$f = 0.0032 + \frac{0.221}{N_{Re}^{0.237}} \tag{E9.9D}$$

$$f = 0.0032 + \frac{0.221}{N_{Re}^{0.237}} = 0.0032 + \frac{0.221}{43,354.9^{0.237}} = 0.0207$$

Therefore, we can see that the values are similar to what was obtained using Moody's diagram (from Example 9.8).

It is also critical to know that the values of the friction factors will continuously change once the pipe becomes operational and can vary widely over time. Due to sustained momentum of the fluid particles, the walls get scoured and may become rougher resulting in increased friction factors and increased major losses. Sometimes, however, the pipes may even become temporarily smoother due to the biofouling that may damp out the effects of the surface roughness. Therefore, it is recommended to build in a factor of safety while choosing the pumps for fluid transport.

Once the value of the friction factor, f is estimated, we can use the Darcy–Weisbach equation to determine the head loss due to the pipe (F_{major}) as we have seen in the preceding examples. Additionally, the fluid transport systems often consist of several pipes of different diameters, and hence the major losses are additive, and the total loss can be expressed as

$$F_{major,total} = \sum_{all\ pipes} \frac{flv^2}{2gD} \tag{9.58}$$

9.7   Minor Losses

While the major losses in pipes are the result of the pipe themselves, the fittings and other accessories individually create local disturbances in the flow and result in frictional losses. For example, when a flow of fluid encounters a Tee-junction, the water abruptly changes the flow direction creating a loss of energy. In general, a piping system will consist of several such fittings and each of such fittings will contribute to the overall head loss. When the pipes are very long (>25 m), the effects of the fittings are usually small. However, in some cases when several fittings are present and/or their effects are large enough, they need to be included in the calculations. In general, the minor loss for each fitting is defined as

$$F_{minor} = K \frac{v^2}{2g}$$ (9.59)

where F_{minor} is the minor loss (in meters) of head due to the given fitting, K is the dimensionless loss coefficient for a given fitting, and v is the velocity of the flow (m/s).

The approximate values of K for the most common fittings employed in the industry are summarized in Table 9.1. It is always best to contact the specific manufacturer to obtain the exact K values for any fittings employed in the piping system. When more

Fitting type	Configuration	K
Tee	Branched-flanged	1.0
	Branched-threaded	2.0
	Inline-flanged	0.2
	Inline-threaded	0.9
Valve	Gate-25% open	15
	Gate-50% open	2.1
	Gate-75% open	0.26
	Gate-100% open	0.15
	Globe-100% open	10
	Ball-100% open	0.05
Sudden expansion	Expanding from d to D	$1.05\left(\dfrac{D^2-d^2}{D^2}\right)^2$
Sudden contraction	Contracting from D to d	$K = 0.5 - 0$ for $\left(\dfrac{d}{D}\right)^2 = 0 - 1.0$ Use v_2 for calculation of minor loss
Elbow	Regular-flanged	0.3
	Regular-threaded	1.5
	45° flanged	0.2
	45° threaded	0.4
Water meter	7.0	

TABLE 9.1 Loss Coefficients for Standard Pipe Fittings

than one fitting is present in a pipeline, the overall effect is additive, and the total minor loss is the sum of all individual loss from each fitting. Therefore,

$$F_{minor,total} = \sum_{all\ fittings} K \frac{v^2}{2g}$$

(9.60)

When all the major and minor losses are included, the energy equation can now be expressed as

$$\frac{P_1}{\gamma} + \frac{v_1^2}{2g} + h_1 + W_P - F_{major,total} - F_{minor,total} = \frac{P_2}{\gamma} + \frac{v_2^2}{2g} + h_2$$

(9.61)

$$\frac{P_1}{\gamma} + \frac{v_1^2}{2g} + h_1 + W_P - \sum_{all\ pipes} \frac{flv^2}{2gD} - \sum_{all\ fittings} K \frac{v^2}{2g} = \frac{P_2}{\gamma} + \frac{v_2^2}{2g} + h_2$$

(9.62)

Example 9.10 (Aquacultural Engineering—Total head loss in a piping system):   A PVC piping network consists of 10-m long pipe of 2.5 cm diameter that abruptly expands to 5 cm in diameter for 10 m and abruptly reduces to its original diameter for another 25 m before discharging into a tank via a fully open globe valve ($K = 10$) at a rate of 100 LPM. Determine the total head losses due to the pipe and the fittings by assuming a constant friction factor of 0.01 (Fig. 9.15).

Step 1: Schematic
The problem is depicted in Fig. 9.15.

Step 2: Given data
Length of the first pipe, $l_1 = 10$ m; Diameter of the first pipe, $D_1 = 2.5$ cm; Length of the second pipe, $l_2 = 10$ m; Diameter of the second pipe, $D_2 = 5$ cm; Length of the third pipe, $l_3 = 25$ m; Diameter of the third pipe, $D_3 = 2.5$ cm; K value for globe valve = 10.0; One sudden expansion and one sudden contraction; Discharge through the pipe, $Q = 100$ LPM $= \dfrac{100}{1000 \times 60} = 1.67 \times 10^{-3} \dfrac{m^3}{s}$

Step 3: Calculations
The major losses and minor losses are additive. Therefore,

$$Total\ head\ losses = \sum_{all\ pipes} \frac{flv^2}{2gD} + \sum_{all\ fittings} K \frac{v^2}{2g}$$

(E9.10A)

$$Total\ head\ losses = \frac{f_1 l_1 v_1^2}{2gD_1} + \frac{f_2 l_2 v_2^2}{2gD_2} + \frac{f_3 l_3 v_3^2}{2gD_3}$$

$$+ \left(1.05\left(\frac{D^2 - d^2}{D^2}\right)^2\right)_{sudden\ expansion} \left(K\frac{v_3^2}{2g}\right)_{sudden\ contraction}$$

$$+ \left(K\frac{v_3^2}{2g}\right)_{globe\ valve}$$

(E9.10B)

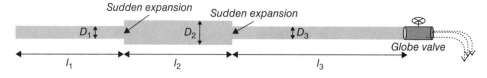

Figure 9.15   Schematic for Example 9.10 depicting the three PVC pipes of different lengths and diameters connected together.

$$\text{Total head losses} = \frac{0.01 \times 10 \times 3.39^2}{2 \times 9.81 \times 0.025} + \frac{0.01 \times 10 \times 0.85^2}{2 \times 9.81 \times 0.05}$$

$$+ \frac{0.01 \times 25 \times 3.39^2}{2 \times 9.81 \times 0.025} + \underbrace{\left(1.05 \times \left(\frac{0.05^2 - 0.025^2}{0.05^2}\right)^2\right)}_{\text{sudden expansion}}$$

$$+ \underbrace{\left(0.4 \times \frac{0.85^2}{2 \times 9.81}\right)}_{\text{sudden contraction}} + \underbrace{\left(10 \times \frac{3.39}{2 \times 9.81}^2\right)}_{\text{globe valve}}$$

Total head losses = 15.01 m

9.8   Flow through Pipes in Series and Parallel

In several practical situations, pipelines might have to be extended from time to time. For example, when the water demand in an animal housing operation, municipal, or a commercial enterprise increases, the existing pipelines are connected with additional pipes in series or parallel configurations. The newer pipes may be of different lengths and diameters. In this section, two most common configurations are analyzed.

To analyze pipes connected in series we will use the combination of mass and energy conservation equations. Let us consider three pipes of different diameters (D) and lengths (l), friction factors (f), connected in series (Fig. 9.16).

The pipes are considered long enough that the minor losses are negligible. From the figure, we can see that the total flow rate in all the pipes will remain the same (mass conservation). The head loss, h_L in each pipe, is different. However, the total head loss between the points (the entrance [A] and exit [B] points) is the summation of all the head losses in all three pipes. Therefore, we can write the following equations

$$Q_{A \text{ to } B} = Q_1 = Q_2 = Q_3 \qquad \text{Mass conservation} \qquad (9.63)$$

$$h_{L\,(A \text{ to } B)} = h_{L,1} + h_{L,2} + h_{L,3} \qquad \text{Energy conservation} \qquad (9.64)$$

$$h_{L\,(A \text{ to } B)} = \frac{f_1 l_1 v_1^2}{2gD_1} + \frac{f_2 l_2 v_2^2}{2gD_2} + \frac{f_3 l_3 v_3^2}{2gD_3} \qquad (9.65)$$

Using Eqs. (9.63)–(9.65) we can determine the total head and power requirements (see examples at the end of this section).

Example 9.11 (Agricultural Engineering–Pipes in series):   Three concrete pipes ($\varepsilon = 0.02$ m) are connected in series to transport 3000 LPM water at 25°C from a pumping station to a specific delivery point 3 km away. The first pipe is 1500-m long with 0.5-m diameter followed by a 500-m long pipe of 0.7-m diameter. The third pipe is of 0.3-m diameter and 1000-m long. Determine the total major losses and the total pressure drop experienced by the piping system. Use the viscosity of water and density at 25°C as 9×10^{-4} Pa·s and 1000 kg/m³ and ignore minor losses (Fig. 9.17).

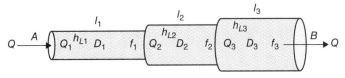

Figure 9.16   Schematic for pipes in series wherein the discharge remains the same and the head losses are additive.

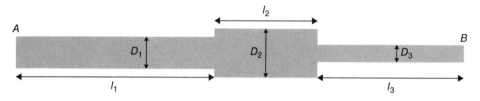

FIGURE 9.17   Schematic for pipes in series arrangement described in Example 9.11.

Step 1: Schematic
The problem is described in Fig. 9.17.

Step 2 Given data
Length of the first pipe, $l_1 = 1500$ m; Diameter of the first pipe, $D_1 = 0.5$ m; Length of the second pipe, $l_2 = 500$ m; Diameter of the second pipe, $D_2 = 0.70$ m; Length of the third pipe, $l_3 = 1000$ m; Diameter of the third pipe, $D_3 = 0.30$ m

The pipes are in series and therefore the major losses are additive. We will calculate the major losses due to each pipe and determine the total major loss.

$$F_{\text{major }(A \text{ to } B)} = \frac{f_1 l_1 v_1^2}{2gD_1} + \frac{f_2 l_2 v_2^2}{2gD_2} + \frac{f_3 l_3 v_3^2}{2gD_3} \tag{E9.11A}$$

The values of f_1, f_2, f_3 can be determine via Moody's diagram or via Swamme and Jain equation or Haaland equation. For this problem, we will employ Swamee and Jain equation as follows:

Pipe length (m)	D (m)	$N_{\text{Re}}(\times 10^5)$	$\dfrac{\varepsilon}{D}$	$v\left(\dfrac{m}{s}\right)$	f^*	F_{major} (m)
1500	0.5	1.41	0.04	0.254	0.064	0.64
500	0.7	1.01	0.0285	0.13	0.056	0.034
1000	0.3	2.35	0.0667	0.70	0.082	6.99
Total						**7.67**

$$^*f = 0.25\left\{\log\left|\frac{\left(\dfrac{e}{D}\right)}{3.7} + \frac{5.74}{N_{\text{Re}}^{0.9}}\right|\right\}^{-2}$$

The total major loss is calculated as **7.67 m of head.**
Now to determine the pressure drop, we will apply the complete Bernoulli's equation between points A and B

$$\frac{P_A}{\gamma} + \frac{v_A^2}{2g} + y_A + W_P - \sum_{\text{all pipes}} \frac{flv^2}{2gD} - \sum_{\text{all fittings}} K\frac{v^2}{2g} = \frac{P_B}{\gamma} + \frac{v_B^2}{2g} + y_B \tag{E9.11B}$$

Noting that $y_A = y_B$ (*same height*), $W_P = 0$ (*no pump*), and $\displaystyle\sum_{\text{all fittings}} K\frac{v^2}{2g} = 0$ (*ignoring minor losses*), we obtain

$$\frac{P_A}{\gamma} + \frac{v_A^2}{2g} - \sum_{\text{all pipes}} \frac{flv^2}{2gD} = \frac{P_B}{\gamma} + \frac{v_B^2}{2g} \tag{E9.11C}$$

or

$$\frac{(P_A - P_B)}{\gamma} = \sum_{\text{all pipes}} \frac{flv^2}{2gD} + \frac{(v_B^2 - v_A^2)}{2g} \qquad \text{(E9.11D)}$$

$$\frac{(P_A - P_B)}{1000 \times 9.81} = 7.67 + \frac{(0.70^2 - 0.254^2)}{2 \times 9.81} \qquad \text{(E9.11E)}$$

$$P_A - P_B = \Delta P = 75,501 \text{ Pa}$$

Power requirement, $P_W = \Delta P \times Q = 75,501 \times 0.05 = 3775$ W

Example 9.12 (Aquacultural Engineering—Pipes in series):   Seawater is to be pumped from the estuary to a large outdoor aquacultural facility located 400 away from the water source and is connected via three PVC pipes ($\varepsilon = 0.0015$ mm) in series. The first pipe is 100-m long and is of 30 cm while the second pipe is of 15 cm in diameter, but 150 m in length. The third pipe has a length of 150 m and 50 cm in diameter. Determine the required pumping power for transporting 100 L/min of sea water with a density of 1040 kg/m^3 and an absolute viscosity of 9.5×10^{-4} Pa·s. Further, if the pump operates 4 hours per day, determine the monthly pumping costs incurred by the hatchery (Fig. 9.18).

Step 1: Schematic
The problem is described in Fig. 9.18.

Step 2: Given data
Length of the first pipe, $l_1 = 100$ m; Diameter of the first pipe, $D_1 = 0.3$ m; Length of the second pipe, $l_2 = 150$ m; Diameter of the second pipe, $D_2 = 0.15$ m; Length of the third pipe, $l_3 = 150$ m; Diameter of the second pipe, $D_3 = 0.50$ m; Required flow rate, $Q = 1,000 \ \dfrac{\text{L}}{\text{min}}$; Density of seawater, $\rho = 1040 \ \dfrac{\text{kg}}{\text{m}^3}$; Viscosity, $\mu = 9.5 \times 10^{-4}$ Pa·s; Pipe (all pipes) roughness, $\varepsilon = 0.0015$ mm

To be determined: Pumping power and cost.

Step 3: Calculations
From the schematic, we can see that the seawater will encounter two types of losses: major losses from each of the three pipes and minor losses due to sudden contraction between pipe 1 and pipe 2 and sudden expansion between pipe 2 and pipe 3. We will evaluate the sum of these losses and calculate the pressure drop, pumping power, and costs.

The first step is to calculate the Reynolds number in all pipes to verify the flow regime.

$$\text{Reynolds number (Pipe 1), } N_{Re,1} = \frac{4Q\rho}{\pi D\mu} = \frac{4 \times \left(\dfrac{1000}{1000 \times 60}\right) \times 1040}{3.14 \times 0.3 \times 9.5 \times 10^{-4}} = 77,500$$

$$\text{Reynolds number (Pipe 2), } N_{Re,2} = \frac{4Q\rho}{\pi D\mu} = \frac{4 \times \left(\dfrac{1000}{1000 \times 60}\right) \times 1040}{3.14 \times 0.15 \times 9.5 \times 10^{-4}} = 155,000$$

$$\text{Reynolds number (Pipe 3), } N_{Re,3} = \frac{4Q\rho}{\pi D\mu} = \frac{4 \times \left(\dfrac{1000}{1000 \times 60}\right) \times 1040}{3.14 \times 0.50 \times 9.5 \times 10^{-4}} = 46,500$$

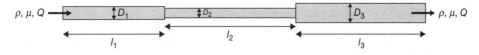

Figure 9.18 Schematic of PVC pipe system connected in series for transporting seawater as described in Example 9.12.

The smallest Reynolds number is still greater than 2100. Therefore, the flow is turbulent through the system, and we can use Colebrook's equation and/or Moody's plot to estimate the friction factors.

Loss in pipe 1

The relative roughness, $\dfrac{\varepsilon}{D} = \dfrac{0.0015/1000}{0.3} = 0.000005$, $N_{\mathrm{Re},1} = 77,500$

Using the values of relative roughness, $\dfrac{\varepsilon}{D}$ and Reynolds number, $N_{\mathrm{Re},1}$ we obtain, $f_1 = 0.019$

Now, the major loss in pipe 1,

$$F_{\mathrm{major},1} = \frac{f_1 l_1 v_1^2}{2gD_1} = \frac{0.025 \times 100 \times \left(\dfrac{\left(\dfrac{1000}{1000 \times 60} \right)}{\left(3.14 \times \dfrac{0.3^2}{4} \right)} \right)^2}{2 \times 9.81 \times 0.3} = 0.017 \text{ m}$$

Similarly, the minor loss due to sudden contraction (between pipe 1 and pipe 2) is expressed as

$$F_{\mathrm{minor},1} = K \frac{v_2^2}{2g} \qquad \text{(E9.12A)}$$

where $K = 0.33$ (see Table 9.1) and v_2 is the velocity in pipe 2 $\left(v_2 = \dfrac{Q}{A_2} = \dfrac{\left(\dfrac{100}{1000 \times 60} \right)}{\left(3.14 \times \dfrac{0.15^2}{4} \right)} = 0.94 \text{ m/s} \right)$.

$$F_{\mathrm{minor},1} = 0.335 \times \frac{0.94^2}{2 \times 9.81} = 0.015 \text{ m}$$

After performing the remaining calculations, all the major and minor losses are tabulated as follows:

Pipe	N_{Re}	$\dfrac{\varepsilon}{D}$	f	Head loss	
				F_{major} (m) $\dfrac{flv^2}{2gD}$	F_{minor} (m) $K\dfrac{v^2}{2g}$
1	77,500	5×10^{-6}	0.019	0.017	
2	155,000	1×10^{-5}	0.016	0.726	
3	46,500	3×10^{-6}	0.021	0.002	
1–2 (sudden contraction)					0.015
2–3 (sudden expansion)					0.87
Total losses (m)				**0.746**	**0.88**

Sometimes, when there is a need to reduce the total head loss, the pipes are connected in parallel. Let us consider a simple parallel arrangement as shown in Fig. 9.19. To analyze this network, we can use the same mass and energy balance equations, of course with appropriate modifications.

From the figure, we see that the flow splits into three sections that flow through pipes 1–3 and rejoin at the exit (B). Therefore, as per mass conservation,

$$Q_{A \text{ to } B} = Q_1 + Q_2 + Q_3 \quad \text{Mass conservation} \qquad (9.66)$$

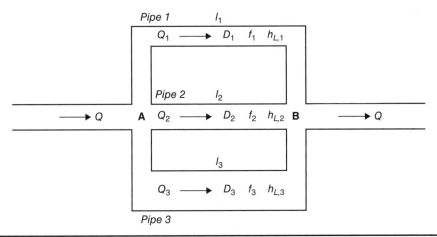

FIGURE 9.19 Schematic of pipes connected in series wherein the head loss remains constant for all the pipes and the discharges are additive.

Similarly, the head loss between the entry and exit points is the same for all the pipes resulting in

$$h_{L\,(A\ to\ B)} = h_{L,1} = h_{L,2} = h_{L,3} \quad \text{Energy conservation} \tag{9.67}$$

$$h_L = \frac{f_1 l_1 v_1^2}{2gD_1} = \frac{f_2 l_2 v_2^2}{2gD_2} = \frac{f_3 l_3 v_3^2}{2gD_3} \tag{9.68}$$

Example 9.13 (Environmental Engineering—Pipes in parallel): Three parallel pipes transport wastewater from a treatment tank to a waste reservoir. The pipes are made of the same material and are of the same length with the diameter ratios of 1:1.5:1.75. If the smallest diameter pipe is expected to carry 100 L/min, determine the total discharge into the reservoir (Fig. 9.20).

Step 1: Schematic
The problem is described in Fig. 9.20.

Step 2: Given Data
Flow rate in the smallest diameter pipe, $Q_1 = 100\dfrac{L}{min} = \dfrac{100}{(1000 \times 60)} = 1.6 \times 10^{-3}\ \dfrac{m^3}{s}$; Ratio of diameters, $D_1{:}D_2{:}D_3 = 1{:}1.5{:}1.75$; Lengths of the pipes, $l_1 = l_2 = l_3$; Friction factors in the pipes, $f_1 = f_2 = f_3$; Density of dairy waste, $\rho = 1020\ \dfrac{kg}{m^3}$; Viscosity of the dairy wastewater, $\mu = 0.001$ Pa·s

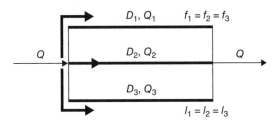

FIGURE 9.20 Schematic of parallel pipe system described in Example 9.13.

Step 3: Calculations

We can write the mass and energy balance equations for a parallel pipe network as

$$Q_{Total} = Q_1 + Q_2 + Q_3 \tag{E9.13A}$$

$$h_L = \frac{f_1 l_1 v_1^2}{2gD_1} = \frac{f_2 l_2 v_2^2}{2gD_2} = \frac{f_3 l_3 v_3^2}{2gD_3} \tag{E9.13B}$$

We will first solve the energy balance and substitute the values in the mass balance to obtain the total discharge. We have been given that $l_1 = l_2 = l_3$, $f_1 = f_2 = f_3$, and $D_1{:}D_2{:}D_3 = 1{:}1.5{:}1.75$. Therefore, Eq. (E9.13B) becomes

$$\frac{v_1^2}{D_1} = \frac{v_2^2}{D_2} = \frac{v_3^2}{D_3} \tag{E9.13C}$$

Expressing all the velocities as Q/A or $\dfrac{Q}{\left(\pi D^2/4\right)}$, we obtain

$$\frac{\left[\dfrac{Q_1}{\left(\pi D_1^2/4\right)}\right]^2}{D_1} = \frac{\left[\dfrac{Q_2}{\left(\pi D_2^2/4\right)}\right]^2}{D_2} = \frac{\left[\dfrac{Q_3}{\left(\pi D_3^2/4\right)}\right]^2}{D_3} \tag{E9.13D}$$

or

$$\frac{Q_1^2}{D_1^5} = \frac{Q_2^2}{D_2^5} = \frac{Q_3^2}{D_3^5} \tag{E9.13E}$$

It is also given that, $Q_1 = 1.6 \times 10^{-3} \dfrac{m^3}{s}$ and $D_1{:}D_2{:}D_3 = 1{:}1.5{:}1.75$ (i.e., $D_2 = 1.5D_1$ and $D_3 = 1.75D_1$). Substitution of all the values in Eq. (E9.13E) will result in

$$Q_2 = 0.012 \frac{m^3}{s} \text{ and } Q_3 = 0.0273 \frac{m^3}{s}$$

Therefore, the total discharge,

$$Q_{Total} = Q_1 + Q_2 + Q_3 = (1.6 \times 10^{-3} + 0.012 + 0.0273) \frac{m^3}{s}$$

$$Q_{Total} = 0.0416 \frac{m^3}{s} = 2500.68 \text{ LPM}$$

Example 9.14 (Bioprocessing Engineering—Pipes in series and parallel):   A shell and tube heat exchanger (HX) is being used in a bioenergy facility. The HX consists of 80 copper tubes (average roughness = 0.001 mm) of 15-mm internal diameter and 25-m long carrying a glycol solution whose viscosity is 0.00845 Pa·s and a density of 1050 kg/m³. The tubes are placed in smooth aluminum shell whose entrance and exit sections are 2-m long and 0.8-m diameter each. Determine the pressure drop for a targeted flow rate of 120 LPM (Fig. 9.21).

Step 1: Schematic

The problem is described in Fig. 9.21.

Step 2: Given Data

Diameter, $D_1 = 1.5$ cm; Length, $l_1 = 25$ m; Diameter, $D_2 = 0.8$ m; Length, $l_2 = 2$ m; Diameter, $D_3 = 0.8$ m; Length, $l_3 = 2$ m

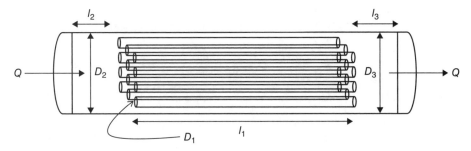

FIGURE 9.21 Schematic of a shell and tube heat exchanger with parallel copper tubes described in Example 9.14.

Area of tube 1, $A_1 = \pi \dfrac{D_1^2}{4} = 3.14 \times \dfrac{0.015^2}{4} = 1.76 \times 10^{-4}\,\text{m}^2$; Area of tube 2, $A_2 = \pi \dfrac{D_2^2}{4} = 3.14 \times \dfrac{0.8^2}{4} =$

$0.50\,\text{m}^2$; Area of tube 3, $A_3 = \pi \dfrac{D_3^2}{4} = 3.14 \times \dfrac{0.8^2}{4} = 0.50\,\text{m}^2$; Density of glycol solution, $\rho = 1050 \dfrac{\text{kg}}{\text{m}^3}$;

Flow rate, $Q = 120 \dfrac{\text{L}}{\text{min}} = \dfrac{120}{(1000 \times 60)} = 0.002\,\dfrac{\text{m}^3}{\text{s}}$; Viscosity of glycol solution, $\mu = 0.00845\,\text{Pa·s}$

To be determined: Pressure drop and the pumping power.

Assumptions: The flow is fully developed, incompressible, and steady.

The liquid flows through the entrance region first and subsequently through the 80 small copper tubes and finally through the exit region. Note that the 80 tubes are parallel and hence they have the same head loss (or major loss) carrying a flow of one-eightieth of the total flow per pipe. Therefore, this problem may be viewed as three pipe systems in series. Additionally, the entrance and exit sections are identical in size and therefore will have the same head loss. We will calculate the head loss in the parallel tubes and the entrance/exit regions and then determine the total head loss by summation.

Head loss in small tubes

$$N_{\text{Re},1} = \frac{\rho v D}{\mu} = \frac{\rho \left(\frac{Q}{A} \right) D}{\mu} = \frac{1050 \times \left(\frac{0.002}{80 \times 1.76 \times 10^{-4}} \right) \times 0.015}{0.00845} = 263.8 < 2100$$

The head loss can be calculated via the Darcy–Weisbach equation as follows:

$$F_{\text{major},1} = h_{F,1} = \frac{f_1 l_1 v_1^2}{2 g D_1} \tag{E9.14A}$$

The value of friction factor, f_1 is calculated using $f = \dfrac{64}{N_{\text{Re}}}$ and Reynolds number ($N_{\text{Re}} = 263.8$).

$$f = 0.24\,\text{m}$$

$$F_{\text{major},1} = h_{F,1} = \frac{f_1 l_1 v_1^2}{2 g D_1} = \frac{0.24 \times 25 \times 0.14^2}{2 \times 9.18 \times 0.015} = 0.41\,\text{m}$$

Head loss in entrance section

The Reynolds number can be determined using the following equation:

$$N_{Re,2} = \frac{4Q\rho}{\pi D_2 \mu} = \frac{4 \times 0.002 \times 1050}{3.14 \times 0.8 \times 0.00845} = 395.73 < 2100 \ (\text{Laminar})$$

The head loss (or the major loss) in the entrance section can be calculated via the same Darcy–Weisbach equation as follows:

$$F_{major,2} = h_{F,2} = \frac{f_2 l_2 v_2^2}{2gD_2} \tag{E9.14B}$$

The friction factor in the entrance section can be determined by the equation $f = \dfrac{64}{N_{Re}}$, because the flow is laminar.

$$f = \frac{64}{N_{Re}} = \frac{64}{395.73} = 0.161$$

$$F_{major,2} = h_{F,2} = \frac{f_2 l_2 v_2^2}{2gD_2} = \frac{0.161 \times 2 \times 0.0039^2}{2 \times 9.81 \times 0.8} = 3.26 \times 10^{-7} \ \text{m} \approx 0 \ (\textit{negligible} = \textit{head loss})$$

Because, the velocity, $v_2 = \dfrac{Q}{A_2} = \dfrac{0.002}{0.50} = 0.0039$ m/s

The major loss in the exit section is the same as that of the entrance section which is 3.26×10^{-7} m ≈ 0 (*negligible = head loss*).

Total major loss, $F_{major,total} = \sum F_{major} = 0.41 + 0 + 0 = 0.41$ m $= h_L$

Pressure drop, $\Delta P = h_L \rho g = 0.41 \times 1050 \times 9.81 \approx 4252.6$ Pa

9.9   Equivalent Pipe

We can also express the entire network of the pipes as a single pipe capable of providing an equivalent discharge with an equivalent head loss. For example, let us consider three pipes connected in series as shown in Fig. 9.22.

Let us imagine that there is an equivalent pipe of a certain length (l_e) and diameter (D_e) that has the same head loss and can provide the same discharge as the series pipe network. Now, we can express the discharges and head losses in all the pipes (neglecting minor losses) as

$$Q_e = Q_1 = Q_2 = Q_3 \tag{9.69}$$

$$h_{L,e} = h_{L,1} + h_{L,2} + h_{L,3} \tag{9.70}$$

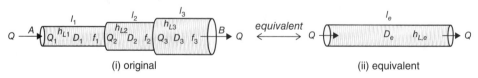

FIGURE 9.22   Various pipes connected in series (i) can be converted into a single pipe (ii) with an equivalent head loss and discharge.

$$h_{L,e} = \frac{f_1 l_1 v_1^2}{2gD_1} + \frac{f_2 l_2 v_2^2}{2gD_2} + \frac{f_3 l_3 v_3^2}{2gD_3} \qquad (9.71)$$

Expressing the velocity in each pipe as $\frac{Q}{A}$, will result in

$$\frac{f_e l_e \left(\dfrac{Q_e}{\dfrac{\pi D_e^2}{4}}\right)^2}{2gD_e} = \frac{f_1 l_1 \left(\dfrac{Q_1}{\dfrac{\pi D_1^2}{4}}\right)^2}{2gD_1} + \frac{f_2 l_2 \left(\dfrac{Q_2}{\dfrac{\pi D_2^2}{4}}\right)^2}{2gD_2} + \frac{f_3 l_3 \left(\dfrac{Q_3}{\dfrac{\pi D_3^2}{4}}\right)^2}{2gD_3} \qquad (9.72)$$

Using the relation $Q_e = Q_1 = Q_2 = Q_3$ and after cancelling out the common terms on both sides will yield

$$\frac{f_e l_e}{D_e^5} = \frac{f_1 l_1}{D_1^5} + \frac{f_2 l_2}{D_2^5} + \frac{f_3 l_3}{D_3^5} \qquad (9.73)$$

In several cases when the pipes are made of the same materials and the flow is turbulent, the friction factors will be similar.

$$f_e \approx f_1 \approx f_2 \approx f_3 \qquad (9.73)$$

Therefore, Eq. (9.73) will reduce to

$$\frac{l_e}{D_e^5} = \frac{l_1}{D_1^5} + \frac{l_2}{D_2^5} + \frac{l_3}{D_3^5} \qquad (9.74)$$

Equation (9.74) is also known as Dupuit equation for compound pipes. Knowing the diameters and the lengths of the existing pipes, one can calculate the equivalent pipe length for a given equivalent diameter.

Example 9.15 (Agricultural Engineering—Equivalent pipe): Three concrete pipes are connected in series. The first pipe is 50 cm in diameter and 200-m long. The second and third pipes are 500 m (75-cm diameter) and 700 m (25-cm diameter), respectively. Convert the system into (a) an equivalent length of 30-cm diameter and (b) equivalent diameter of a 1000-m long concrete pipe (Fig. 9.23).

Step 1: Schematic
The problem is described in Fig. 9.23.

Step 2: Given data
Diameter of the first pipe, $D_1 = 50$ cm; Length of the first pipe, $l_1 = 200$ m; Diameter of the second pipe, $D_2 = 75$ cm; Length of the second pipe, $l_2 = 500$ m; Diameter of the third pipe, $D_1 = 25$ cm; Length of the third pipe, $l_1 = 700$ m

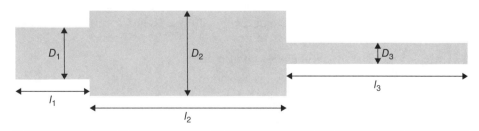

Figure 9.23 Schematic of the pipe system connected in series described in Example 9.15.

To be determined: Equivalent length for a given diameter and an equivalent diameter for a given length.

Assumptions: (1) The flow is fully developed, incompressible, and steady and (2) the friction factors are the same in all pipes.

Step 3: Calculations
We will use the equivalent pipe equation to determine the length and diameter, respectively.

$$\frac{l_e}{D_e^5} = \frac{l_1}{D_1^5} + \frac{l_2}{D_2^5} + \frac{l_3}{D_3^5} \tag{E9.15A}$$

For an equivalent diameter of 30 cm,

$$\frac{l_e}{0.3^5} = \frac{200}{0.5^5} + \frac{500}{0.75^5} + \frac{700}{0.25^5} \tag{E9.15B}$$

Solving for l_e

$$l_e = 1762.49 \text{ m}$$

The series network will have a head loss equivalent to a head loss in a 1762.49-m long pipe of 0.3-m diameter.

To determine an equivalent diameter for a 1000-m long pipe,

$$\frac{1000}{D_e^5} = \frac{200}{0.5^5} + \frac{500}{0.75^5} + \frac{700}{0.25^5} \tag{E9.15C}$$

Solving for D_e

$$D_e = 0.267 \text{ m}$$

The series network will have a head loss equivalent to a head loss in a 1000-m long pipe of 0.267-m diameter.

Practice Problems for the FE Exam

1. A Reynolds number of 11,000 in a circular PVC pipe of 30-cm diameter is considered
   a. Laminar.
   b. Turbulent.
   c. Transition.
   d. Fully developed.

2. For a steady-state, incompressible, fully developed laminar flow in a pipe of radius R, the maximum velocity occurs
   a. When the shear stress is minimum.
   b. When the shear stress is maximum.
   c. At the wall of the pipe.
   d. None of the above.

3. For a steady-state, incompressible, fully developed laminar flow in a 10-cm diameter pipe the average velocity occurs at

   a. The wall of the pipe.

   b. 3 cm from the walls of the pipe.

   c. 4 cm from the center of the pipe.

   d. The center of the pipe.

4. The discharge through a 100-m long horizontal pipe of 10-cm diameter carrying soybean oil $\left(\rho = 930 \dfrac{kg}{m^3} \text{ and } \mu = 70 \text{ mPa} \cdot \text{s} \right)$ with a pressure drop of 58 kPa is approximately

   a. 24 L/s.

   b. 30 L/s.

   c. 2.4 L/s.

   d. 0.24 L/s.

5. In Problem 4, the power requirement assuming a 70% pump efficiency is

   a. 3.5 kW.

   b. 10 kW.

   c. 5 kW.

   d. 2 kW.

6. Honey (density = 1420 kg/m^3 and kinematic viscosity = 0.01 m^2/s) is pumped at 200 LPM through a 20-m long pipe of 7.5-cm diameter using a pump. Assuming that the flow is laminar, the expected head loss across the pipe is nearly

   a. 0.7 m.

   b. 1 m.

   c. 0.5 m.

   d. None of the above.

7. The major loss experienced by a 200-m long pipe of 7.5-cm diameter ($f = 0.001$) carrying 1000 LPM is closest to

   a. 1.8 m.

   b. 1 m.

   c. 2 m.

   d. 2.5 m.

8. The minor loss for a fitting attached to a 2.5-cm diameter pipe carrying 50 LPM of water with a loss coefficient, $K = 9.0$ is

   a. 0.9 m.

   b. 1 m.

   c. 1.3 m.

   d. None of the above.

9. Which of the following is true for pipes connected in series?

   a. The total flow is the sum of individual flows in each pipe.

   b. The total head loss is the sum of individual head losses in each pipe.

   c. The total head loss in each pipe remains the same.

   d. None of the above.

10. Which of the following is true for pipes connected in parallel?

    a. The total flow is the sum of individual flows in each pipe.

    b. The total head loss is the sum of individual head losses in each pipe.

    c. Both a and b.

    d. None of the above.

Practice Problems for the PE Exam

1. A 100-m long, 10-cm diameter CI pipe is transporting a lubricant at 25°C (density = 720 kg/m^3 and viscosity = 0.3 Pa·s). If the pressure drop in the pipe was measured to be 70 kPa, calculate the maximum viscous force and the velocity through the pipe.

2. Diluted bio-oil (density = 800 kg/m^3 and a viscosity of 0.6 Pa·s) from a pyrolysis unit is being pumped in 30-m long pipe of 10-cm diameter using a centrifugal pump at 30°C. If the measured pressure drop was found to be 64 kPa, what will be the discharge through the pipe? Also, determine the power drawn by the pump assuming a 75% efficiency.

3. An engineer decides to transfer oil (27°C) from the production hall to the stocking room which is 50 m away. She chooses a smooth horizontal stainless-steel pipe of 10-cm diameter. Assuming the density and viscosity of the oil as 950 kg/m^3 and 120 mPa·s, respectively, calculate the pumping power needed to transport 1000 LPM oil from the production hall to the stock room.

4. In a pipe network, three CI pipes are connected in series such that a 25-m long pipe of 2.5-cm diameter abruptly expands to 7.5 cm in diameter for 20 m and abruptly reduces to 5 cm diameter for another 25 m before discharging into a tank via a fully open globe valve (K = 10) at a rate of 250 LPM. Determine the total head losses due to the pipe and the fittings by assuming a constant friction factor of 0.05.

5. In a bioenergy facility, ethanol (density = 800 kg/m^3 and viscosity = 0.9 mPa·s) at 25°C is stored in a large cylindrical tank whose bottom is connected to a 5-cm diameter, 10-m long pipe. What should be the level of ethanol in the tank to facilitate a minimum flow of 500 LPM at all times?

6. In a water distribution system, 0.5 m^3/s of water at 20°C is to be pumped between two reservoirs 1000 m apart. Four concrete (roughness = 0.0003 m) pipes were laid in series as listed in Table P9.1.

Pipe No	Length (m)	Diameter (m)
1	300	0.3
2	400	0.2
3	100	0.25
4	200	0.35

TABLE P9.1   Dimensions of the Concrete Pipes
Connected in Series Described in Problem 6

What will be power required by the pump to transport the water assuming a 80% pump efficiency? If the pump is operated 2 hours a day, determine the annual cost of pumping at 20 cents/kWh. Assume the density and viscosity of water to be 1000 kg/m³ and 0.001 Pa·s.

7. Four identical parallel pipes made are used to transport water from a treatment tank to a storage reservoir. The diameters of the pipes are 10 cm, 18 cm, 20 cm, and 30 cm, respectively. Assuming that the discharge through the 18-cm diameter pipe is 500 LPM, how much water is transported from the treatment tank to the storage reservoir in a 24-h period?

8. Three parallel pipes transport wastewater from a treatment tank to a waste reservoir. The pipes are made of the same material and are of the same length with the diameter ratios of 1:1.5:1.75. If the smallest diameter pipe is expected to carry 100 L/min, determine the total discharge into the reservoir.

9. In a water distribution system, four PVC pipes are connected in series. The first pipe is 75 cm in diameter and 100-m long. The second, third, and the fourth pipes are 350 m (50-cm diameter), 700 m (40-cm diameter), and 600 m (30-cm diameter), respectively. Convert the pipe network into (a) an equivalent length of 60-cm diameter and (b) an equivalent diameter of a single 1100-m long pipe.

CHAPTER 10

Pumps and Fans

Chapter Objectives

The content presented in this chapter will provide students with the background needed to

- Understand the common types of pumps and fans and their working mechanisms.
- Size and select a suitable pump needed for transporting fluid for a given situation.
- Employ scaling laws to predict the performance of pumps and fans.

10.1   Motivation

The purpose of a pump or a fan is to impart energy to a fluid. Pumps and fans are routinely used in all water and air treatment and distribution, agricultural and industrial operations, HVAC systems, automobiles, and even in our personal computers. Engineers are frequently tasked to provide recommendations on selection of pumps and predict their performance metrics such as flow rate (discharge), power requirement, and available pressure head. Therefore, engineers should be equipped with the knowledge related to the working of common types of pumps and fans and procedures for the selection and scaling. The purpose of this chapter is to familiarize the student with the basic concepts of pumps and fans and provide mathematical background to integrate pumps and fans into a given fluid transport problem.

10.2   Fluid Moving Equipment

Depending on the type of fluid being transported, fluid moving equipment can be divided into two broad categories, namely pumps and fans. Pumps move liquids while fans move gasses. Regardless of this classification, the mathematical analysis of pumps and fans is similar. Depending on the mode of operation, pumps are generally divided into two categories: rotodynamic pressure pumps and positive displacement pumps. Similarly, fans are also categorized into centrifugal, axial, and mixed flow fans.

   Rotodynamic pumps (RDPs) involve a rotating impeller that imparts kinetic energy to the fluid being pumped. The most common types of RDPs are **centrifugal pumps** and **axial pumps**. **Mixed flow pumps** are an offshoot of RDPs that combine the features of centrifugal and axial pumps to provide head and flow rates in the intermediate range.

10.3   Centrifugal Pump

In its simple configuration, a centrifugal pump consists of a spiral-shaped casing (also called a volute) that contains an impeller connected to a motor. A sample schematic of a centrifugal pump is shown in Fig. 10.1.

The impeller is the heart of a centrifugal pump. During the rotation of the impeller, a low-pressure regime is created around the eye of the impeller that causes the liquid from the inlet to rush toward the eye. The liquid element is then subjected to a centrifugal force by the blades of the rotating impeller that imparts kinetic energy to the liquid element and causes the liquid to move radially toward the outer edge of the volute. Finally, as the fluid travels in the gradually expanding volute, its velocity is decreased (due to gradually increasing area) causing an increase in the pressure of the liquid at the outlet of the pump (Fig. 10.1). In an alternate design, also called turbine design, a diffuser is placed around the impeller. The vanes of the diffuser are designed such that the area between the vanes is gradually increased from the base of the vanes to the edge of the vanes. This increase in the area will decrease the velocity and increase the pressure of the fluid. Centrifugal pumps are the most used pumps in the industry and are available in many configurations (Fig. 10.2).

10.4   Axial Pump

In an axial pump, the liquid flows in a straight path, along the axis of the impeller (Fig. 10.3). The impeller in an axial pump is equipped with a twist in the middle that acts like a propeller in an airplane. The rotating impeller causes a small pressure differential between the upstream and the downstream of the impeller blades which moves the fluid forward. Although axial pumps are not equipped to add significant pressure energy to the liquid, they can provide high discharge and are very appropriate in agricultural applications where significant quantities of water are needed.

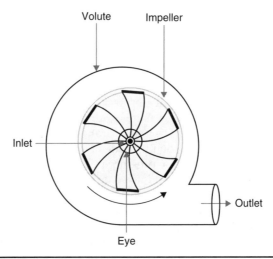

Figure 10.1   A general schematic of a common centrifugal pump.

Figure 10.2 An image of a popular centrifugal water pump (Gorman-Rupp 6500 series) used in agricultural, construction, food and bioprocessing, wastewater industries. (Image courtesy of Gorman-Rupp Mansfield Division. Reprinted with permission.)

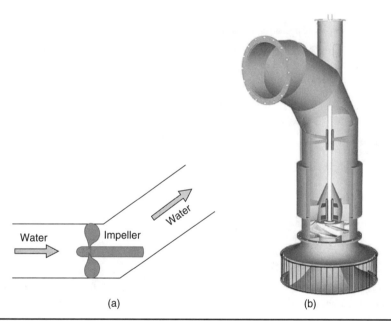

Figure 10.3 A simple schematic of an axial pump (a) and an image of a popular industrial vertical axial pump (b) manufactured by ETEC, Colombia. (Image courtesy of ETEC, Colombia. Reprinted with permission.)

Example 10.1 (Agricultural Engineering–Pump efficiency): A 10-kW centrifugal pump is being used to pump water from a 45-m deep well in a soybean farm. Estimate the discharge from the pump if the efficiency of the pump is 78% (Fig. 10.4).

Step 1: Schematic
The problem is depicted in Fig. 10.4.

Step 2: Given data
Rated power of the pump, $P_{BHP} = 10$ kW; Required head, $W_p = 45$ m; Pump efficiency, $\eta = 78\%$

To be determined: Pump discharge.

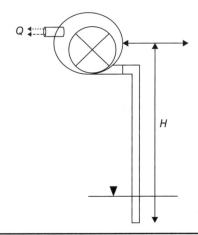

Figure 10.4   Schematic of the centrifugal pump for Example 10.1 to lift water from a well.

Assumptions: The performance and the efficiency of the pump remain constant.

Approach: We will directly use the equation for the power to determine the discharge from the pump.

Step 3: Calculations
The equation for the power of the pump is

$$P_{WHP} = \gamma Q W_P \qquad\qquad\qquad\text{(E10.1A)}$$

We also know that

$$P_{WHP} = \eta P_{BHP} \qquad\qquad\qquad\text{(E10.1B)}$$

Therefore,

$$\eta P_{BHP} = \gamma Q W_P \qquad\qquad\qquad\text{(E10.1C)}$$

$$Q = \frac{\eta P_{BHP}}{\gamma W_P} = \frac{\eta P_{BHP}}{\rho g W_P} = \frac{0.78 \times 10,000}{9.81 \times 1000 \times 45} = 0.0177 \ \frac{m^3}{s} \approx 1060.1 \ \text{LPM}$$

10.5   Pump Specific Speed

As engineers we are required to choose a type of pump for a given application. This is where the concept of specific speed will be helpful. It is defined as the speed of a hypothetical geometrically similar pump capable of moving $1 \ \frac{m^3}{s}$ of a liquid against a vertical head of 1 m. This can mathematically be expressed as

$$N_{Sp} = \frac{NQ^{0.5}}{(gH)^{0.75}} \qquad\qquad\qquad\text{(10.1)}$$

where N is the rotational speed (rad/s), Q is the flow rate $\left(\dfrac{m^3}{s}\right)$, g is the gravity $\left(\dfrac{m}{s^2}\right)$, and H is the head of the fluid (m).

N_{Sp} is a **dimensionless quantity** whose value will provide information on the type of pump that will be needed for a given application. Consider, for example, a centrifugal pump that usually provides a high head and small discharge. This will result in a smaller value of specific speed relative to a specific speed of an axial pump.

Type of the Pump	Range of N_{Sp}
Axial	≥3.5
Mixed	1.6–3.4
Centrifugal	≤1.5

TABLE 10.1 Typical Range of Specific Speed for Commonly Used Pumps in Industry (Data from Cengel and Cimbala)

The ranges for the specific speeds for centrifugal, axial, and mixed pumps are provided in Table 10.1.

However, in the United States, a different version of Eq. (10.1) is used that is not technically nondimensional. It is expressed as

$$N_{Sp} = \frac{NQ^{0.5}}{H^{0.75}} \tag{10.2}$$

with $\left(\frac{L^{0.75}}{T^{1.5}}\right)$ as units and Q, N, and H are expressed as GPM, RPM, and feet, respectively.

Example 10.2 (Selection of pump type): Water from a reservoir is to be pumped to a field at 2000 LPM against a head of 1 m (Fig. 10.5). If the impeller speed is limited to 1500 rpm, what type of pump is recommended?

Step 1: Schematic
The problem is described in Fig. 10.5.

Step 2: Given data
Discharge, $Q = 2000$ LPM; Head, $H = 1$ m; Impeller speed, $N = 1500$ rpm

To be determined: Pump type.

Approach: We will determine the specific speed from which we can decide on the type of the pump.

Step 3: Calculations
The specific speed for a pump can be determined as

$$N_{Sp} = \frac{NQ^{0.5}}{(gH)^{0.75}} \tag{E10.2A}$$

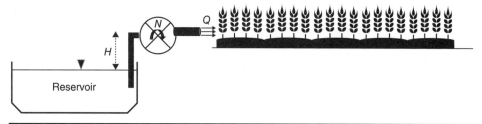

FIGURE 10.5 The schematic of the problem described in Example 10.2.

Substituting the data in the above equation, we obtain

$$N_{Sp} = \frac{\left(\dfrac{1500 \times 2 \times \pi}{60}\right) \times \left(\dfrac{2000}{1000 \times 60}\right)^{0.5}}{(9.81 \times 1)^{0.75}}$$

$$N_{Sp} = 5.17 \tag{E10.2B}$$

Therefore, an axial pump (from Table 10.1) would be appropriate for this situation.

10.6 Matching a Pump for a Given System

Once the type of pump is selected, the next most important task of an engineer is to determine pump parameters (power, impeller diameter, and efficiency) for a given system. This is accomplished by matching the pump system curve and the pump performance curve, whose procedure is described in this section.

10.7 Pump System Curve

The mathematical relationship between the required pump head and the flow rate for a system under consideration is called the pump system curve (PSC). The PSC is the first step in pump selection for a given situation. A simple numerical example will clarify the concept of PSC. Consider two water bodies at different levels (tank 2 is 10 m above tank 1) connected with pipes and relevant fittings (Fig. 10.6). The goal here is to transfer water from tank 1 to tank 2 using a pump. To determine the required pump head (W_p), we will apply energy equation between locations A and B as follows:

$$\frac{P_A}{\gamma} + \frac{v_A^2}{2g} + h_A + W_P - \sum_{\text{all pipes}} \frac{f l v_B^2}{2gD} - \sum_{\text{all fittings}} K \frac{v_B^2}{2g} = \frac{P_B}{\gamma} + \frac{v_B^2}{2g} + h_B \tag{10.3}$$

Considering that both tanks are exposed to atmospheric pressure ($P_A = P_B = 1$ atm), velocity at location A is very small ($v_A \approx 0$), and the location B is 10 m above location A ($h_B - h_A = 10$ m), we can rewrite the above equation in terms of the required pump head as

$$W_P = \frac{v_B^2}{2g} + h_B - h_A + \sum_{\text{all pipes}} \frac{f l v_B^2}{2gD} + \sum_{\text{all fittings}} K \frac{v_B^2}{2g} \tag{10.4}$$

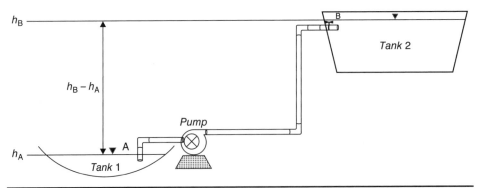

FIGURE 10.6 Schematic of the pump system moving water between two tanks.

Now for the sake of discussion, let us assume that the pipe is 5 cm in diameter and the total length of the pipe is 50 m with a friction factor of 0.01. Similarly, from Fig. 10.6, we can see that there are three threaded elbow fittings ($K = 1.5$) and one globe valve ($K = 10$) connected to the pipe carrying the water. Substituting these values and expressing the velocity as $v = \dfrac{Q}{A} = \dfrac{4Q}{(\pi \times 0.05^2)} = \dfrac{Q}{0.0019}$, we obtain

$$W_P = \frac{Q^2}{0.0019^2 \times 2 \times 9.81} + 10 + \frac{0.01 \times 50 \times Q^2}{0.0019^2 \times 2 \times 9.81 \times 0.05}$$

$$+ ((3 \times 1.5) + 10) \times \frac{Q^2}{0.0019^2 \times 2 \times 9.81} \tag{10.5}$$

or

$$W_P = 10 + 337{,}459.6 \times Q^2 \tag{10.6}$$

$$W_P = \underset{\substack{\downarrow \\ \text{Elevation head}}}{10} + \underbrace{337{,}459.6 \times Q^2}_{\substack{\downarrow \\ \text{Velocity and frictional head}}}$$

Equation (10.6) is the system equation, and the plot of the system equation for a chosen range of discharges is called the **pump system curve**. Let us choose target discharges between 0 and 600 LPM and the corresponding pump head required is plotted (both in LPM and m^3/s) as the system curve in Fig. 10.7.

10.7.1 Pump Performance Curve

Like the PSC, the plot between the available pump head and the flow rate for a given pump is called the pump performance curve (PPC). The PPC curves are determined by the pump manufacturers for each pump for a given rotational speed and the diameter of the impeller. A PPC essentially summarizes important information about a given **pump performance** (discharge, power consumption, and efficiency) for a given head. A typical PPC is depicted (Fig. 10.8) that consists of three curves. The first curve is the

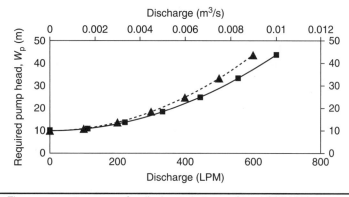

FIGURE 10.7 The pump system curve for discharge between 0 and 600 LPM.

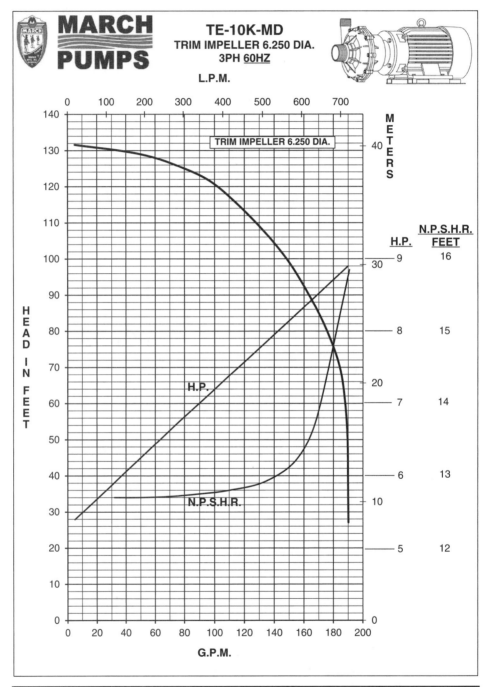

FIGURE 10.8 Pump performance curve for a 6.25-in. impeller TE-10K-MD pump as measured and provided by the manufacturer. (Image courtesy of March Pumps, IL, USA. Reprinted with permission.)

discharge versus head curve that provides the ability of the pump to move the fluid at a given head. The second curve is the amount of energy (HP) drawn by the impeller at a given head and flow rate. The third curve is the net positive suction head required (NPSHR), an important design parameter for the pump, which will be discussed in a separate section.

10.8   Pump Matching and Selection

Once we collect the PSC and PPC, we select the pump by superimposing the PSC and PPC and locate the operating point as the point of intersection of the curves such that the $W_\text{p} = H_\text{p}$ (Fig. 10.9). To explain this, the discharge vs head data (PPC) for the TE-10K-MD pump (5.25-in. diameter impeller) is replotted and superimposed with the PSC (from Fig. 10.7) as shown in Fig. 10.9. From the plot, we can see that the operating point occurs at an operating flow rate of about 530 LPM at an operating head of about 35 m.

10.9   Net Positive Suction Head (NPSH)

During a centrifugal pump's operation, the impeller region in the pump experiences low pressure. When the pressure in the impeller region approaches or drops below the vapor pressure of the fluid (at the given temperate), the fluid begins to boil and experiences the formation of bubbles. This phenomenon is called cavitation. Further, as the fluid crosses the impeller, its pressure increases significantly causing the fluid to change its state and the bubbles begin to collapse rapidly. These rapidly collapsing bubbles interact with the impeller and cause significant surface erosions resulting in permanently damaging the impeller. As an engineer, one must always size the pump correctly to ensure sufficient pressure in the suction pipe to avoid cavitation.

The net positive suction head is the minimum head needed for the liquid to flow through the impeller without causing cavitation within the pump. It is defined using the following equation:

$$\text{NPSH} = \frac{P_\text{s}}{\gamma} + \frac{v_\text{s}^2}{2g} - \frac{P_\text{v}}{\gamma} \qquad (10.7)$$

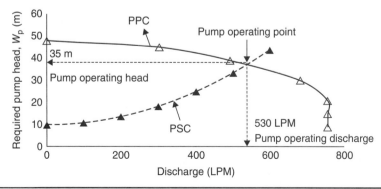

Figure 10.9   The superimposition of PSC and PPC will yield the pump operating point.

where $\dfrac{P_s}{\gamma}$ and $\dfrac{v_s^2}{2g}$ represent the pressure head and the velocity head at the suction of the pipe (at the inlet of the pipe), and $\dfrac{P_v}{\gamma}$ is the vapor pressure head at the given water temperature.

One of the tasks of the pump manufacturer is to evaluate a given pump under various laboratory conditions that simulate the field conditions that the pump will be subjected to. The data collected is transformed into a practical parameter called net positive suction head required (NPSHR). When matching a pump to a piping system and choosing a flow rate, one must ensure that the calculated NPSH is always greater than NPSHR. The following example will illustrate the concept of NPSH and NPSHR.

Example 10.3 (Bioprocessing Engineering—Determination of NPSH): Ethanol (density = 800 kg/m³, viscosity = 0.001 Pa·s, and vapor pressure = 5870 Pa) in an industrial facility is being pumped from a large tank via 6.25-in. diameter impeller centrifugal pump (March pump, TE-10K-MD) at 530 LPM to an overhead tank. If the vertical distance between the ethanol tank (suction height) and the pump is 4 m, do you foresee a possibility of cavitation assuming total major and minor losses during the transportation of ethanol to be 3 m.

Step 1: Schematic
The problem is illustrated in Fig. 10.10.

Step 2: Given data
The density of ethanol, $\rho = 800\ \dfrac{\text{kg}}{\text{m}^3}$; The viscosity of ethanol, $\mu = 0.001$ Pa·s; Vapor pressure of ethanol, $P_v = 5870$ Pa; Pump specifications: March pump, TE-10K-MD with a 6.25-in. diameter impeller; Flow rate, $Q = 530$ LPM; Vertical (suction height) distance between the pump and the tank, $z = 4$ m; Major and minor losses, $F_{\text{major}} + F_{\text{minor}} = 3$ m

To be determined: The possibility of cavitation.

Assumptions: The flow is incompressible and steady state.

Approach: We will determine the NPSH available for this system and compare with the NPSHR for the pump at 530 LPM. If NPSH > NPSHR, we can conclude that the pump will not experience cavitation.

Step 3: Calculations
From Fig. 10.10, we see that location 1 represents ethanol surface and location 2 represents the pump inlet, where cavitation might occur. Applying Bernoulli's equation between locations 1 and 2 will result in

$$\frac{P_1}{\gamma} + \frac{v_1^2}{2g} + z_1 - F_{\text{major}} - F_{\text{minor}} = \frac{P_2}{\gamma} + \frac{v_2^2}{2g} + z_2 \tag{E10.3A}$$

Noting that $P_1 = P_{\text{atm}}$, $v_1 \approx 0$, $z_1 = 0$ (datum level), $P_2 = P_s$, $v_2 = v_s$, and $z_2 - z_1 = z$, the equation now becomes

$$\frac{P_{\text{atm}}}{\gamma} - F_{\text{major}} - F_{\text{minor}} - z = \frac{P_s}{\gamma} + \frac{v_s^2}{2g} \tag{E10.3B}$$

Now, we also know that the NPSH is expressed as

$$\text{NPSH} = \frac{P_s}{\gamma} + \frac{v_s^2}{2g} - \frac{P_v}{\gamma} \tag{E10.3C}$$

Combining the above two equations will yield

$$\text{NPSH} = \frac{P_{\text{atm}}}{\gamma} - \frac{P_v}{\gamma} - z - F_{\text{major}} - F_{\text{minor}} \tag{E10.3D}$$

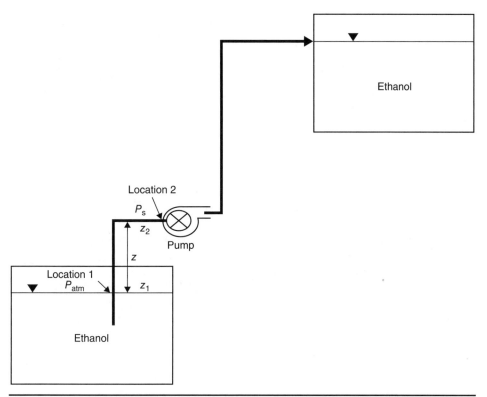

FIGURE 10.10 Schematic for Example 10.3 in which ethanol is being pumped to an overhead tank.

Now substituting the data into the above equation,

$$\text{NPSH} = \frac{101,325}{800 \times 9.81} - \frac{5870}{800 \times 9.81} - 4 - 3 = 5.16 \text{ m}$$

The **NPSHR** for this pump at **530 LPM** is given from the pump performance curve as 13 ft or about **3.96 m.**

Because NPSH > NPSHR, we can conclude that the pump will not undergo cavitation at 530 LPM.

Example 10.4 (Aquacultural Engineering—Minimum length of the suction pipe): Brackish water (density = 1020 kg/m³) is being pumped from a creek into an overhead tank of a shrimp hatchery using a TE-10K-MD pump (5.25-in. diameter impeller) at 300 LPM through a 100-m long, 10-cm diameter pipe ($f = 0.05$). Determine the maximum height the pump can be placed above the creek to avoid cavitation.

Step 1: Schematic
The problem is described in Fig. 10.11.

Step 2: Given data
Pump specifications: TE-10K-MD pump (5.25 inches diameter impeller); Density of the brackish water, $\rho = 1020$ kg/m³; Flow rate, $Q = 300$ LPM; Length of the pipe, $L = 100$ m; Diameter of the pipe, $D = 10$ cm; Friction factor, $f = 0.05$

To be determined: The maximum height between the pump and the creek surface to avoid cavitation, z.

Assumptions: The flow is incompressible and steady state. The minor losses in the pipe are ignored.

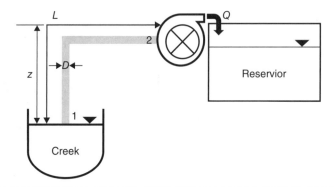

FIGURE 10.11   Schematic for pump system described in Example 10.4.

Approach: We will combine the NPSHR, and the Bernoulli's equation to determine the maximum height of the pump above the creek's surface.

Step 3: Calculations
Using Fig. 10.11, we can write the Bernoulli's equation between 1 and 2 as follows:

$$\frac{P_1}{\gamma} + \frac{v_1^2}{2g} + z_1 - F_{major} = \frac{P_2}{\gamma} + \frac{v_2^2}{2g} + z_2 \qquad \text{(E10.4A)}$$

Noting that $P_1 = P_{atm}$, $v_1 \approx 0$, $z_1 = 0$ (datum level), $P_2 = P_s$, $v_2 = v_s$, and $z_2 - z_1 = z$, the equation now becomes

$$\frac{P_{atm}}{\gamma} - F_{major} - z = \frac{P_s}{\gamma} + \frac{v_s^2}{2g} \qquad \text{(E10.4B)}$$

From the definition of NPSH, the right-hand part of the above equation becomes

$$\frac{P_s}{\gamma} + \frac{v_s^2}{2g} = \text{NPSH} + \frac{P_v}{\gamma} \qquad \text{(E10.4C)}$$

Combining the above equation, we obtain

$$\frac{P_{atm}}{\gamma} - F_{major} - z = \text{NPSH} + \frac{P_v}{\gamma} \qquad \text{(E10.4D)}$$

$$z = \frac{P_{atm}}{\gamma} - \frac{P_v}{\gamma} - F_{major} - \text{NPSH} \qquad \text{(E10.4E)}$$

Based on the pump performance curve, the minimum net positive suction head required for a flow rate of 300 LPM is 3.5 m.

Now, we will determine the extent of major losses as follows:

$$F_{major} = \frac{flv^2}{2gD} \qquad \text{(E10.4F)}$$

Therefore,

$$F_{major} = \frac{fl\left(\dfrac{Q}{A}\right)^2}{2gD} = \frac{fl\left(\dfrac{4Q}{3.14 \times D^2}\right)^2}{2gD} \qquad \text{(E10.4G)}$$

Substituting the data into the above equation will yield

$$F_{major} = \frac{0.05 \times 100 \times \left(\dfrac{4 \times 300 \times 0.001}{60 \times 3.14 \times 0.1^2}\right)^2}{2 \times 9.81 \times 0.1} = 1.03 \text{ m}$$

We can now determine the value of z using Eq. (E10.4E) as

$$z = \frac{101,325}{1020 \times 9.81} - \frac{3170}{1020 \times 9.81} - 1.03 - 3.5$$

The maximum height between the suction of the pump and the surface of the water is 5.27 m.

10.10   Scaling of Pumps

The performance metrics of a pump, that is, the discharge (Q), head (H), and the required power (P_{BHP}) depend on two pump input variables, that is, the impeller diameter and the impeller speed of the pump. For the same impeller speed, if its diameter is increased, the values of Q, H, and P_{BHP} will vary as summarized in Table 10.2. Similarly, if the diameter is maintained constant and the impeller speed is increased, the output parameters change accordingly (Table 10.2).

We can combine the effects of the change in impeller diameter and the impeller speed, and express the pump design laws as follows:

$$\frac{Q_1}{Q_2} = \left(\frac{D_1}{D_2}\right)^3 \frac{N_1}{N_2} \tag{10.11}$$

$$\frac{H_1}{H_2} = \left(\frac{D_1}{D_2}\right)^2 \left(\frac{N_1}{N_2}\right)^2 \tag{10.12}$$

$$\frac{P_{BHP,1}}{P_{BHP,2}} = \frac{\rho_1}{\rho_2}\left(\frac{D_1}{D_2}\right)^5 \left(\frac{N_1}{N_2}\right)^3 \tag{10.13}$$

Constant Speed, but Diameter Varied	Constant Diameter, but Speed Varied	Eqn #
$\dfrac{H_1}{H_2} = \left(\dfrac{D_1}{D_2}\right)^2$	$\dfrac{H_1}{H_2} = \left(\dfrac{N_1}{N_2}\right)^2$	10.8
$\dfrac{Q_1}{Q_2} = \left(\dfrac{D_1}{D_2}\right)^3$	$\dfrac{Q_1}{Q_2} = \dfrac{N_1}{N_2}$	10.9
$\dfrac{P_{BHP1}}{P_{BHP2}} = \dfrac{\rho_1}{\rho_2}\left(\dfrac{D_1}{D_2}\right)^5$	$\dfrac{P_{BHP1}}{P_{BHP2}} = \dfrac{\rho_1}{\rho_2}\left(\dfrac{N_1}{n_2}\right)^3$	10.10

TABLE **10.2**   The Mathematical Relationships Summarizing the Dependence of Impeller Diameter and Impeller Speed on the Pump Head, Discharge, and the Power Consumption

Example 10.5 (Bioprocessing Engineering—Pumping of tomato puree):   In a food processing facility, tomato puree is being pumped using an existing pump having an impeller size of 25 cm at 1500 rpm. The pump is capable of a head of 120 m and consumes 5.5 kW of energy. Considering a projected demand, it was decided to replace the current impeller with a larger one of 37.5 cm and operate at 1000 rpm to enhance the performance (Fig. 10.12). You were approached for the advice. What would be your recommendation?

Step 1: Schematic
The problem is depicted in Fig. 10.12.

Step 2: Given data
Let us denote the current pump configuration with 1 and the future with 2; Current impeller size, $D_1 = 25$ cm; Current impeller speed, $N_1 = 1500$ rpm; Current pumping head, $H_1 = 120$ m; Current energy consumption, $P_{BHP,1} = 5.5$ kW; Proposed impeller size, $D_2 = 37.5$ cm; Proposed impeller speed, $N_2 = 1000$ rpm

To be determined: Effect on pump performance metrics.

Assuming the density of the tomato puree remains constant, we can use the pump scaling laws to determine the effect of larger impeller size and speed on Q, H, and the power consumption.

Effect on available pump head (H)

$$\frac{H_1}{H_2} = \left(\frac{D_1}{D_2}\right)^2 \left(\frac{N_1}{N_2}\right)^2 \tag{E10.5A}$$

Because we are dealing with ratios, the unit conversions will cancel out and therefore we can use the data in given units

$$\frac{120}{H_2} = \left(\frac{25}{37.5}\right)^2 \times \left(\frac{1500}{1000}\right)^2 \tag{E10.5B}$$

$$H_2 = 120 \text{ m (same as before)}$$

Effect on pump capacity (discharge) (Q)

$$\frac{Q_1}{Q_2} = \left(\frac{D_1}{D_2}\right)^3 \frac{N_1}{N_2} \tag{E10.5C}$$

$$\frac{Q_1}{Q_2} = \left(\frac{25}{37.5}\right)^3 \times \frac{1500}{1000} \tag{E10.5D}$$

$$\frac{Q_2}{Q_1} = 2.25 \tag{E10.5E}$$

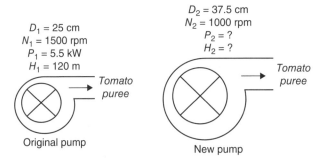

FIGURE 10.12   Schematic for Example 10.5 to determine the effects of changing impeller diameter and speed on pump performance metrics.

Effect on power consumption (P)

$$\frac{P_{BHP,1}}{P_{BHP,2}} = \frac{\rho_1}{\rho_2}\left(\frac{D_1}{D_2}\right)^5\left(\frac{N_1}{N_2}\right)^3 \qquad\qquad (E10.5F)$$

$$\frac{5.5}{P_{BHP,2}} = 1.0\left(\frac{25}{37.5}\right)^5 \times \left(\frac{1500}{1000}\right)^3 \qquad\qquad (E10.5G)$$

Because $\dfrac{\rho_1}{\rho_2} = 1$

$$P_{BHP,2} \approx 12.37 \text{ kW}$$

The larger impeller will increase the discharge 2.25 times and provides the same head. It consumes 11.25 kW of power even though it is operated at a lower speed of 1000 rpm.

Example 10.6 (Environmental Engineering–Transport of manure slurry): In a confined animal feeding operation, 100 L/min of manure slurry is pumped via a centrifugal pump with an impeller diameter of 20 cm against a head of 25 m. What should be the diameter of a new impeller if 500 L/min manure is to be moved using the old impeller speed? Also determine the new head the larger impeller can provide, and the additional power needed to operate the new pump (Fig. 10.13).

Step 1: Schematic
The problem is depicted in Fig. 10.13.

Step 2: Given data
Let us denote the current pump configuration with 1 and the new with 2; Current impeller size, $D_1 = 20$ cm; Current pumping head, $H_1 = 25$ m; Current impeller speed, $N_1 =$ New impeller speed, N_2; Current flow rate, $Q_1 = 100$ LPM; New flow rate required, $Q_2 = 500$ LPM

To be determined: The new impeller size, new head, and the additional power requirement.

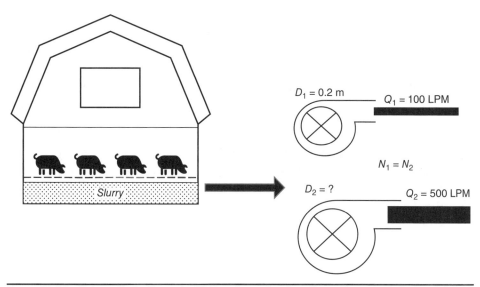

Figure 10.13 Schematic for Example 10.6 to determine the effect of increased discharge on pump size and power requirement.

Step 3: Calculations

Assuming the density of the manure remains constant, we can use the pump scaling laws to determine the size of new impeller needed, available head, and the power consumption.

New diameter (D_2)

$$\frac{Q_1}{Q_2} = \left(\frac{D_1}{D_2}\right)^3 \frac{N_1}{N_2} \tag{E10.6A}$$

$$\frac{100}{500} = \left(\frac{20}{D_2}\right)^3 \times 1 \tag{E10.6B}$$

$$D_2 \approx 34.2 \text{ cm}$$

New available head (H_2)

$$\frac{H_1}{H_2} = \left(\frac{D_1}{D_2}\right)^2 \left(\frac{N_1}{N_2}\right)^2 \tag{E10.6C}$$

$$\frac{25}{H_2} = \left(\frac{20}{34.2}\right)^2 \times 1 \tag{E10.6D}$$

$$H_2 = 73.1 \text{ m}$$

Additional power required (P_{BHP2})

$$\frac{P_{BHP,1}}{P_{BHP,2}} = \frac{\rho_1}{\rho_2}\left(\frac{D_1}{D_2}\right)^5 \left(\frac{N_1}{N_2}\right)^3 \tag{E10.6E}$$

$$\frac{P_{BHP,1}}{P_{BHP,2}} = 1 \times \left(\frac{20}{34.2}\right)^5 \times 1 \tag{E10.6F}$$

$$\frac{P_{BHP,2}}{P_{BHP,1}} = 14.62$$

$$\text{Additional power required} = \frac{P_{BHP,2} - P_{BHP,1}}{P_{BHP,1}} \times 100 = \left(\frac{P_{BHP,2}}{P_{BHP,1}} - 1\right) = 1362\% \text{ increase}$$

10.11   Pumps in Series and Parallel

In some situations, we might have to enhance the pressure head of the pumping system. Imagine pumping to a large multistory building or moving a liquid from several hundreds of feet from under the ground. In such cases, one approach is to connect identical pumps in series as shown in Fig. 10.14. Similar to the pipes in series (Chap. 9), we can apply the conservation of mass and energy to quantify the performance of the new pumps in series system.

$$Q_T = Q_A = Q_B \quad \text{Conservation of mass} \tag{10.14}$$

$$H_T = H_A + H_B \quad \text{Conservation of energy} \tag{10.15}$$

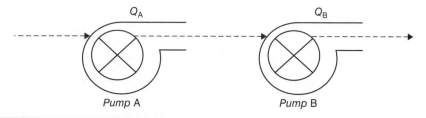

FIGURE 10.14 Two pumps in series will increase the available head for the same discharge.

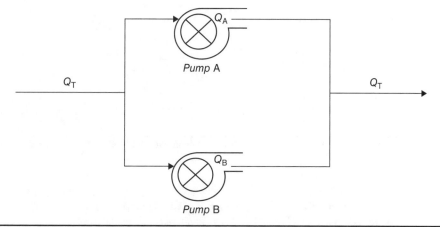

FIGURE 10.15 Two pumps in parallel will increase the discharge for the available head.

In other cases, when we require a significantly higher amount of discharge, we connect similar pumps in parallel (Fig. 10.15). It is to be noted that the head will remain constant while the discharge is increased. Therefore, we can write the energy and mass balance equations as follows:

$$Q_T = Q_A + Q_B \quad \text{Conservation of mass} \tag{10.16}$$

$$H_T = H_A = H_B \quad \text{Conservation of energy} \tag{10.17}$$

It is of practical importance to look at the pump performance curve when the pumps are connected in series and parallel. The following example will illustrate the effect of pumps in series and parallel.

Example 10.7 (General Engineering—All Concentrations):   For the following data provided for a given pump, construct a pump performance curve when two such pumps are placed in series and parallel (Fig. 10.16).
  [Head (m), Q (LPM)] = [25, 0], [22, 100], [15, 250], [10, 300], [5, 350].

Step 1: Schematic
The problem is depicted in Fig. 10.16.

Step 2: Given data
Impeller diameter, $D = 17.8$ cm; Impeller speed, $N = 1150$ rpm; Discharge versus head is also provided

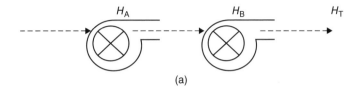

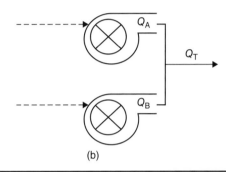

FIGURE 10.16 Schematic for Example 10.7 in which two identical pumps are arranged in series (a) and parallel (b).

To be determined: Performance curves for two identical pumps in series and parallel.

Assumptions: The discharge and efficiency of both the pumps are identical.

Approach: We will apply the conservation of the mass and energy and determine the new head and discharge for the series and parallel configurations.

Step 3: Calculations
Let us call the two pumps A and B. For the pumps in series, $Q_T = Q_A = Q_B$ and $H_T = H_A + H_B = 2H_A$. The new discharge and head for pumps in series are calculated (Table 10.1).

Similarly, for the pumps in parallel, $Q_T = Q_A + Q_B = 2Q_A$ and $H_T = H_A = H_B = H_A$, the new head and discharge for pumps in parallel are calculated (Table 10.1). The original and the new pump performance curves are plotted in Fig. 10.17.

Original Head	Original Discharge	Two Pumps in Series New Pump Head	New Pump Discharge	Two Pumps in Parallel New Pump Head	New Pump Discharge
H (m)	Q (LPM)	H_T (m)	Q_T (LPM)	H_T (m)	Q_T (LPM)
25	0	50	0	25	0
22	100	44	100	22	200
15	250	30	250	15	500
10	300	20	300	10	600
5	350	10	350	5	700

TABLE 10.1 Summary of the Original and New Head for Two Pumps in Series (Fig. 10.16a) and New Discharge for Two Pumps in Parallel (Fig. 10.16b)

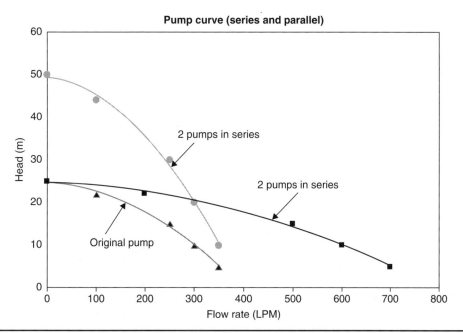

FIGURE 10.17 The original and new head and discharges for two pumps in series.

10.12 Multistage Pumps

Another option to significantly increase the pressure head of a pump is to accommodate multiple impellers in the same pump casing (Fig. 10.18). Such an arrangement is called a multistage pump. In principle, this is somewhat similar to pumps in series. However, unlike pumps in series configurations, which have multiple motors, multistage pumps are operated by a single motor. Therefore, multistage pumps offer higher efficiencies than pumps in series configurations that suffer cumulative losses due to the presence of several motors. In addition, the compactness of multistage pumps allows for their use in space-limited applications. Despite these advantages, multistage pumps,

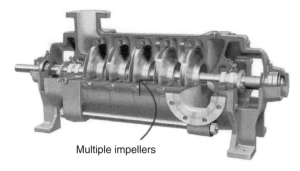

Multiple impellers

FIGURE 10.18 A sectional view of a multistage centrifugal pump. (Image courtesy of Castle Pumps, UK. Reprinted with permission.)

due to smaller clearances between the impellers, and are not suitable for liquids with higher solid content.

10.13   Reciprocating Pumps

As specialized pumps, reciprocating pumps find applications in situations where pre-determined and accurate flow rates at high pressures are desired. They are also useful in moving viscous liquids. Therefore, several industries including food processing, oil and gas, and pharmaceutical operations employ reciprocating pumps. When compared to centrifugal pumps, they are usually more efficient. A simple schematic of a recipro-cating pump is depicted in Fig. 10.19. The pump contains a cylindrical chamber that encloses a piston capable of moving back and forth via a connecting rod. The connect-ing rod is connected to an electrically powered crank that converts the rotational motion into a reciprocating motion. Therefore, when the crack rotates half a turn the connecting rod will pull the piston back resulting in suction formation in the cylinder. During the suction cycle, the one-way suction valve is opened and the fluid rushes in due to the pressure differential. Subsequently, in the next cycle (also called as delivery stroke), when the crank completes a full turn ($\alpha = 360°$), the piston moves forward and pushes the fluid (positive pressure) through the one-way discharge valve into the delivery pipe. When this process keeps repeating at high rotational speeds, fluid delivery at high pressures can be obtained.

10.14   Discharge, Power, and Slip

One complete revolution of the crank will move the piston back (suction) and forth (delivery) and moves the fluid over the actual length of the cylinder (one stroke). If A is the area of the cross section and the length of the cylinder is L, the volume of the fluid

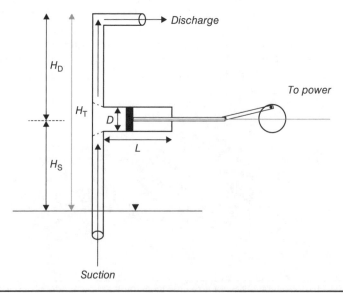

Figure 10.19   A simplified schematic of a single acting reciprocating pump.

moved is $A \times L$. Now if the crank is rotating at N rpm, the maximum theoretical flow rate, Q_T may be expressed as

$$Q_T = \frac{A\,(m^2)L(m)N(rpm)}{60\,(s)} \qquad (10.18)$$

$$Q_T = \frac{ALN}{60} = \frac{\pi D^2 LN}{4 \times 60} = 0.013 D^2 LN \qquad (10.19)$$

Once the flow rate is calculated, the power can be calculated as

$$P_{WHP} = \gamma Q_T H \qquad (10.20)$$

It should, however, be noted that in practice the actual measured flow rate, Q is less than the maximum theoretical flow rate, Q_T due to leakages in the pump. Therefore, the differential between the maximum theoretical flow rate, Q_T and actual measured flow rate, Q is called the slip (S) of a reciprocating pump and is defined as

$$S = Q_T - Q \qquad (10.21)$$

Example 10.8 (Bioprocessing Engineering—Pumping of a slurry):   A reciprocating pump is employed to pump a food slurry with a density of 1200 kg/m^3 from an underground tank to a storage tank located 20 m above. If the piston of the pump is 10 cm in diameter with a stroke length of 40 cm, determine the maximum possible discharge at 200 rpm. Additionally, if the measured flow rate was 500 LPM, estimate the slip of the pump (Fig. 10.20). At this flow rate, what will be the power requirement?

Step 1: Schematic
The problem is depicted in Fig. 10.20.

Step 2: Given data
Total head, $H = 20$ m; Rotational speed, $N = 200$ rpm; Length of the stroke, $L = 40$ cm; Diameter of the piston, $D = 10$ cm; Measured flow rate, $Q = 500$ LPM $= 0.01\,\dfrac{m^3}{s}$

To be determined: Maximum possible flow rate (Q_T), slip (S), and the power requirement (Q_{WHP}).

Step 3: Calculations
For a single acting pump, the maximum possible flow rate (Q_T) is given by

$$Q_T = 0.013 D^2 LN = 0.013 \times (0.1)^2 \times 0.4 \times 200 \qquad (E10.8A)$$

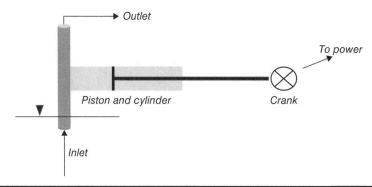

FIGURE 10.20   A schematic for the single acting reciprocating pump for Example 10.8.

$$Q_T \approx 628 \text{ LPM} = 0.0104 \, \frac{m^3}{s}$$

However, the actual flow rate that was measured was 500 LPM or about $0.0083 \, \frac{m^3}{s}$. Therefore, the slip, S can be calculated as

$$S = Q_T - Q = 0.0104 - 0.0083 = 0.0004 \, \frac{m^3}{s} = 124 \text{ LPM or about } 20\%$$

Now the power can be calculated as

$$P_{WHP} = \gamma Q_T H = \rho g Q_T H = 1200 \times 9.81 \times 0.0104 \times 20 \approx 2.46 \text{ kW}$$

10.15   Double Acting Reciprocating Pump

When the suction and delivery actions occur concurrently, such arrangement is called double effect or double acting pump (Fig. 10.21). When the crank rotates by 180°, the piston moves backward and creates a suction pressure on the left half of the cylinder (zone 1). This will open the suction value and the fluid fills up the cylinder due to the pressure differential. Simultaneously, the piston rod creates a positive pressure on the fluid collected on the right half of the cylinder (zone 2) resulting in opening the delivery valve and subsequently discharge into the delivery line. In the next cycle, the crank completes the remaining half of the rotation, during which, the piston moves forward, this creating a suction in zone 2 and positive pressure in zone 1. Thus, after one complete revolution of the crank (1 stroke), the fluid will have moved a distance of 2L units

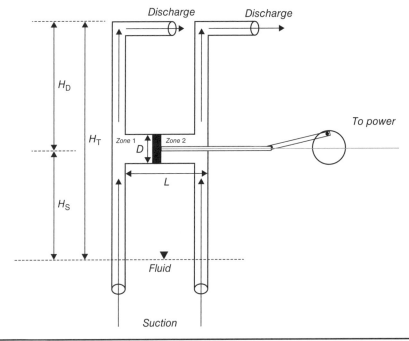

Figure 10.21   A simplified schematic of a double acting reciprocating pump.

and (after neglecting the area of the piston rod) the flow rate is twice that of the single acting pump and is expressed as

$$Q_T = \frac{A(2L)N}{60} = \frac{\pi D^2(2L)N}{4 \times 60} = 0.026D^2LN \tag{10.22}$$

Example 10.9 (Agricultural Engineering—Pumping from deep well):   It was proposed to use a double acting reciprocating pump rated for 5 kW to pump water from a deep underground well to the 2 m above ground level. If the expected flow rate was 1000 LPM, determine the required rotational speed of the pump with a piston diameter of 10 cm and a stroke length of 30 cm (Fig. 10.22). What will be limiting head the pump can provide at this rotational speed?

Step 1: Schematic
The problem is depicted in Fig. 10.22.

Step 2: Given data
Power of the pump, $P = 5$ kW; Flow rate, $Q = 1000$ LPM $\approx 0.0167\ \frac{m^3}{s}$; Length of the stroke, $L = 30$ cm; Diameter of the piston, $D = 10$ cm

To be determined: The rotational speed, N and the total head, H.

Assumptions: The flow will be steady state and the pump performance remains constant.

Approach: We will use the equation for the double acting reciprocating pump to solve for the crank's rotational speed and the total head provided by the pump.

Step 3: Calculations
For a double acting pump, the maximum possible flow rate (Q_T) in terms of rotational speed is given by

$$Q_T = 0.026D^2LN$$

$$\frac{1000}{(1000 \times 60)} = 0.026 \times 0.1^2 \times 0.3 \tag{E10.9A}$$

$$N = 213.67 \approx 214 \text{ rpm}$$

The maximum height the water can be lifted to will be

$$P_{WHP} = \gamma Q_T H$$

$$5000 = 1000 \times 9.81 \times \left(\frac{1000}{1000 \times 60}\right) \times H \tag{E10.9B}$$

$$H = 30.58 \text{ m (including the 2 m above the ground)}$$

The pump can lift the water from 28.58 m under the ground.

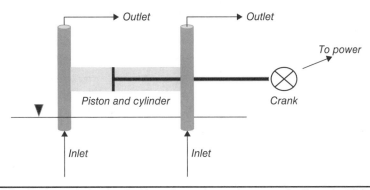

Figure 10.22   A schematic for the double acting reciprocating pump for Example 10.9.

10.16   Airlift Pumps

Airlift pumps are one of the simplest and ancient pumps. It is believed to have been developed by Carl Löscher, a German engineer in the late 1700s. Airlift pumps have no mechanical and moving parts and, therefore, they are highly suited for moving corrosive liquids that would otherwise damage the impellers in the conventional pumps.

In its simplest design, an airlift pump consists of a pipe whose bottom is connected to a source of compressed air (Fig. 10.23). Because of buoyancy, the air-water mixture rises, moves up, and eventually flows out through the outlet.

Another application of airlift pump is in aquaculture where the water must always contain dissolved oxygen above the recommended limit (usually greater than 6 ppm). Pure gaseous oxygen can be used in lieu of air to increase the concentration gradient and the subsequent mass transfer of oxygen to water.

10.17   Fans

Fluid moving equipment that moves a gas is called a fan. Fans typically move gases at low pressure (e.g., exhaust fan in our homes). When a fan moves a gas at very high pressure (albeit at low flow rate), it is commonly called a compressor. Similarly, a blower is simply a fan that can move a gas at moderate pressure and flow rate. Fans can be either of axial type (Fig. 10.24a) capable of moving large quantities of a gas at low pressure or centrifugal type (Fig. 10.24b) which moves gases at high pressure. Like pumps, fans can be arranged in series to increase the head or in parallel (Fig. 10.25) to increase the discharge.

From a functional perspective, a fan is basically a pump, and therefore most of the mathematical principles pertaining to the pumps are equally valid for fans. We will demonstrate the concepts of fans via the following examples.

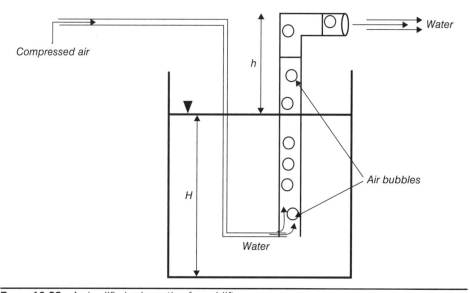

Figure 10.23   A simplified schematic of an airlift pump.

(a)   (b)

FIGURE 10.24   Images of popular industrial wall axial fan (a) and Series 41 Fiberglass Backward Curved Centrifugal Fan (b). (Image courtesy of Cincinnati Fans and Hartzell, USA. Reprinted with permission.)

FIGURE 10.25   Image of hog barn equipped with several fans in parallel. (Image courtesy of Drs. Brett Ramirez and Steven Hoff, Iowa State University Agricultural and Biosystems Engineering. Reprinted with permission.)

Example 10.10 (Agricultural Engineering—Grain drying): Freshly harvested grain is being dried using three identical centrifugal fans in series. It is also given that each fan moves against a head of 10 m and has a specific speed of 1.4. What should be the fan speed needed for this task if the fan system is required to move 1 m³/s of dry air. Also estimate the total power needed by the fan assuming a mechanical efficiency of 65% (Fig. 10.26).

Step 1: Schematic
The problem is depicted in Fig. 10.26.

Step 2: Given data
Three fans (A, B, and C) in series, pressure head for each fan, $H_A = H_B = H_C$, Fan specific speed, $N_{Sp} = 1.4$; Flow rate, $Q = 1$ m³/s; Fan efficiency, $\eta = 65\%$

To be determined: Impeller speed of the fan, N.

Assumptions: The flow is steady state and the discharge and efficiency of all three fans are identical. In addition, we may assume that the average density of air for the entire drying process is 1.18 kg/m³.

Approach: We will determine the impeller speed using the fan specific speed and subsequently determine the power requirement.

Step 3: Calculations
Because the fans are in series, the total head (H_T) is the summation of each pressure head. Therefore,

$$H_T = H_A + H_B + H_C = 30 \text{ m} \tag{E10.10A}$$

We can now determine the impeller speed from the specific speed as follows:

$$N_{Sp} = \frac{NQ^{0.5}}{(gH_T)^{0.75}} \tag{E10.10B}$$

Substituting the given data into the above equation,

$$1.4 = \frac{N \times 1^{0.5}}{(9.81 \times 30)^{0.75}} \tag{E10.10C}$$

We obtain, $N = 99.47 \dfrac{\text{rad}}{\text{s}}$ or 949 rpm

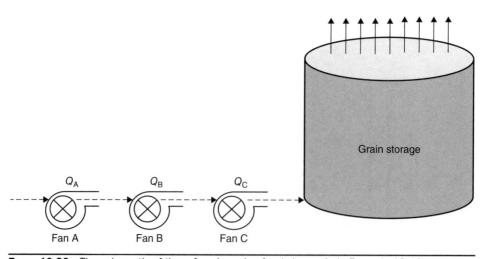

FIGURE 10.26 The schematic of three fans in series for drying grain in Example 10.10.

The total brake horsepower requirement for all the three pumps in series is calculated as

$$P_{BHP} = \frac{\rho g Q H_T}{\eta}$$

After substituting the given data into the above equation, we obtain

$$P_{BHP} = \frac{1.18 \times 9.81 \times 1 \times 30}{0.65} = 534.26 \text{ watts}$$

Example 10.11 (Bioprocessing Engineering–Cooling of a gasifier):   A centrifugal fan (for cooling a gasifier) operating at 1200 rpm equipped with a 31-cm diameter impeller provides 3 m³/s of air against a head of 4 m. It was desired to increase the flow rate to 6.5 m³/s. The available options are to either use a larger fan and maintain the same fan speed or increase the fan speed but keep the fan size the same (Fig. 10.27). What will be your approach?

Step 1. Schematic
The problem is described in Fig. 10.27.

Step 2: Given data
Original discharge from the fan, $Q_{original} = 3$ m³/s; Original diameter of the fan impeller, $D_{original} = 0.31$ m; Original speed of the fan impeller, $N_{original} = 1200$ rpm; Original head for the fan, $H_{original} = 4$ m; New discharge from the fan, $Q_{new} = 6.5$ m³/s

To be determined: Effect of increasing the fan impeller diameter or fan impeller speed.

Assumptions: The dimensional similarity is valid. The fluid density remains constant.

Approach: We will use the fan scaling laws (same as pump scaling laws and analyze each option).

Step 3: Calculations
Let us first analyze the effect of using a larger fan size. The new impeller diameter is calculated from the fan laws as follows:

$$\frac{Q_{new}}{Q_{original}} = \frac{N_{new}}{N_{original}} \left(\frac{D_{new}}{D_{original}} \right)^3 \qquad \text{(E10.11A)}$$

Because the impeller speed remains the same, we can determine the new diameter needed for the target flow rate of 6.5 m³/s as follows:

$$\frac{Q_{new}}{Q_{original}} = \left(\frac{D_{new}}{0.31} \right)^3 \qquad \text{(E10.11B)}$$

$$D_{new} = 40.1 \text{ cm}$$

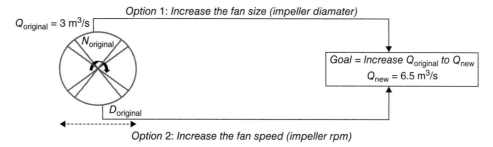

Option 1: Increase the fan size (impeller diamater)

$Q_{original} = 3$ m³/s

$N_{original}$

$D_{original}$

Goal = Increase $Q_{original}$ to Q_{new}
$Q_{new} = 6.5$ m³/s

Option 2: Increase the fan speed (impeller rpm)

Figure 10.27   The schematic of the fan in Example 10.11 whose discharge is to be increased.

The larger impeller fan will also be able to handle an increased pressure head

$$\frac{H_{new}}{H_{original}} = \left(\frac{N_{new}}{N_{original}}\right)^2 \left(\frac{D_{new}}{D_{original}}\right)^2$$ (E10.11C)

$$\frac{H_{new}}{H_{original}} = \left(\frac{0.401}{0.31}\right)^2$$ (E10.11D)

$$H_{new} = 1.66 H_{original} \approx 6.7 \text{ m}$$

However, an increase in the fan size will also increase the power requirement as follows:

$$\frac{BHP_{new}}{BHP_{original}} = \frac{\rho_{new}}{\rho_{original}} \left(\frac{N_{new}}{N_{original}}\right)^3 \left(\frac{D_{new}}{D_{original}}\right)^5$$ (E10.11E)

Because the impeller speed and the fluid density remain the same, the new power requirement is

$$\frac{BHP_{new}}{BHP_{original}} = \left(\frac{0.401}{0.31}\right)^5$$ (E10.11F)

$$BHP_{new} \approx 3.63 \text{ } BHP_{original}$$

Therefore, using a larger diameter fan will consume about 3.63 times more power than the original fan.

Now, we will look at the effect of increasing the fan speed but keep the fan size the same. The new impeller speed is calculated from the fan laws as follows:

$$\frac{Q_{new}}{Q_{original}} = \frac{N_{new}}{N_{original}} \left(\frac{D_{new}}{D_{original}}\right)^3$$ (E10.11G)

Substituting the given data into the above equation,

$$\frac{6.5}{3} = \frac{N_{New}}{1200}$$ (E10.11H)

$$N_{new} = 2600 \text{ rpm}$$

The faster fan will also be able to handle an increased pressure head

$$\frac{H_{new}}{H_{original}} = \left(\frac{N_{new}}{N_{original}}\right)^2 \left(\frac{D_{new}}{D_{original}}\right)^2$$ (E10.11I)

$$\frac{H_{new}}{H_{original}} = \left(\frac{2600}{1200}\right)^2$$ (E10.11J)

$$H_{new} = 4.69 \text{ } H_{original} \approx 18.78 \text{ m}$$

However, this will also increase the power requirement as follows:

$$\frac{BHP_{new}}{BHP_{original}} = \frac{\rho_{new}}{\rho_{original}} \left(\frac{N_{new}}{N_{original}}\right)^3 \left(\frac{D_{new}}{D_{original}}\right)^5$$ (E10.11K)

$$\frac{BHP_{new}}{BHP_{original}} = \left(\frac{2600}{1200}\right)^3$$ (E10.11L)

$$BHP_{new} = 10.17 \text{ } BHP_{original}$$

Therefore, increasing the impeller speed from 1200 rpm to 2600 rpm will consume 10.17 times more power than the original fan.

Practice Problems for the FE Exam

1. A pump always increases the:

   a. flow rate.

   b. pressure.

   c. flow velocity.

   d. None of the above.

2. The maximum head, a pump of 5.1 kW of water horsepower can provide while pumping 200 LPM of water is closet to:

   a. 150 m.

   b. 156 m.

   c. 125 m.

   d. 100 m.

3. To avoid cavitation, one must ensure that:

   a. NPSH > NPSHR.

   b. NPSH = NPSHR.

   c. NPSH < NPSHR.

   d. None of the above.

4. The old impeller of a pump was replaced with a new impeller whose diameter is 30% larger. If the impeller spins at the same speed as before, the new discharge from the pump will:

   a. increase by about 120%.

   b. decrease by about 120%.

   c. remain the same.

   d. increase by 30%.

5. In the new impeller in the above problem, the increase in the power requirement will be nearly:

   a. 4 times the original power.

   b. 3 times the original power.

   c. 1.8 times the original power.

   d. None of the above.

6. In the above problem, the increase in the pressure head due to the larger impeller is nearly:

   a. 30%.

   b. 170%.

   c. 70%.

   d. None of the above.

7. The specific speed for a water pump with an impeller speed of 2000 rpm working against a head of 10 m while pumping 100 LPM is closest to:

   a. 0.67.

   b. 0.1.

   c. 0.3.

   d. None of the above.

8. Three pumps each capable of moving 175 LPM against a head of 10 m are arranged in series. The total flow of all the pumps together is:

   a. 325 LPM.

   b. 1750 LPM.

   c. 525 LPM.

   d. 175 LPM.

9. In Problem 8, the total head provided by all the pumps together will be:

   a. 30 m.

   b. 10 m.

   c. 1750 m.

   d. None of the above.

10. A fan used to dry the bed of manure in a barn is found to draw 1.25 HP of power. It was decided to double the fan speed to shorten the drying time. The new power requirement will be:

    a. 20 HP.

    b. 2.5 HP.

    c. 5 HP.

    d. 10 HP.

Practice Problems for the PE Exam

1. Determine the power requirement for an axial pump for discharging 2000 LPM of water against a head of 1.5 m assuming a 65% efficiency.

2. 8000 liters of biodiesel (density = 850 kg/m^3) is to be pumped via a centrifugal pump to a tank located 5 m above the suction level. An existing pump of 1500 W is available. If the efficiency of the pump is 50%, how long will the pump take to complete the task assuming the major and minor losses in the pipe are negligible.

3. It was proposed to pump biodiesel at 200 LPM against a total head of 12 m using a centrifugal pump whose specific speed was given as 0.55. What should be the speed of the impeller needed for this task?

4. A centrifugal pump is pumping seawater (density = 1020 kg/m^3) open to atmosphere through a 15-cm diameter pipe ($f = 0.02$) to a receiving station open

to atmosphere located 100 m away against vertical head of 20 m. Determine the system characteristic curve for this pump.

5. Why is it important to avoid cavitation during pumping? Can cavitation occur while pumping seawater? In addition, could cavitation occur in fans?

6. Determine the NPSH for a pump moving 2000 LPM of water between two streams (at atmospheric level) through a 200-m long pipe ($f = 0.05$) of 20-cm diameter neglecting minor losses. It is also given that the pump is at the same level as that of the water being pumped.

7. A double acting reciprocating pump is used to pump biodiesel (density = 900 kg/m^3) to an overhead tank located 5 m above the source. If the pump has a piston diameter of 15 cm and a stroke length of 40 cm, determine the mass flow rate of the biodiesel at 500 rpm and the power drawn by the pump at a 67% efficiency.

8. A centrifugal pump (impeller diameter = 12.7 cm) operated at 1150 rpm was tested in a laboratory and the following data was collected. Construct a pump performance curve for a geometrically similar pump with a 1700 rpm. Also, determine how much excess power is drawn by the impeller.

[Head (m), Q (LPM)] = [27, 0], [24, 95], [18, 220], [7, 320], [3, 400], [1, 450].

9. A centrifugal fan of 75-cm diameter operating at 1000 rpm moves 5 m^3/s against a head of 2 m. What will be the effect of increasing the impeller diameter to 1 m and speed to 1600 rpm on the fan performance?

CHAPTER 11

Fundamentals of Mass Transfer

Chapter Objectives

The content presented in this chapter will equip students to

- Estimate the diffusive mass transfer rates for simple two-component systems.
- Understand the mathematical similarities between heat transfer and mass transfer.
- Determine the convective mass transfer coefficients and rates for a given system.

11.1 Motivation

Several natural and engineered processes involve transport of mass of a given species due to concentration gradient. For instance, movement of ammonia and odors from an animal housing operation, drying of foods, and absorption of water by the roots of a plant all involve mass transfer. As engineers, we are expected to determine the rates of transport of a given species and design systems to optimize the mass transfer rates. Fortunately, because of the similarity of the transport mechanisms, the mathematical treatment of mass transfer problems is similar to those of heat transfer and we can therefore directly extend the concepts developed in heat transfer to mass transfer problems. The goal of this chapter is to introduce the basic principles of mass transfer and appreciate the mathematical similarities between heat and mass transfer.

11.2 Mathematical Description of Mass Transfer (Fick's Law of Diffusion)

To illustrate the idea of mass transfer, let us consider a room with people. Now, we will spray a perfume at one end of the room (Fig. 11.1) forming a perfume cloud. As we can imagine, the molecules of the perfume will travel from the location where it was sprayed (perfume cloud) to all the locations in the room due to the concentration gradient. In addition, the molecules will travel faster if the available area between the perfume cloud and the people is larger. Further, it is also easy to recognize that the people nearest to the perfume cloud will smell it faster than the people sitting farther away.

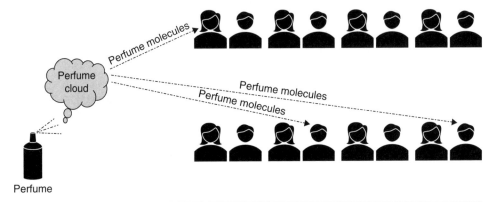

FIGURE 11.1 Simplified illustration of the diffusional mass transfer mechanism wherein perfume molecules travel from the source (perfume cloud) to the destination (people in the room) due to concentration gradient.

Therefore, if we define $\dot{m}$ as the rate of mass transfer of these perfume molecules, our observations can be mathematically summarized as

$$\dot{m} \propto \text{Concentration gradient} \tag{11.1}$$

$$\dot{m} \propto \text{Area available for trasnport} \tag{11.2}$$

$$\dot{m} \propto \frac{1}{\text{distance}} \tag{11.3}$$

If we repeat this experiment with different molecular species (perfume vs. after-shave vs. acetone [a major ingredient in nail polish remover]), we will observe the same trend (Eqs. (11.1)–(11.3)). However, it is important to note that the rate of mass transfer will be different for each species in a given medium (air in this case) suggesting that the mass transfer rate also strongly depends on the inherent transport property of the given species in a medium. That inherent transport property is called mass diffusivity or diffusion coefficient or simply diffusivity and is denoted by D_{12}. It is noted that the subscript 12 represents the diffusion of mass from species 1 to species 2 (or the medium).

Now combining the above ideas, the rate of mass transfer via diffusion for a given species 1, is expressed as

$$\dot{m}_D = -\frac{\text{Diffusivity}_{12} \times \text{Area} \times \text{Concentration gradient}}{\text{Distance}} \tag{11.4}$$

or

$$\dot{m}_D = -D_{12} A \frac{dC_1}{dx} \tag{11.5}$$

where $\dot{m}_D$ is the rate of mass transfer via diffusion $\left(\dfrac{\text{kg}}{\text{s}} \text{ or } \dfrac{\text{kmol}}{\text{s}}\right)$, D_{12} is the mass diffusivity of species 1 into species 2 $\left(\dfrac{\text{m}^2}{\text{s}}\right)$, dC_1 is the concentration gradient $\left(\dfrac{\text{kg}}{\text{m}^3} \text{ or } \dfrac{\text{kmol}}{\text{m}^3}\right)$, and dx is the transport distance (m). The negative sign is included to account for the negative concentration gradient (similar to Fourier's law of heat conduction).

The above equation is called the famous Fick's first law of diffusion.

It is important to understand that the processes of mass transfer occur in liquids and solids as well. For example, let us consider a container with water. Now, if we gently add a few drops of red food color, we can see that the food color will begin to travel from the top of the container to the bottom (the color of the water column will gradually change from colorless to red). Given enough time (usually several hours), the entire water will turn red.

Mass transfer between different phases is also possible. For instance, consider a rubber balloon that is filled with helium. At the beginning the balloon is lighter than air and therefore tends to rise. After several hours, the helium molecules (gas) *"escape"* (they are actually transported via the wall of the rubber [solid] to the outside atmosphere, while the atmospheric gases [nitrogen and oxygen] enter the balloon via the same wall thereby establishing the concentration [and pressure] equilibrium resulting in dropping the balloon to the floor). This type of mass transfer is called **gas-solid** mass transfer. Several other examples of various inter-phase mass transfer can be seen in our daily lives and the students are strongly encouraged to look out for these processes.

In Eq. (11.5) the concentration is expressed on a mass basis. However, concentrations are also expressed on a mass fraction (w) or mole fraction (y) basis. Fick's first law therefore is also expressed as

$$\dot{m}_D = -\rho D_{12} A \frac{dw_1}{dx} \text{ (on a mass fraction basis)} \tag{11.6}$$

$$\dot{m}_D = -C D_{12} A \frac{dy_1}{dx} \text{ (on a mass fraction basis)} \tag{11.7}$$

where $\rho \left(\frac{kg}{m^3} \right)$ is the density and $C \left(\frac{kmol}{m^3} \right)$ is the total concentration of the species.

11.3 The Concept of Mass Diffusivity (D)

The mass diffusivity is an important inherent property of a given material. It governs the speed at which mass will diffuse from one species to another. It is numerically equivalent to the rate of mass transfer that occurs due to a unit concentration gradient separated by a unit distance across a unit area of cross section. Therefore,

$$\text{when } A = 1 \text{ unit} = \Delta C = \Delta x$$

Equation (11.5) becomes

$$\dot{m}_D = D_{12} \text{ (ignoring the negative sign)} \tag{11.8}$$

Careful inspection of Eq. (11.5) reveals that the unit of mass diffusivity is given by m^2/s, which is the same as the kinematic viscosity or momentum diffusivity (v) and thermal diffusivity (α). Therefore, each of these properties can be thought of playing similar roles in their respective transports.

The values of mass diffusivities are measured via systematic experiments. Some of the representative data on mass diffusivity are presented in Tables 11.1. The mass diffusivities increase with an increase in temperature and decrease with an increase in pressure as described in the following equation:

$$D_{1-2} \propto \frac{T^{1.5}}{P} \tag{11.9}$$

Species 1	Species 2	Temperature (°C)	$D_{12}(\times 10^5)$
Air	Ammonia	25	2.6
Air	Ethanol	25	1.2
Air	Oxygen	25	2.1
Air	Water vapor	25	2.5
Oxygen	Ammonia	20	2.5
Oxygen	Water vapor	25	2.5
Carbon dioxide	Water vapor	25	1.6
Species 1	**Species 2**	**Temperature (°C)**	$D_{12}(\times 10^9)$
Ammonia	Water	22	1.6
Carbon dioxide	Water	25	2.0
Chlorine	Water	22	1.4
Ethanol	Water	15	1.0
Glucose	Water	25	0.69
Oxygen	Water	25	2.4
Methane	Water	20	1.5
Species 1	**Species 2**	**Temperature (°C)**	$D_{12}(\times 10^{10})$
Carbon dioxide	Rubber (natural)	25	1.1
Oxygen	Rubber (natural)	25	2.1
Hydrogen	Iron	25	0.0026
Helium	Pyrex	20	0.000045
Carbon	γ-Iron	500	0.00005
Zinc	Copper	500	4×10^{-8}

TABLE 11.1   Mass Diffusivities of Dilute Gases, Liquids, and Solids at 1 atm (data compiled from Cengel & Ghajar, Heat and Mass Transfer: Fundamentals and Applications, Fifth Edition. McGraw Hill, 2016. Reprinted with permission.)

Several biological and agricultural engineering situations involve the transport of moisture into the air. Fortunately, Marerro and Manson (1972) developed a simple yet very useful relationship to determine the mass diffusivity of water in air for a temperature range of 7–177°C.

$$D_{\text{water–air}} = 1.87 \times 10^{-10} \frac{T^{2.072}}{P} \frac{\text{m}^2}{\text{s}} \tag{11.10}$$

where T is the temperature (K) and P is the pressure (atm).

Example 11.1   Determine the mass diffusivity of water in air in Denver, Colorado (0.88 atm) on a given summer day (35°C).

Step 1: Given data
Pressure in Denver, CO, $P = 0.88$ atm; Temperature, $T = 35$°C

To be determined: Mass diffusivity of water in air, $D_{\text{water in air}}$.

Assumptions: The temperature and the pressure remain constant.

Approach: We will use the Marerro and Manson correlation to determine the mass diffusivity of water in air.

Step 2: Calculations
The mass diffusivity can be determined as below:

$$D_{\text{water-air}} = 1.87 \times 10^{-10} \frac{T^{2.072}}{P} \tag{E11.1A}$$

Substituting the given data into Eq. (E11.1A) we obtain

$$D_{\text{water-air}} = 1.87 \times 10^{-10} \frac{(35+273)^{2.072}}{0.88} = 3.04 \times 10^{-5} \frac{\text{m}^2}{\text{s}}$$

11.4 Similarity with Heat Transfer

By now you must have realized that the diffusion equation (Fick's law) is remarkably similar to that of the conduction heat transfer equation (Fourier's law). Just like heat conduction, mass also moves from high concentration to low concentration (until an equilibrium is attained), and its rate of transport depends on the area and the inverse of the distance. Therefore, the mathematical equations describing heat conduction and diffusional mass transfer are similar (Fig. 11.2) and expressed in general terms as below:

$$\dot{q} = -kA \frac{dT}{dx} \quad \text{(Fourier's law of heat conduction)} \tag{11.11}$$

$$\dot{m} = -D_{12} A \frac{dC}{dx} \quad \text{(Fick's law of mass transfer)} \tag{11.12}$$

11.5 One-Dimensional Steady-State Diffusional Mass Transfer

Now we shall analyze the mass transfer for standard geometric configurations. We will first begin with a rectangular section and extend the same principles to circular and cylindrical sections.

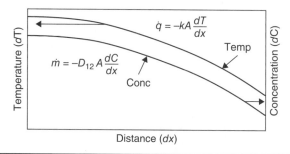

FIGURE 11.2 Graphical representation of conduction heat transfer and diffusional mass transfer.

11.5.1   Mass Transfer through Rectangular Section

Imagine a simple rectangular section (a plain wall that we call as species "j") of thickness "a" and an area "A" subjected to a concentration of a given species "i" on either side. For the sake of discussion, we will also assume that the left face of the wall is exposed to a higher concentration ($C_{i,L}$) while the right side is exposed to a lower concentration ($C_{i,R}$) (Fig. 11.3).

Now consider a small elemental section of thickness dx across which a concentration gradient dC_i exists due to movement of species i through the section. Now, if we assume a constant diffusivity of species i into species j, D_{ij}, we can mathematically express the mass transfer as

$$\dot{m}_i = -D_{ij} A \frac{dC_i}{dx} \tag{11.13}$$

The above equation can be solved by rearranging and applying the appropriate boundary conditions:

$$\frac{\dot{m}_i}{D_{ij}A} dx = -dC_i \tag{11.14}$$

Integrating and applying the limits $x = 0$, $C_i = C_{i,L}$ and $x = a$, $C_i = C_{i,R}$

$$\frac{\dot{m}_1}{D_{ij}A} \int_0^a dx = -\int_{C_{i,L}}^{C_{i,R}} dC_i \tag{11.15}$$

After simplifying, we obtain

$$\dot{m}_i = \frac{D_{ij}A(C_{i,L} - C_{i,R})}{a} \tag{11.16}$$

The above equation may also be written in terms of mass transfer resistance form (Eq. 11.17). The numerator in Eq. (11.17) is viewed as the driving force that facilitates the transfer of mass from the left face of the wall to the right face of the wall, while the denominator is the opposing force (i.e., resistance) to the mass transfer that depends on the wall thickness, area, and the diffusivity of the species being transported.

$$\dot{m}_i = \frac{(C_{i,L} - C_{i,R})}{\left(\dfrac{a}{D_{ij}A}\right)} = \frac{\text{Driving force}}{\text{Resistance force}} \tag{11.17}$$

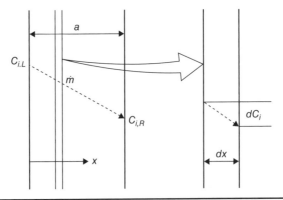

Figure 11.3   Schematic of a plain wall subjected to a concentration gradient of species i.

Example 11.2 (Bioprocessing Engineering—Moisture transport through a composite): A dairy manufacturer packs 700 kg of milk powder (4% moisture) in a square box made of a plastic composite material of dimension 1 m × 1 m × 1 m and 1-cm thick and stores such boxes in a warehouse maintained at 15°C with a relative humidity of 50%. How long can the boxes be safely stored if the maximum moisture content cannot exceed 5% assuming a diffusivity of water into plastic composite is $7 \times 10^{-7} \frac{m^2}{s}$ and a relative humidity inside the box as 5%?

Step 1: Schematic
The problem is described in Fig. 11.4.

Step 2: Given data
Total surface area available for mass transfer = 6 m²; Thickness of the box = 1 cm; Relative humidity outside the box (RH_o) = 50%; Relative humidity inside the box (RH_i) = 5%; Temperature, T = 15°C; Diffusivity of water into composite $\left(D_{H_2O-composite} = 7 \times 10^{-7} \frac{m^2}{s} \right)$; Initial moisture content (MC_i) = 4%; and final (maximum) moisture content (MC_{final}) = 5%

To be determined: Safe storage time.

Assumptions: (a) Ideal gas laws are valid, (b) the milk powder is hygroscopic and absorbs all the water, (c) the concentration of moisture in the air is the same as that of on the surface of the box, (d) saturated conditions exist, and (e) the universal gas constant, $R = 8.31 \frac{m^3 Pa}{mol\ K}$.

Total mass of water gained by the milk powder is 5% of 700 kg – 4% of 700 kg = 7 kg.

This is the total water that can be safely transported to minimize the loss of quality of the dry milk. Now, let us determine the rate of water transfer into the box using the following equation:

$$\dot{m}_{H_2O} = \frac{D_{H_2O-composite} A (C_{H_2O,o} - C_{H_2O,i})}{a} \qquad \text{(E11.2A)}$$

We can estimate the concentrations of water inside and outside the box using the ideal gas law and steam tables at 15°C as follows:

$$P_{H_2O,o} = RH_o \times P_{sat} = 0.5 \times 1.70 = 0.85 \text{ kPa}$$

Similarly,

$$P_{H_2O,i} = RH_i \times P_{sat} = 0.05 \times 1.70 = 0.085 \text{ kPa}$$

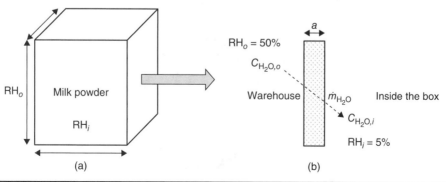

FIGURE 11.4 Schematic of the box containing the milk powder (a) subjected to a differential relative humidity resulting in mass transfer of water from the warehouse into the box (b).

From the ideal gas law,

$$PV = nRT \qquad \text{(E11.2B)}$$

$$\frac{n}{V} = C = \frac{P}{RT} \qquad \text{(E11.2C)}$$

Therefore,

$$\frac{n_{H_2O,o}}{V} = C_{H_2O,o} = \frac{P_{H_2O,o}}{RT} \qquad \text{(E11.2D)}$$

$$C_{H_2O,o} = \frac{0.85}{8.31 \times (273+15)} = 3.55 \times 10^{-4} \frac{\text{kmol}}{\text{m}^3}$$

Or, on a mass basis,

$$C_{H_2O,o} = 3.55 \times 10^{-4} \frac{\text{kmol}}{\text{m}^3} \times 18 \frac{\text{kg}}{\text{kmol}} = 6.39 \times 10^{-3} \frac{\text{kg}}{\text{m}^3}$$

Similarly,

$$C_{H_2O,i} = \frac{0.085}{8.31 \times (273+15)} = 3.55 \times 10^{-5} \frac{\text{kmol}}{\text{m}^3}$$

Or, on a mass basis

$$C_{H_2O,i} = 6.39 \times 10^{-4} \frac{\text{kg}}{\text{m}^3}$$

Now,

$$\dot{m}_{D,H_2O} = \frac{7 \times 10^{-7} \times 6 \times (6.39 \times 10^{-3} - 6.39 \times 10^{-4})}{0.01}$$

$$\dot{m}_{D,H_2O} = 2.41 \times 10^{-6} \frac{\text{kg}}{\text{s}}$$

Thus, the maximum storage time for the boxes for limiting the transport of 7 kg of water will be = $\frac{7}{2.55 \times 10^{-6}} = 2.89 \times 10^6$ s or 33.52 days.

11.6   Mass Transfer through Spherical Section

Now, let us choose a spherical-shaped body with inner and outer radii as r_i and r_o, respectively (Fig. 11.5) that is subjected to a concentration gradient of C_i and C_o such that $C_i > C_o$.

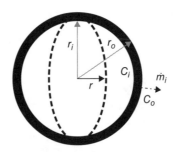

FIGURE 11.5   Schematic of a spherical shell subjected to a concentration gradient of species i.

Using Fick's law, we express the rate of mass transfer at a distance r from the center as

$$\dot{m}_i = -D_{ij} A \frac{dC}{dr} \tag{11.18}$$

After expressing the area of the sphere as $A = 4\pi r^2$, the above equation becomes

$$\dot{m}_i = -D_{ij} 4\pi r^2 \frac{dC}{dr} \tag{11.19}$$

$$\frac{\dot{m}_i}{4\pi D_{ij}} \left(\frac{dr}{r^2} \right) = -dC \tag{11.20}$$

Integrating, applying the limits $r = r_i$, $C = C_1$ and $r = r_o$, $C = C_o$ will yield

$$\frac{\dot{m}_i}{4\pi D_{ij}} \int_{r_i}^{r_o} \left(\frac{dr}{r^2} \right) = -\int_{C_i}^{C_o} dC \tag{11.21}$$

which after simplification becomes

$$\dot{m}_i = \frac{4\pi r_i r_o D_{ij}(C_i - C_o)}{r_o - r_i} \tag{11.22}$$

The equation can also be expressed in terms of mass transfer resistance as follows:

$$\dot{m}_{D,i} = \frac{(C_i - C_o)}{\left(\dfrac{r_o - r_i}{4\pi r_i r_o D_{ij}} \right)} = \frac{\text{Driving force}}{\text{Resistance force}} \tag{11.23}$$

Example 11.3 (Agricultural Engineering—Oxygen loss from a storage tank): Oxygen is stored under a pressure of 4 atm at 27°C in a copper spherical tank of 3-m diameter and 4-cm thick. If the solubility of concentration of oxygen in copper is 0.0003 $\dfrac{\text{kmol}}{\text{m}^3\text{atm}}$, determine the mass of oxygen lost over a 30-day period assuming the diffusivity of oxygen in copper at 27°C to be $5 \times 10^{-11} \dfrac{\text{m}^2}{\text{s}}$.

Step 1: Schematic
The problem is shown in Fig. 11.6.

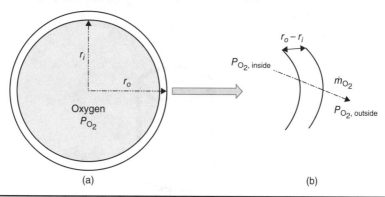

FIGURE 11.6 Schematic of the spherical shell containing the pressurized oxygen (a) subjected to a differential concentration gradient resulting in mass transfer of oxygen from within the shell to the outside the shell (b).

Step 2: Given data

Inner radius (r_i) of the sphere = 1.5 m; Outer radius (r_o) of the sphere = 1.54 m;

Pressure of oxygen in the shell ($P_{O_2,i}$) = 4 atm; Pressure of oxygen outside the shell ($P_{O_2,o}$) = 0.21 atm;

Temperature, $T = 27°C$; Diffusivity of oxygen into copper $\left(D_{O_2-Cu} = 5 \times 10^{-11} \dfrac{m^2}{s} \right)$;

Solubility of oxygen in copper $\left(S_{O_2-Cu} = 0.0003 \dfrac{kmol}{m^3 atm} \right)$.

To be determined: Total loss of oxygen in 30 days.

Step 3: Calculations

Assumptions: (a) Ideal gas laws are valid and (b) the pressure of oxygen outside the tank is 0.21 atm.

The rate of diffusional mass transfer of oxygen through the copper walls of the spherical shell can be determined by the following governing equation:

$$\dot{m}_{O_2} = \frac{4\pi r_i r_o D_{O_2-Cu}\left(C_{O_2,i} - C_{O_2,o} \right)}{r_o - r_i} \tag{E11.3A}$$

Expressing concentrations in terms of solubility and pressure will convert the governing equation into

$$\dot{m}_{O_2} = \frac{4\pi r_i r_o D_{O_2-Cu} S_{O_2-Cu}\left(P_{O_2,i} - P_{O_2,o} \right)}{r_o - r_i} \tag{E11.3B}$$

Substituting the values in the above equation will result in

$$\dot{m}_{O_2} = \frac{4 \times 3.14 \times 1.5 \times 1.54 \times 5 \times 10^{-11} \times 0.0003 \times (4 - 0.21)}{1.54 - 1.5} = 4.12 \times 10^{-11} \frac{kmol}{s}$$

In terms of mass loss, it will be $4.12 \times 10^{-11} \dfrac{kmol}{s} \times 32 \dfrac{kg}{kmol} = 1.32 \times 10^{-9} \dfrac{kg}{s}$

Therefore, in 30 days, the total loss of oxygen from the shell is $1.32 \times 10^{-9} \dfrac{kg}{s} \times 60 \times 60 \times 24 \times 30 \ s = 0.0034 \ kg = 3.4 \ g$.

11.7   Mass Transfer through Cylindrical Section

The above-described approach can also be extended to a cylindrical section shown in Fig. 11.7 with inner and outer radii as r_i and r_o, respectively that is subjected to a concentration gradient of C_i and C_o such that $C_i > C_o$.

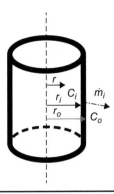

FIGURE 11.7   Schematic of a cylinder subjected to a concentration gradient of species i.

The rate of mass transfer ($\dot{m}_{D,i}$) for species, i can be written using Fick's law as follows:

$$\dot{m}_i = -D_{ij} A \frac{dC}{dr} \tag{11.24}$$

or

$$\dot{m}_i = -D_{ij} 2\pi rL \frac{dC}{dr} \tag{11.25}$$

$$\frac{\dot{m}_i}{2\pi LD_{ij}} \frac{dr}{r} = dC \tag{11.26}$$

Integrating, applying the limits $r = r_i$, $C = C_i$ and $r = r_o$, $C = C_o$ will yield

$$\frac{\dot{m}_i}{2\pi LD_{ij}} \int_{r_i}^{r_o} \frac{dr}{r} = -\int_{C_i}^{C_o} dC \tag{11.27}$$

which after simplification and rearrangement becomes

$$\dot{m}_i = \frac{2\pi LD_{ij}(C_i - C_o)}{\ln\left(\dfrac{r_o}{r_i}\right)} \tag{11.28}$$

Or in terms of resistance,

$$\dot{m}_i = \frac{(C_i - C_o)}{\left(\dfrac{\ln\left(\dfrac{r_o}{r_i}\right)}{2\pi LD_{ij}}\right)} = \frac{\text{Driving force}}{\text{Resistance force}} \tag{11.29}$$

Example 11.4 (Environmental Engineering—Diffusional mass transfer through a pipe): A manure-to-biogas facility is pumping biogas (a mixture of 70% methane and 30% carbon dioxide by weight) in 1-cm thick PVC pipe of 25-cm inner radius under a pressure of 2 atm at an average temperature of 25°C (Fig. 11.8). Estimate the rate of loss of methane and carbon dioxide from the pipe per unit length, assuming the diffusivities of methane and carbon dioxide at 25°C to be 2×10^{-13} and 3×10^{-12}, respectively.

Step 1: Schematic
The problem is shown in Fig. 11.8.

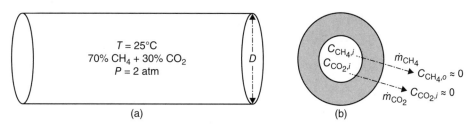

(a) (b)

FIGURE 11.8 Schematic of the PVC pipe containing the pressurized biogas (methane and carbon dioxide) (a) subjected to a differential concentration gradients resulting in mass transfer of methane and carbon dioxide from within the pipe to the outside environment (b).

Step 2: Given data

Inner radius (r_i) of the pipe = 25 cm; Outer radius (r_o) of the pipe = 26 cm; Length, L = 1 m;

Total pressure in the pipe (P_T) = 2 atm = 202.6 kPa; Weight percentage of methane = 70%;

Weight percentage of carbon dioxide = 30%; Temperature, T = 25°C;

Diffusivity of methane into PVC $\left(D_{CH_4-PVC} = 2 \times 10^{-13} \dfrac{\text{m}^2}{\text{s}} \right)$; Diffusivity of carbon dioxide into PVC

$\left(D_{CO_2-PVC} = 3 \times 10^{-12} \dfrac{\text{m}^2}{\text{s}} \right)$

To be determined: Rate of mass transfer of methane and carbon dioxide through the walls of the PVC pipe.

Assumptions: (a) Ideal gas laws are valid and (b) the concentrations of methane and carbon dioxide outside the pipe are negligible.

Step 3: Calculations

Mass Transfer Rate of Methane

The rate of mass transfer of methane can be determined by the following equation.

$$\dot{m}_{CH_4} = \frac{2\pi L D_{CH_4-PVC}\left(C_{CH_4,i} - C_{CH_4,o}\right)}{\ln\left(\dfrac{r_o}{r_i}\right)} \qquad (E11.4A)$$

To solve the above equation, we need to estimate the concentration of methane. We are given that methane is 70% by weight and carbon dioxide is 30% by weight. Therefore, the mass fraction of methane is

$$x_{methane,i} = \frac{\text{mass of methane}}{\text{mass of methane} + \text{mass of carbon dioxide}} \qquad (E11.4B)$$

Symbolically,

$$x_{CH_4,i} = \frac{m_{CH_4}}{m_{CH_4} + m_{CO_2}} = \frac{70}{70+30}$$

Converting mass into moles, we obtain

$$y_{CH_4,i} = \frac{\left(70/16\right)}{\left(70/16\right) + \left(30/44\right)} = 0.865$$

The partial pressure of methane in the pipe,

$$P_{CH_4,i} = y_{CH_4} \times P_T \qquad (E11.4C)$$

Substituting the relevant values in the above equation will yield

$$P_{CH_4,i} = 0.865 \times 202.6 \text{ kPa} = 175.28 \text{ kPa}$$

Now, we will use the ideal gas law to determine the concentration of methane gas in the pipe.

$$P_{CH_4,i} V = n_{CH_4} RT \qquad (E11.4D)$$

$$\frac{n_{CH_4,i}}{V} = C_{CH_4,i} = \frac{P_{CH_4,i}}{RT} \qquad (E11.4E)$$

$$C_{CH_4,i} = \frac{P_{CH_4,i}}{RT} = \frac{175.28}{8.314 \times (273+25)} = 0.07 \frac{\text{kmol}}{\text{m}^3}$$

Expressing in terms of mass concentration,

$$C_{CH_4,i} = 0.07 \frac{kmol}{m^3} \times 16 \frac{kg}{kmol} = 1.13 \frac{kg}{m^3}$$

Now, we will substitute all the values into Eq. (14.3, Chapter 14) to determine the rate of methane transfer per unit length of the PVC pipe:

$$\dot{m}_{D,CH_4} = \frac{2 \times 3.14 \times 1 \times 2 \times 10^{-13} \times (1.13 - 0)}{\ln\left(\dfrac{0.26}{0.25}\right)} = 3.62 \times 10^{-11} \frac{kg}{s}$$

The diffusional mass transfer rate of methane through the walls of the PVC pipe is determined to be $3.62 \times 10^{-11} \frac{kg}{s}$.

Mass Transfer Rate of Carbon Dioxide

We can repeat the same process to determine the mass transfer rate of carbon dioxide. The relevant steps are summarized as follows:

$$y_{CO_2,i} = \frac{\left(0.3 / 44\right)}{\left(0.7 / 16\right) + \left(0.3 / 44\right)} = 0.135$$

Note: Alternatively, we can calculate the $y_{CO_2,i} = 1 - y_{CH_4,i} = 1 - 0.865 = 0.135$

$$P_{CO_2,i} = y_{CO_2,i} \times P_T = 0.135 \times 202.6 \text{ kPa} = 27.31 \text{ kPa}$$

$$C_{CO_2,i} = \frac{P_{CO_2,i}}{RT} \times MW_{CO_2} = \frac{27.31}{8.314 \times (273 + 25)} \times 44 \frac{kg}{kmol} = 0.485 \frac{kg}{m^3}$$

$$\dot{m}_{CO_2} = \frac{2\pi L D_{CO_2-PVC}\left(C_{CO_2,i} - C_{CO_2,o}\right)}{\ln\left(\dfrac{r_o}{r_i}\right)} = \frac{2 \times 3.14 \times 1 \times 3 \times 10^{-12} \times (0.485 - 0)}{\ln\left(\dfrac{0.26}{0.25}\right)} = 2.33 \times 10^{-10} \frac{kg}{s}$$

A direct comparison of one-dimensional conduction heat transfer and diffusive mass transfer for common geometries is presented in Table 11.2.

Configuration	1-D Heat Transfer	1-D Mass Transfer
Rectangle General form	$\dot{q} = \dfrac{kA(T_L - T_R)}{a}$	$\dot{m}_i = \dfrac{D_{ij}A(C_{i,L} - C_{i,R})}{a}$
Resistance form	$\dot{q} = \dfrac{(T_L - T_R)}{\left(\dfrac{a}{kA}\right)}$	$\dot{m}_i = \dfrac{(C_{i,L} - C_{i,R})}{\left(\dfrac{a}{D_{ij}A}\right)}$
Cylinder General form	$\dot{q} = \dfrac{4\pi r_i r_o k(T_i - T_o)}{r_o - r_i}$	$\dot{m}_i = \dfrac{4\pi r_i r_o D_{ij}(C_i - C_o)}{r_o - r_i}$
Resistance form	$\dot{q} = \dfrac{(T_i - T_o)}{\left(\dfrac{r_o - r_i}{4\pi r_i r_o k}\right)}$	$\dot{m}_i = \dfrac{(C_i - C_o)}{\left(\dfrac{r_o - r_i}{4\pi r_i r_o D_{ij}}\right)}$

TABLE 11.2 Summary of Equations for One-Dimensional Heat and Mass Transfer in General and Resistance Forms

Configuration	1-D Heat Transfer	1-D Mass Transfer
Cylinder General form	$$\dot{q} = \dfrac{2\pi Lk(T_i - T_o)}{\ln\left(\dfrac{r_o}{r_i}\right)}$$	$$\dot{m}_i = \dfrac{2\pi LD_{ij}(C_i - C_o)}{\ln\left(\dfrac{r_o}{r_i}\right)}$$
Resistance form	$$\dot{q} = \dfrac{(T_i - T_o)}{\left(\dfrac{\ln\left(\dfrac{r_o}{r_i}\right)}{2\pi Lk}\right)}$$	$$\dot{m}_i = \dfrac{(C_i - C_o)}{\left(\dfrac{\ln\left(\dfrac{r_o}{r_i}\right)}{2\pi LD_{ij}}\right)}$$

TABLE 11.2   Summary of Equations for One-Dimensional Heat and Mass Transfer in General and Resistance Forms (*Continued*)

11.8   One-Dimensional Mass Transfer through Composite Sections

It is common to encounter situations where multiple layers of dissimilar materials of different thickness and diffusivities are joined together to form composite materials. For example, the cardboard cartons used for packaging foods are lined with plastic layers to minimize transfer of oxygen and moisture within the carton. Similarly, the walls in our houses are fabricated by joining vapor barriers, insulation, drywalls, and sometimes a layer of bricks. In such cases, we can simply extend the ideas from thermal resistance network to determine the mass transfer.

Let us consider a rectangular composite section made of three different materials, viz, 1, 2, and 3 subjected to a concentration gradient (of species j) of $C_{j,L} - C_{j,R}$ on the left and right sides of the section as shown in Fig. 11.9. Let the thicknesses and the

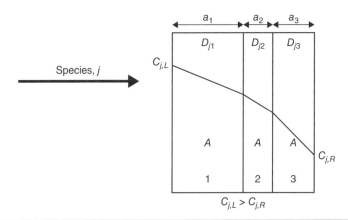

FIGURE 11.9   Schematic of one-dimensional diffusional mass transfer through a composite section subjected to a concentration gradient.

diffusivities be a_1, a_2, a_3 and D_{j1}, D_{j2}, D_{j3}, respectively. The area of all the sections is the same, which we will denote as A.

Because these sections are connected in series, we can simply apply the thermal resistance in series concepts that we discussed in heat transfer. Therefore, the rate of mass transfer across the composite wall in terms of diffusional resistance (R_D) can be written as

$$\dot{m}_{D,j} = \frac{C_{j,L} - C_{j,R}}{\dfrac{a_1}{D_{j1}A} + \dfrac{a_2}{D_{j2}A} + \dfrac{a_3}{D_{j3}A}} \tag{11.30}$$

$$\dot{m}_D = \frac{C_{j,L} - C_{j,R}}{R_{D,1} + R_{D,2} + R_{D,3}} = \frac{\text{Driving force}}{\text{Resistance 1} + \text{Resistance 2} + \text{Resistance 3}} \tag{11.31}$$

or

$$\dot{m}_{D,j} = \frac{\Delta C}{R_{D,\text{Total}}} \tag{11.32}$$

11.9   One-Dimensional Unsteady-State Mass Transfer

In some situations, mass transfer becomes transient (unsteady) and a simple case of one-dimensional unsteady-state mass transfer is expressed as

$$\frac{\partial C_1}{\partial t} = D_{12} \frac{\partial^2 C}{\partial x^2} \tag{11.33}$$

where C_1 is the concentration of the given species 1, D_{12} is the mass diffusivity of species 1 into species 2, x is the thickness of the section (in the x-direction), and t represents the time.

We can immediately see that the above equation is very similar to the one-dimensional transient heat conduction equation. Therefore, we can also use the same approaches to solve for the concentrations at any given time and location within the system. Drawing from heat transfer, we can define three nondimensional parameters for mass transfer (MT) as follows:

Biot number for mass transfer: $\qquad N_{\text{Bi,MT}} = \dfrac{h_{\text{MT}} L_c}{D_{ij}}$ $\qquad\qquad$ (11.34)

Fourier number for mass transfer: $\qquad N_{\text{Fo,MT}} = \dfrac{D_{ij} t}{L_c^2}$ $\qquad\qquad$ (11.35)

Concentration ratio for mass transfer: $\qquad \theta_{\text{MT}} = \dfrac{C_t - C_\infty}{C_i - C_\infty}$ $\qquad\qquad$ (11.36)

The temperature, time, and the mass transfer coefficients for any transient state mass transfer problems can be analyzed using the same graphical and analytical approximation procedures described in Chap. 4.

11.10   Convection Mass Transfer

Similar to convection heat transfer, convection mass transfer occurs whenever there is a flow of fluid between two systems separated by a concentration gradient. To understand this clearly, let us consider the same perfume example that we used in diffusional mass transfer. In the previous case, there was no flow (motion) of fluid (air) and the transport of the perfume molecules from the source (perfume cloud) to the nostrils of people was via random molecular movement due to diffusion. Now imagine a fan blowing air in the room resulting in the movement of the air between the perfume cloud and the people. The moving air significantly enhances the transport of the perfume molecules in the room. Such flow-assisted diffusional mass transfer is called convection mass transfer.

If we observe carefully, convection mass transfer is ubiquitous. The reason we mix sugar in our coffee is to initiate a fluid flow to enhance the transport of sugar to all locations in the cup (to attain a uniform sugar concentration in the coffee cup). Similarly, before the advent of clothes drier, the wet clothes were hung on a line which allowed flow of air for several hours that resulted in the evaporation (transport of moisture from the wet clothes to dry air) of water from the clothes. Other examples include transfer of oxygen from atmosphere to water in open waterbodies and movement of air pollutants from the source to the surroundings.

It is easy to understand the physical process of convection mass transfer. However, it is to be noted that it is extremely difficult to mathematically describe the process of mass transfer. Nonetheless, due to the remarkable physical similarities between convection heat transfer and convection mass transfer, we will use Newton's law of cooling ($\dot{q} = hA(T_s - T_\infty)$) as the fundamental basis and extend its definition to the rate of convection mass transfer. Therefore, we can write the rate of convection mass transfer as

$$\dot{m}_C = h_{MT} A (C_{i,s} - C_{i,\infty}) \tag{11.37}$$

where $\dot{m}_C$ is the rate of convection mass transfer $\left(\dfrac{kg}{s}\right)$ (commonly called mass transfer rate in Biological and Agricultural Engineering literature), A is the mass transfer area (m^2), h_{MT} is the mass transfer coefficient $\left(\dfrac{m}{s}\right)$, $C_{i,s}$ and $C_{i,\infty}$ are the concentrations of the species $\left(\dfrac{kg}{m^3}\right)$ at the surface and a location far away.

11.11   The Concept of the Mass Transfer Coefficient (h_m)

If we study Eq. (11.37), we observe that the concentrations of any given species, i and the area subjected to the mass transfer, A can easily be measured. What makes the mass transfer analysis complicated is the calculation of the mass transfer coefficient. We need to realize that, unlike area and concentration, the mass transfer coefficient depends on the flow regime (laminar or turbulent), physical properties of the system (roughness and characteristic length), fluid properties (density, viscosity, diffusion coefficient), the system configuration (cylinder vs. sphere), and the angle of the flow. Mathematically,

$$h_{MT} = f(v, \mu, \rho, l_c, D_{mn}, \varepsilon, \text{flow regime, geometry, system configuration}) \tag{11.38}$$

Therefore, analytically determining the values of the mass transfer coefficient is quite a task. Hence, we rely on alternative approaches such as dimensional analysis to express mass transfer coefficient as a function of certain variables that can be either measured or estimated as we did in convection heat transfer. To aid in our analysis, we will also define a couple of *dimensionless numbers* as follows.

11.12   Schmidt Number (N_{Sc})

Imagine a convective mass transfer situation (such as movement of ammonia gas through the air from a poultry house to the environment far away from the source) in which the fluid with a viscosity μ and density ρ is transporting a species with a diffusion coefficient of D_{mn}. The Schmidt number is defined as the ratio of the fluid viscosity (or momentum diffusivity) to the diffusion coefficient (or mass diffusivity) and is expressed as

$$N_{Sc} = \frac{\mu}{\rho D_{mn}} = \frac{v}{D_{mn}} = \frac{\text{diffusion of momentum}}{\text{diffusion of mass}} = \frac{\text{momentum diffusivity}}{\text{mass diffusivity}} \quad (11.39)$$

Another way to understand the physical meaning of the Schmidt number is to use the boundary layer theory. Imagine the same fluid (described before) flowing over a solid body (assumed to be a flat plate for illustration purposes). As the flow progresses, there will be two boundary layers in laminar flow, one associated with the concentration (diffusivity) and the other related to the fluid flow (i.e., viscosity). The Schmidt number compares the relative thicknesses of the concentration boundary layer (CBL) that is represented by the mass diffusivity and the velocity boundary layer (VBL) that is represented by the fluid viscosity (Fig. 11.10). In this circumstance, depending on the relative thickness, there exist three distinct possibilities.

Case (1): VBL-dominated regime (VBL > CBL): When the transfer of momentum is significantly greater than the transfer of mass, the thickness of the VBL will be greater than the thickness of the CBL. In other words, the value of the Schmidt number will be large. The implication of this will be that the moving fluid is easily able to transfer the momentum but not the mass of the species that the fluid is carrying. Revisiting the perfume example, we can conclude that a large Schmidt number will result in decreased rate of transport of the perfume molecules. Mass transfer in liquid–liquid systems are gas–liquid systems generally possess large Schmidt numbers in the range of greater than 100.

Case (2): CBL-dominated regime (CBL > VBL): Now, let us reverse the process described in case (1). When the rate of transfer of mass is larger than the rate of transfer of momentum, the thickness of CBL will be greater than the thickness of VBL resulting in a small value of the Schmidt number. Smaller Schmidt numbers will allow for faster mass transfer and therefore are highly desired in catalytic chemical reactions where the reactants need to reach the catalyst surface. Small Schmidt numbers are usually associated with gaseous systems and range between 0.1 and 4.

Case (3): Normal regime (VBL ≈ CBL): In several situations, the rates of momentum (the thickness of the VBL) and mass transfers (the thickness of CBL) are similar. This means that the value of the Schmidt number is 1.0 or close to 1.0. The implication of this condition is that mass and momentum transfer is optimum and is usually encountered in methane–air, oxygen–air, carbon dioxide–air mass transfer at room temperatures.

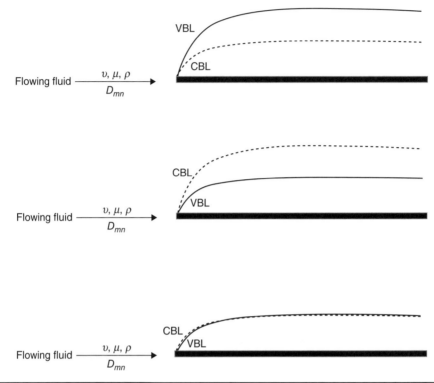

FIGURE 11.10 Illustration of the Schmidt number as a function of CBL and VBL for a flow on a flat plate.

By now, we recognize that the Schmidt number (N_{Sc}) is analogous to the Prandtl number (N_{Pr}) that we extensively used in convection heat transfer.

Example 11.5 (General Engineering–All concentrations): Estimate the Schmidt number for a mass transfer in a two-component system consisting of (a) methane in air, (b) oxygen in water, and (c) ethanol in water at 20°C (Fig. 11.11).

Step 1: Schematic
The problem is described in Fig. 11.11.

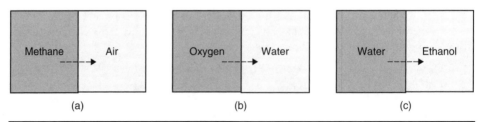

FIGURE 11.11 Schematic for the two-component systems described in Example 11.5.

Step 2: Given data

Temperature, $T = 25°C$. At 25°C, the relevant properties of the given species are tabulated as follows:

Property (units)	Methane	Air	Oxygen	Ethanol	Water
Viscosity ($\times 10^5$ Pa·s)	1.1	1.82	2.04	110	89
Density $\left(\dfrac{kg}{m^3}\right)$	0.66	1.2	1.3	790	1000
Diffusivity $\left(\times 10^9 \dfrac{m^2}{s}\right)$	2100 (in air)	—	2.4 (in water)	1.1 (in water)	—

To be determined: Schmidt numbers for the given two-component systems.

Assumptions: Constant temperature and pressure conditions.

Step 3: Calculations

The equation for determination of Schmidt number is

$$N_{Sc} = \frac{\mu}{\rho D_{12}} = \frac{\nu}{D_{12}} \tag{E11.5A}$$

(a) Methane in air: The Schmidt number for methane–air system is

$$N_{Sc} = \frac{\mu}{\rho D_{12}} = \frac{1.1 \times 10^{-5}}{0.66 \times 2100 \times 10^{-9}} = 7.9$$

(b) Oxygen in water: The Schmidt number for oxygen–water system is

$$N_{Sc} = \frac{\mu}{\rho D_{12}} = \frac{2.04 \times 10^{-5}}{1.3 \times 2.4 \times 10^{-9}} = 6538.4$$

(c) Ethanol in water: The Schmidt number for ethanol–water system is

$$N_{Sc} = \frac{\mu}{\rho D_{12}} = \frac{110 \times 10^{-5}}{790 \times 1.1 \times 10^{-9}} = 1265.2$$

11.13   Sherwood Number (N_{Sh})

We have defined the convection mass transfer as diffusional mass transfer coupled with the bulk flow. In this context, Sherwood number is defined as the ratio of convection mass transfer rate to the diffusive mass transfer rate. Mathematically, Sherwood number is expressed as

$$N_{Sh} = \frac{\text{Convection mass transfer rate}}{\text{Diffuional mass transfer rate}} = \frac{\dot{m}_C}{\dot{m}_D} = \frac{h_{MT} A(C_{i,s} - C_{i,\infty})}{\left(D_{12} A(C_{i,s} - C_{i,\infty})\middle/ a\right)} \tag{11.40}$$

$$N_{Sh} = \frac{h_{MT} a}{D_{12}} \tag{11.41}$$

When expressed in terms of a general characteristic length L_c, the Sherwood number

$$N_{Sh} = \frac{h_{MT} L_c}{D_{12}} \tag{11.42}$$

It is to be noted that the Sherwood number quantifies the effect of bulk flow on mass transfer. When there is no fluid flow, the mass transfer would occur purely via diffusion alone. The flow of fluid on the other hand will increase the overall mass transfer of the system. Therefore, the Sherwood number represents the actual increase in mass transfer due to the flow. For example, a Sherwood number of 100 represents a 100-fold increase in mass transfer (compared to diffusion alone).

Example 11.6 (Bioprocessing Engineering—Increase in mass transfer due to mixing): Pure oxygen (99%) is being injected in an aerobic reactor of 30-cm diameter containing a 50-L dilute solution of microbial culture maintained at 30°C (Fig. 11.12). The reactor is agitated at 300 rpm and previous experience with the reactor mixing conditions suggested a convective mass transfer coefficient of $0.000017 \frac{m}{s}$ at 30°C. Comment on the effect of mixing on the transfer of oxygen into the water if the diffusivity of oxygen in water at 30°C is $2.6 \times 10^{-9} \frac{m^2}{s}$.

Step 1: Schematic
The problem is described in Fig. 11.12.

Step 2: Given data
Diameter of the reactor, $D = L_c = 0.3$ m; Volume of the culture, $V = 50$ L; Temperature, $T = 30°C$; Rate of agitation, $N = 300$ rpm; Convective mass transfer coefficient at 30°C, $h_{MT} = 0.000017$ m/s; mass diffusivity of oxygen in water at 25°C, $D_{Ox\,in\,water} = 2.6 \times 10^{-9} \frac{m^2}{s}$

To be determined: Quantitative effect of agitation.

Assumptions: Constant temperature and pressure conditions within the reactor.

Approach: We can determine the Sherwood number from the given data, which can provide us with the effect of agitation on the mass transfer of oxygen in the water.

Step 3: Calculations
The Sherwood number can be determined via

$$N_{Sh} = \frac{h_{MT} L_c}{D_{12}} \tag{E11.6A}$$

Using the given data, we obtain

$$N_{Sh} = \frac{0.000017 \times 0.3}{2.6 \times 10^{-9}} = 1961.5$$

Therefore, the agitation of the reactants at 300 rpm is enhancing the mass transfer by 1961.5 times (relative to nonagitated system in which mass transfer occurs only via diffusion).

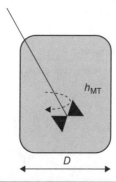

FIGURE 11.12 Schematic of the agitated aerobic microbial reactor described in Example 11.6.

Additionally, because convection mass transfer involves fluid flow (for both free and forced convection) we will use the Grashof number (N_{Gr}) and Reynolds number (N_{Re}) that we defined in convection heat transfer

$$N_{Gr} = \frac{\rho^2 g L_c^3}{\mu^2} \frac{\Delta\rho}{\rho} = \frac{g L_c^3}{v^2} \frac{\Delta\rho}{\rho} \tag{11.43}$$

$$N_{Re} = \frac{\rho v L_c}{\mu} = \frac{v L_c}{v} \tag{11.44}$$

It is noted that in mass transfer applications the Grashof number is modified to include the mass concentration gradient.

In certain cases, we are also interested in the relative magnitudes of heat and diffusional mass transfer. Therefore, we will define another nondimensional number called the Lewis number (N_{Le}) to compare the thermal and mass diffusivities as follows:

$$N_{Le} = \frac{N_{Sc}}{N_{Pr}} = \frac{\left(\dfrac{\mu}{\rho D_{12}}\right)}{\left(\dfrac{\mu}{\rho\alpha}\right)} = \frac{\alpha}{D_{12}} = \frac{\text{Thermal diffusivity}}{\text{Mass diffusivity}} \tag{11.45}$$

11.14 Determination of Mass Transfer Coefficient (h_m)

One of the most important tasks of an engineer is to determine the mass transfer coefficient (and subsequently the mass transfer rate) for a given system. As described earlier, we will employ the Buckingham's π theorem to develop a mathematical relation between several nondimensional variables that govern the mass transfer process for a forced convection. At this time, we will not worry about the system geometry as we are mainly interested in the more general relationship. Thus, we express the mass transfer coefficient as

$$h_{MT} = f(\rho,\ \mu,\ L_c,\ v,\ D_{12}) \tag{11.46}$$

The above function has 6 variables and 3 primary dimensions (mass [M], length [L], and time [T]) (Table 11.3).

# of variables, $n = 6$; # of primary dimensions, $m = 3$; and therefore the number of nondimensional parameters (π terms) $= n - m = 6 - 3 = 3$.

Now, choosing ρ, μ, and L_c as repeating variables and v, D_{12}, and h_{MT} as nonrepeating variables, we write the three π terms as

$$\pi_1 = \rho^a \mu^b L_c^c v \tag{11.47}$$

$$\pi_2 = \rho^a \mu^b L_c^c D_{12} \tag{11.48}$$

$$\pi_3 = \rho^a \mu^b L_c^c h_{MT} \tag{11.49}$$

Expressing each π term in its fundamental dimensions, we obtain

$$\pi_1 = M^0 L^0 T^0 = (ML^{-3})^a (ML^{-1}T^{-1})^b (L)^c (LT^{-1}) \tag{11.50}$$

Variable	Unit	Dimension
Density (ρ)	$\dfrac{\text{kg}}{\text{m}^3}$	ML^{-3}
Viscosity (μ)	Pa · s	$ML^{-1}T^{-1}$
Characteristic length (L_c)	m	L
Velocity (v)	$\dfrac{\text{m}}{\text{s}}$	LT^{-1}
Diffusivity (D_{12})	$\dfrac{\text{m}^2}{\text{s}}$	L^2T^{-1}
Mass transfer coefficient (h_{MT})	$\dfrac{\text{m}}{\text{s}}$	LT^{-1}

TABLE 11.3 List of Variables and Their Primary Dimensions Employed in Convective Mass Transfer

Equating the exponents

$$0 = a + b \tag{11.51}$$

$$0 = -3a - b + c + 1 \tag{11.52}$$

$$0 = -b - 1 \tag{11.53}$$

Solving will yield,

$$a = 1,\ b = -1,\ \text{and } c = 1 \tag{11.54}$$

Substituting the values of a, b, and c in Eq. (E11.47, Chapter 11) will result in

$$\pi_1 = \frac{\rho v L_c}{\mu} = N_{\text{Re}} \tag{11.55}$$

Similarly, we repeat the same process for solving Eqs. (14.48) and (14.49) to obtain π_2 and π_3

$$\pi_2 = \rho^a \mu^b L_c^c D_{12} \tag{11.56}$$

$\pi_2 = M^0 L^0 T^0 = (ML^{-3})^a (ML^{-1}T^{-1})^b (L)^c (L^2 T^{-1})$ resulting in $a = 1$, $b = -1$, and $c = 0$

$$\pi_2 = \frac{\rho D_{12}}{\mu} = \frac{D_{12}}{v} = \frac{1}{N_{\text{Sc}}} \tag{11.57}$$

Since π_2 is a dimensionless parameter, the inverse of π_2 is also nondimensionless. Therefore,

$$\pi_2^* = \pi_2^{-1} = \frac{v}{D_{12}} = N_{\text{Sc}} \tag{11.58}$$

Further,
$$\pi_3 = \rho^a \mu^b L_c^c h_{MT} \tag{11.59}$$

$$\pi_2 = M^0 L^0 T^0 = (ML^{-3})^a (ML^{-1}T^{-1})^b (L)^c (LT^{-1}) \tag{11.60}$$

resulting in
$$a = 1, \; b = -1, \; \text{and} \;\; c = 1 \tag{11.61}$$

$$\pi_3 = \frac{\rho L_c h_{MT}}{\mu} \tag{11.62}$$

Again, due to the nondimensional nature of the π terms, we can define a new π term, π_3^* as a ratio of π_3 and π_2

$$\pi_3^* = \frac{\pi_3}{\pi_2} = \frac{\dfrac{\rho L_c h_{MT}}{\mu}}{\dfrac{\rho D_{12}}{\mu}} \tag{11.63}$$

After simplification,

$$\pi_3^* = \frac{L_c h_C}{D_{12}} = N_{Sh} \tag{11.64}$$

Finally, following the general equation of convection mass transfer coefficient, we can express the nondimensional π terms as

$$\pi_3^* = f(\pi_2^*, \pi_1) \tag{11.65}$$

In other words, we can relate Sherwood number as a function of Reynolds and Schmidt numbers

$$N_{Sh} = f(N_{Re}, N_{Sc}) \tag{11.66}$$

or

$$N_{Sh} = K N_{Re}^{n_1} N_{Sc}^{n_2} \tag{11.67}$$

where K, n_1, and n_2 are constants that depend on the flow regime, system geometry, and system configuration and are experimentally determined. Table 11.4 summarizes the experimentally determined mathematical relationships between Sherwood, Reynolds, and Schmidt numbers for forced mass convection in some of the commonly encountered situations.

If the mass transfer process occurs via free convection, the mathematical relationship to describe convection mass transfer is expressed as

$$N_{Sh} = K N_{Gr}^{n_1} N_{Sc}^{n_2} \tag{11.68}$$

and the values of K, n_1, and n_2 for a specific geometry can be found in literature.

Situation	Condition	Correlation
Flow over a cylinder	$N_{Re}N_{Sc} > 0.2$	$N_{Sh} = 0.3 + \dfrac{0.62N_{Re}^{\frac{1}{2}}N_{Sc}^{\frac{1}{3}}}{\left[1 + \left(\dfrac{0.4}{N_{Sc}}\right)^{\frac{2}{3}}\right]^{\frac{1}{4}}}\left\{1 + \left(\dfrac{N_{Re}}{282,000}\right)^{\frac{5}{8}}\right\}^{\frac{4}{5}}$
Flow over a flat plate	When $N_{Re} = 500,000 - 700,000$ (turbulent) and $N_{Pr} = 0.6 - 60$	$N_{Sh} = 0.037N_{Re}^{0.8}N_{Sc}^{\frac{1}{3}}$
Flow within smooth pipe (laminar)	$N_{Re} < 2300$	$N_{Sh} = 3.66$
Flow within smooth pipe (turbulent)	$N_{Re} > 10,000$ $0.7 < N_{Sc} < 160$	$N_{Sh} = 0.023N_{Re}^{0.8}N_{Sc}^{0.4}$

TABLE 11.4   Correlations for the Estimation of Convective Mass Transfer Coefficients for Various Geometrical Configurations

Example 11.7 (Environmental Engineering—Calculation of mass transfer):   Carbon dioxide is added to a long pipe of 50-cm diameter carrying $4\,\dfrac{m^3}{s}$ of air at 25°C (Fig. 11.13). Determine the mass transfer coefficient of carbon dioxide in air if it is given that the mass diffusivity of carbon dioxide in air at 25°C is $1.6 \times 10^{-5}\,\dfrac{m^2}{s}$ and the properties of air at 25°C as viscosity $= 1.8 \times 10^{-5}\,Pa \cdot s$, density $= 1.2\,\dfrac{kg}{m^3}$. Also, determine the mass transfer flux of carbon dioxide in air assuming that $1\,\dfrac{g}{m^3}$ of carbon dioxide is injected at the source.

Step 1: Schematic
The problem is described in Fig. 11.13.

Step 2: Given data
Diameter of the pipe, $D = L_c = 0.5\,m$; Flow rate of air, $Q = 4\,\dfrac{m^3}{s}$; Temperature, $T = 25°C$; Diffusivity of carbon dioxide in air at 25°C, $D_{CO_2\ in\ air} = 1.6 \times 10^{-5}\,\dfrac{m^2}{s}$; Viscosity of air at 25°C, $\mu = 1.8 \times 10^{-5}\,Pa \cdot s$; Density of air at 25°C, $\rho = 1.2\,\dfrac{kg}{m^3}$; Concentration of carbon dioxide at the injection port, $C_{CO_2,injection} = 1\,\dfrac{g}{m^3}$

To be determined: Mass transfer coefficient and the rate of mass transfer of carbon dioxide in air in the pipe.

Assumptions: The temperature and the pressure in the pipe remain constant and the flow is fully developed. The original concentration of carbon dioxide in air is negligible ($C_{CO_2,air} = 0$).

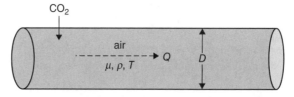

FIGURE 11.13   Schematic of the pipe described in Example 11.7 wherein carbon dioxide is mixed with flowing air.

Approach: This is a forced convection problem. Therefore, we will determine the Reynolds number first to check the validity of the flow regime. Subsequently, we will use the appropriate correlation and determine the Sherwood number, mass transfer coefficient, and the rate of mass transfer.

Step 3: Calculations

The Reynolds number can be calculated using Eq. (8.10, Chapter 8) (based on the flow rate):

$$N_{Re} = \frac{4Q\rho}{\pi D \mu} \tag{E11.7A}$$

Substituting the given data into the above equation, we obtain

$$N_{Re} = \frac{4 \times 4 \times 1.2}{3.14 \times 0.5 \times 1.8 \times 10^{-5}} = 6.794 \times 10^5$$

Now, the Schmidt number is calculated as

$$N_{Sc} = \frac{\nu}{D_{mn}} = \frac{\mu}{\rho D_{mn}} \tag{E11.7B}$$

Again, using the given data, we obtain

$$N_{Sc} = \frac{1.8 \times 10^{-5}}{1.2 \times 1.6 \times 10^{-5}} = 0.93$$

Because the flow is turbulent ($N_{Re} > 10{,}000$) and $0.7 < N_{Sc} < 160$, we can use the following correlation (Table 11.4):

$$N_{Sh} = 0.023 N_{Re}^{0.8} N_{Sc}^{0.4} \tag{E11.7C}$$

Substituting the values of N_{Re} and N_{Sc} into the above equation, we obtain

$$N_{Sh} = 0.023 \times (6.79 \times 10^5)^{0.8} \times (0.93)^{0.4} = 1038$$

or

$$N_{Sh} = 1038 = \frac{l_c h_{MT}}{D_{mn}} = \frac{0.5 \times h_{MT}}{1.6 \times 10^{-5}} \tag{E11.7D}$$

or the mass transfer coefficient, $h_{MT} = 0.033 \, \frac{m}{s}$.

The rate of mass flux can be calculated as

$$\dot{m}''_{CO_2} = h_{MT} \left(C_{CO_2,source} - C_{CO_2,air} \right) \tag{E11.7E}$$

$$\dot{m}''_{CO_2} = 0.033 \times (0.001 - 0) = 3.3 \times 10^{-5} \, \frac{kg}{m^2 s}$$

11.15   Similarity with Convection Heat Transfer

Careful inspection of Table 11.4 will inform us that the mathematical relationships to determine the mass transfer coefficients are similar to the equations that we discussed in convection heat transfer. In most cases, we can simply convert heat transfer equation to the mass transfer equation by replacing the Nusselt number with Sherwood number and Prandtl number with Schmidt number as shown in Fig. 11.14.

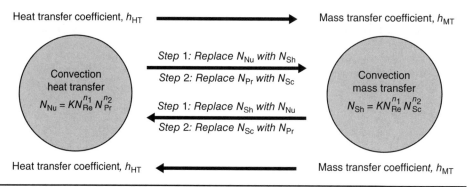

FIGURE 11.14 Flow diagram illustrating the interconversion of heat transfer equations to mass transfer equations.

To illustrate this idea, let us consider the correlation to determine the heat transfer coefficient for a fluid flowing over a cylinder.

$$N_{Nu} = 0.3 + \frac{0.62 N_{Re}^{0.5} N_{Pr}^{0.33}}{\left[1+\left(\dfrac{0.4}{N_{Pr}}\right)^{0.67}\right]^{0.25}} \left\{1 + \left[\frac{N_{Re}}{282,000}\right]^{0.625}\right\}^{0.8} \qquad N_{Re} N_{Pr} > 0.2 \qquad (11.69)$$

Now, imagine that a fluid with a concentration gradient flows over the same cylinder and we are interested in determining the mass transfer coefficient for this system. It turns out that the correlation can be easily modified as shown in Eq. (11.70).

$$N_{Sh} = 0.3 + \frac{0.62 N_{Re}^{0.5} N_{Sc}^{0.33}}{\left[1+\left(\dfrac{0.4}{N_{Sc}}\right)^{0.67}\right]^{0.25}} \left\{1 + \left[\frac{N_{Re}}{282,000}\right]^{0.625}\right\}^{0.8} \qquad N_{Re} N_{Sc} > 0.2 \qquad (11.70)$$

11.15.1   Relation between Heat, Mass, and Momentum Transfer

Now, let us try to understand the parallels between heat, mass, and momentum transfer. For convection heat transfer (Chap. 5), we derived a relation between Nusselt (N_{Nu}), Reynolds (N_{Re}), and Prandtl (N_{Pr}) numbers as $N_{Nu} = f(N_{Re}, N_{Pr})$.

We can now rewrite the above equation into another nondimensional equation famously called the Stanton number (N_{St}). However, because we are specifically dealing with heat transfer (HT), we will use the subscript HT for clarity. Hence,

$$N_{St,HT} = \frac{N_{Nu}}{N_{Re} N_{Pr}} \qquad (11.71)$$

Expressing each of these numbers in their basic form, we obtain

$$N_{St,HT} = \frac{\dfrac{h_{HT} L_c}{k}}{\left(\dfrac{v L_c}{v}\right)\left(\dfrac{v}{\alpha}\right)} = \frac{h_{HT}\alpha}{kv} \qquad (11.72)$$

Similarly, for mass transfer, we have recently seen the relation between Sherwood (N_{Sh}), Reynolds (N_{Re}), and Schmidt (N_{Sc}) as $N_{Sh} = f(N_{Re}, N_{Sc})$, which can also be transformed into the nondimensional Stanton number for mass transfer ($N_{St,MT}$).

$$N_{St,MT} = \frac{N_{Sh}}{N_{Re}N_{Sc}} \tag{11.73}$$

Again, expressing each of these nondimensional numbers in their basic form, we obtain

$$N_{St,MT} = \frac{\dfrac{h_{MT}L_c}{D_{12}}}{\left(\dfrac{vL_c}{v}\right)\left(\dfrac{v}{D_{12}}\right)} = \frac{h_{MT}}{v} \tag{11.74}$$

Combining Eqs. (11.72) and (11.74)

$$\frac{N_{St,HT}}{N_{St,MT}} = \frac{h_{HT}}{h_{MT}}\left(\frac{\alpha}{k}\right) = \frac{h_{HT}}{h_{MT}}\left(\frac{1}{\rho C_p}\right) \tag{11.75}$$

As originally proposed by Reynolds, we can also extend the concepts we learned in fluid mechanics, heat and mass transfer to develop equations relating heat, mass, and momentum transfer. For example, for a flow of fluid within a pipe, the slopes of the velocity, temperature, and concentration profiles represent the coefficients of wall friction $\left(\dfrac{C_f}{2}N_{Re}\right)$, heat transfer ($N_{Nu}$), and mass transfer ($N_{Sh}$), respectively. Assuming the effects of heat, mass, and momentum are the same ($\alpha = D_{12} = v$ or $N_{Pr} = N_{Sc} = N_{Le} = 1$), we can express the Reynolds' analogy as

$$\frac{C_f}{2}N_{Re} = N_{Nu} = N_{Sh} \tag{11.76}$$

$$\frac{C_f}{2} = \frac{N_{Nu}}{N_{Re}} = \frac{N_{Sh}}{N_{Re}} \tag{11.77}$$

$$\frac{C_f}{2} = \frac{N_{Nu}}{N_{Re}N_{Pr}} = \frac{N_{Sh}}{N_{Re}N_{Sc}} \tag{11.78}$$

Because $N_{Pr} = N_{Sc} = 1$

$$C_f = 2N_{St,HT} = 2N_{St,MT} \tag{11.79}$$

$$C_f = 2\frac{h_{HT}\alpha}{kv} = 2\frac{h_{MT}}{v} \tag{11.80}$$

Therefore, determining the value of either of the mass transfer or heat transfer or the wall friction coefficients will provide a complete picture of remaining the transport processes involved in the fluid flow.

However, several practical situations involve dissimilar Prandtl and Schmidt numbers (i.e., $N_{Pr} \neq N_{Sc} \neq 1$). Such cases were extensively studied by Chilton and Colburn independently via experiments and developed a modified Reynolds analogy also called Chilton–Colburn (1934) analogy as follows:

$$C_f = 2N_{St,HT}\, N_{Pr}^{\frac{2}{3}} = 2N_{St,MT}\, N_{Sc}^{\frac{2}{3}} \tag{11.81}$$

$$\frac{N_{St,HT}}{N_{St,MT}} = \frac{h_{HT}}{h_{MT}}\left(\frac{1}{\rho C_p}\right) = \left(\frac{\alpha}{D_{12}}\right)^{\frac{2}{3}} = (N_{Le})^{\frac{2}{3}} \tag{11.82}$$

Because

$$N_{Le} = \frac{\alpha}{D_{12}} \quad \text{(from 11.45)}$$

or

$$\frac{h_{HT}}{h_{MT}} = \rho C_p \left(\frac{\alpha}{D_{12}}\right)^{\frac{2}{3}} \tag{11.83}$$

Example 11.8 (General Engineering): Using the data provided in Example 11.7, determine the heat transfer coefficient and the wall friction in the pipe.

Step 1: Given data
Diameter of the pipe, $D = L_c = 0.5$ m; Flow rate of air, $Q = 4\dfrac{m^3}{s}$; Temperature, $T = 25°C$; Diffusivity of carbon dioxide in air at 25°C, $D_{CO_2 \text{ in air}} = 1.6 \times 10^{-5}\,\dfrac{m^2}{s}$; Viscosity of air at 25°C, $\mu = 1.8 \times 10^{-5}\,Pa \cdot s$; Density of air at 25°C, $\rho = 1.2\,\dfrac{kg}{m^3}$

In addition, we also know the conductivity $\left(k = 0.025\,\dfrac{W}{m \cdot °C}\right)$ and the specific heat $\left(C_p = 1.005\,\dfrac{kJ}{kg\,°C}\right)$ of air at 25°C.

To be determined: The heat transfer coefficient, h_{HT} and the wall friction in the pipe, C_f.

Assumptions: Constant temperature and pressure are maintained within the pipe.

Approach: We have already determined the Schmidt number ($N_{Sc} = 0.93$) and mass transfer coefficient in the previous example (Example 11.7). Therefore, in this problem, we will calculate the Prandtl number. If the Prandtl number is also close to unity, we can directly employ the Reynolds analogy.

Step 2: Calculations
The Prandtl number is determined as follows:

$$N_{Pr} = \frac{v}{\alpha} = \frac{\mu}{\rho \alpha}$$

The value of thermal diffusivity, α for air at 25°C is determined using

$$\alpha = \frac{k}{\rho C_p} = \frac{0.025}{1.2 \times 1.005 \times 1000} = 2.07 \times 10^{-5}\,\frac{m^2}{s}$$

$$N_{Pr} = \frac{v}{\alpha} = \frac{\mu}{\rho \alpha} = \frac{1.8 \times 10^{-5}}{1.2 \times 2.07 \times 10^{-5}} = 0.72$$

Because Prandtl number and Schmidt number are close to unity, we will simply use Reynolds analogy to determine the heat transfer coefficient and wall friction coefficient as under

$$C_f = 2\frac{h_{HT}\alpha}{kv} = 2\frac{h_{MT}}{v}$$

$$C_f = 2 \times \frac{h_{HT} \times 2.07 \times 10^{-5}}{0.025 \times \left(\dfrac{3.14 \times 0.5^2}{4}\right)} = 2 \times \frac{0.033}{\left(\dfrac{3.14 \times 0.5^2}{4}\right)}$$

resulting in $C_f = 0.003$ and $h_{HT} = 39.8 \ \dfrac{W}{m^2 \, ^\circ C}$.

Practice Problems for the FE Exam

1. The mass diffusivity of a gas
   a. Increases with an increase in pressure.
   b. Decreases with an increase in pressure.
   c. Is independent of the pressure.
   d. None of the above.

2. The mass diffusivity of a gas
   a. Increases with an increase in temperature.
   b. Decreases with an increase in temperature.
   c. Increases until up to 110°C and then decreases with a further increase in temperature beyond 110°C.
   d. None of the above.

3. The left face of a 0.1-mm thick composite film $\left(\text{mass diffusivity} = 3 \times 10^{-7} \dfrac{cm^2}{s}\right)$ is subjected to a concentration gradient of sodium chloride of 10 g/L. The diffusional mass transfer flux rate across the film is nearly

   a. $3 \times 10^{-6} \ \dfrac{g}{m^2 s}$.

   b. $3 \times 10^{-6} \ \dfrac{kg}{m^2 s}$.

   c. $1.2 \times 10^{-6} \ \dfrac{kg}{m^2 s}$.

   d. $3 \times 10^{-6} \ \dfrac{kg}{s}$.

4. Pure oxygen is added to flowing air at 35°C (viscosity = 1.9×10^{-5} Pa·s and density = $1.1 \frac{\text{kg}}{\text{m}^3}$). If the diffusion coefficient of oxygen in air at 35°C is $3 \times 10^{-5}\,\text{m}^2/\text{s}$, the value of the Schmidt number is almost

   a. 3.0.

   b. 0.5.

   c. 1.0.

   b. 2.0.

5. Using the data from Problem 4, and assuming that the thermal conductivity and the specific heat of the air as 0.05 W/m·K and 1.01 kJ/kg·K, respectively, the Lewis number now becomes

   a. 0.6.

   b. 0.9.

   c. 1.2.

   d. 0.5.

6. Heat transfer coefficient can be converted into the corresponding mass transfer coefficient by

   a. Replacing the Nusselt number with Sherwood number.

   b. Replacing the Prandtl number with Schmidt number.

   c. Replacing the Nusselt and Prandtl numbers with Sherwood and Schmidt numbers.

   d. None of the above.

7. Oxygen is diffusing into air at 25°C with a mass diffusivity of $2 \times 10^{-5}\,\text{m}^2/\text{s}$. The rate of the diffusional mass transfer rate of oxygen due to a unit concentration gradient, across an unit area, and through a unit thickness is

   a. 2 mol/s.

   b. 5 mol/s.

   c. 1 mol/s.

   d. 2×10^{-5} mol/s.

8. The change in the value of the mass diffusivity due to a 30% increase in the temperature and 10% decrease in the pressure is closest to

   a. 6.5%.

   b. 65%.

   c. 70%.

   d. None of the above.

9. The mass diffusivity of water at 30°C and 1 atm is

   a. $5 \times 10^{-5}\,\text{m}^2/\text{s}$.

   b. $4 \times 10^{-5}\,\text{m}^2/\text{s}$.

c. $2.77 \times 10^{-5} \, \text{m}^2/\text{s}$.

d. $7 \times 10^{-5} \, \text{m}^2/\text{s}$.

10. A 1-mm thick membrane separates two sugar solutions of 50 g/L and 5 g/L at 25°C. If the diffusivity of sugar into the membrane is $3 \times 10^{-13} \, \text{m}^2/\text{s}$, the rate of movement of sugar across the membrane is closest to

   a. $1.35 \times 10^{-8} \, \text{kg/s m}^2$.

   b. $1.9 \times 10^{-8} \, \text{kg/s m}^2$.

   c. $13.5 \times 10^{-8} \, \text{kg/s m}^2$.

   d. $0.135 \times 10^{-8} \, \text{kg/s m}^2$.

11. An underwater 2-mm thick rubber pipe of 5-cm diameter and 30-m length carries 200 g/L salt solution at a flow rate of 350 LPM at 25°C. If the mass diffusivity of the salt into the rubber is $1.2 \times 10^{-13} \, \text{m}^2/\text{s}$, the amount of salt transferred into the water in 90-day period is about

   a. 174 g.

   b. 254 g.

   c. 500 g.

   d. 457 g.

12. For an oxygen–air mixture (mass diffusivity $= 2 \times 10^{-5} \, \text{m}^2/\text{s}$) in a 50-cm diameter pipe at 30°C, the Sherwood number was calculated as 450. The mass transfer coefficient for this system is

   a. 0.008 m/s.

   b. 1.8 cm/s.

   c. 0.01 m/s.

   d. None of the above.

13. A Sherwood number of 400 suggests

   a. The bulk motion of the fluid increases the rate of mass transfer increases by 400 times.

   b. The bulk motion of the fluid decreases the rate of mass transfer by 400 times.

   c. The mass diffusivity increases by 400 times.

   d. The mass diffusivity decreases by 400 times.

14. For a water–air system, the mass diffusivity of water at 30°C and 1 atm is given as $2.77 \times 10^{-5} \, \text{m}^2/\text{s}$. If the conductivity, density, specific heat of the water–air system are 0.025 W/m·°C, 1.15 kg/m³, and 1.007 kJ/kg·°C, respectively, the value of the Lewis number is approximately

   a. 0.8.

   b. 0.75.

   c. 1.

   d. 0.9.

15. For an air–vapor system, the convective transfer coefficient was determined to be $45\dfrac{W}{m^2 \cdot °C}$. If the density and the specific heat of the air–vapor mixture are given as $1.15\ kg/m^3$ and $1.05\ kJ/kg \cdot K$, respectively, the mass transfer coefficient can be approximated as

   a. 0.9 m/s.

   b. 0.09 m/s.

   c. $0.04\ \dfrac{m}{s}$.

   d. 0.01 m/s.

16. Pure hydrogen (3 atm) separated from a gasifier is pumped through a 50-cm diameter and 100-m long cast iron pipe at 27°C. If the solubility and mass diffusivity of hydrogen through cast iron at 27°C are $0.0023\ kmol/m^3 \cdot atm$ and $3 \times 10^{-13}\ m^2/s$, respectively, the rate of loss of hydrogen in a 365-day period through the pipe is approximately

   a. 10 g.

   b. 0.9 g.

   c. 0.4 g.

   d. None of the above.

Practice Problems for the PE Exam

1. Determine the mass diffusivities of water in air for a temperature range of 10°C–55°C and plot them as a function of temperature.

2. In an aquacultural hatchery, 20% salt water (density = $2190\ kg/m^3$) is being pumped through a 0.1-cm thick rubber pipe (ID = 2.5 cm) that is submerged in 1% salt water at 25°C. If the mass diffusivity of salt in rubber at 25°C is $2.7 \times 10^{-13}\ m^2/s$, determine the rate of movement of salt into the fresh water per unit length of the pipe.

3. A 10-cm thick rectangular wall fabricated using a carbon composite separates concentrated sucrose solution (50%) and water. If the density of sucrose is $1700\ kg/m^3$ and its diffusivity through the composite wall is $1.5 \times 10^{-11}\ m^2/s$, determine the mass flux of sucrose through the wall. If the mass loss of sucrose is to be limited to 3 g/day, what should be the thickness of the wall?

4. An 18-cm long cylindrical container of diameter 5-cm filled with ethanol (density = $800\ kg/m^3$) is exposed to air at 30°C at 1 atm. If the average mass diffusivity of ethanol in air at 30°C is given as $0.16\ cm^2/s$, determine the time needed for complete evaporation of ethanol from the container.

5. Pressured oxygen (2 atm) is stored at 25°C in a spherical rubber container (inner diameter = 1 m) of 1 cm thickness. The solubility of oxygen in rubber is given as $0.0035\ kmol/m^3 \cdot atm$. Assuming a mass diffusivity of oxygen through rubber at 25°C as $2 \times 10^{-10}\ m^2/s$, determine the rate of oxygen transfer through the container.

6. A long rectangular 1-cm thick rubber slab is exposed to 1 g/m^3 of carbon dioxide on both sides. Assuming a constant temperature, determine the concentration of carbon dioxide at the center of the slab after 10 days of exposure.

7. Warm air (30% RH) at 45°C is forced at 5 m/s over a wet food assumed to be cylinder in shape (diameter 0.25 m and 0.5-m long). Assuming the density of air and viscosity at 45°C as 1.1 kg/m^3 and 1.9×10^{-5} Pa·s, respectively, determine the mass transfer coefficient.

8. Dry air (10% RH) at 30°C flows over a swimming pool of 10 m × 20 m at 5 km/h. If the average surface water temperature is 20°C, determine the rate of evaporation of water and the rate of heat transferred from air to water.

9. The mass transfer coefficient for a dilute oxygen–air system moving at 2 m/s (density = 1.1 kg/m^3 and specific heat = 1005 J/kg·°C) is 0.02 m/s. Assuming a unit Lewis number, calculate the heat transfer coefficient and the wall friction.

CHAPTER **12**

Introduction to Psychrometrics

Chapter Objectives

After studying the content presented in this chapter, students will be able to

- Determine the properties of air–water mixture using standard formulae.
- Use psychrometric chart to determine the thermodynamic properties of moist air.
- Solve simple numerical problems related to heating, cooling and dehumidification, and mixing of air streams.

12.1 Motivation

As per the American Society of Heating, Refrigerating, and Air-Conditioning Engineers (ASHRAE), psychrometrics is a subset of physics that pertains to the physical and thermodynamic properties of moist air. In our homes, places of work, and automobiles that we travel in, the right amount of water vapor in the air is very critical to our comfort levels. Similarly, many engineering operations, such as drying of food and other biological materials, preservation of artifacts, and design of surgery rooms in hospitals, require constant control and maintenance of right proportions of moisture in the air thereby making the study of psychrometrics important and relevant for engineers.

12.2 Introduction

The air around us is composed of two components: dry air and water vapor. Dry air is the mixture of nitrogen, oxygen, carbon dioxide, and other gases (in very small percentages) whose composition remains constant. For all practical purposes, we consider the entire mixture of dry air as one component with an average molecular weight of 28.97 g/mol, and average specific heat of 1.005 kJ/kg·K. However, the amount of water vapor varies depending on the elevation, season, and even time of the day. As engineers, we are expected to know the properties of air–water mixtures before we begin to design and analyze HVAC systems. Therefore, we will begin with the fundamental definitions and discuss the concepts of psychrometric charts and their applications in biological and agricultural engineering.

12.3   Humidity

Humidity is defined as the amount of water present in a given mass of air. It is similar to the moisture content of a solid material. We typically use two types of humidity when working with air–vapor mixtures: specific humidity and relative humidity.

12.3.1   Specific Humidity (ω)

The mass of water vapor present in a unit mass of dry air is called specific humidity. It is denoted by ω although other symbols such as W, w, and H are also frequently used. Nonetheless, it is mathematically expressed as

$$\omega = \frac{m_{\text{vapor}}}{m_{\text{dry air}}} \tag{12.1}$$

The units of humidity ratio are $\dfrac{\text{kg vapor}}{\text{kg dry air}}$.

Using the ideal gas law, we can express the mass of vapor as

$$P_v V_v = n_v R T_v = \frac{m_v}{MW_v} R T_v \tag{12.2}$$

where P_v, V_v, n_v, R, T_v, m_v, and MW_v are the pressure, volume, number of moles, universal gas constant, temperature, mass, and the molecular weight of water vapor, respectively.

Rearranging the terms, we obtain

$$m_v = \frac{P_v V_v}{R T_v} MW_v \tag{12.3}$$

Similarly, the mass of dry air can be written as

$$m_a = \frac{P_a V_a}{R T_a} MW_a \tag{12.4}$$

Now, substituting Eqs. (12.3) and (12.4) into Eq. (12.1), we obtain

$$\omega = \frac{\dfrac{P_v V_v}{R T_v} MW_v}{\dfrac{P_a V_a}{R T_a} MW_a} \tag{12.5}$$

Knowing that the respective volumes and the temperatures of the air and the vapor are the same, and the molecular weights of air and vapor are 28.97 and 18 g/mol, we obtain

$$\omega = \frac{P_v}{1.606(P_{\text{atm}} - P_v)} \tag{12.6}$$

Example 12.1 (Environmental Engineering):   The air in a poultry barn was measured to be 25°C with a 60% RH. Determine the humidity ratio of the air circulating in the barn.

Step 1: Given data
Temperature, $T = 25°C$; Relative humidity, $RH = 60\%$

To be determined: Humidity ratio, ω.

Assumptions: The pressure in the barn is 1 atm, and the temperature remains constant.

Approach: We will look up the value of the vapor pressure at 25°C for 60% RH and use the humidity ratio equation.

Step 2: Calculations
From steam tables at 25°C for 60% saturation,

$$P_v = 0.6 \times 3.16 \text{ kPa} = 1.896 \text{ kPa}$$

Therefore, the humidity ratio is calculated as

$$\omega = \frac{P_v}{1.6(P_{atm} - P_v)} = \frac{1.896}{1.6 \times (101.5 - 1.896)}$$

$$\omega = 0.011 \ \frac{\text{kg water vapor}}{\text{kg dry air}}$$

12.3.2 Relative Humidity (RH)

Consistent with its name, the relative humidity is the amount of vapor present in the air mass relative to the maximum allowable capacity of air at saturation at a given temperature. The mathematical definition is the ratio of the partial pressure of water to the saturation pressure of water at that temperature expressed on a percentage basis.

$$\text{RH} = \frac{P_v}{P_{v,\text{sat}}} (\%) \tag{12.7}$$

12.4 Saturated Pressure (P_{sat})

Saturated pressure or saturation pressure (or sometimes called saturated vapor pressure) is that vapor pressure when the water vapor reaches a dynamic equilibrium (evaporation equals condensation) with the surface of pure water at any given temperature (Fig. 12.1).

The saturated pressure is highly dependent of temperature. At higher temperatures, the molecules are equipped with higher energies and result in higher saturation pressures. The values of the saturated pressures at various temperatures are readily

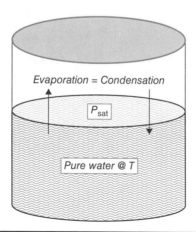

FIGURE 12.1 Schematic to illustrate the concept of saturated pressure that occurs when the rate of water vapor evaporation is balanced by the rate of water vapor condensation.

available in steam tables. Besides, several excellent correlations are also available in the literature. However, due to its simplicity and accuracy, improved Magnus equation developed by Alduchov and Eskridge (1995) can be used to predict the saturation pressure at a given temperature in degree Celsius (°C), which is presented as follows:

$$P_{sat} = 610.94 \exp\left(\frac{17.625T}{T+243.04}\right) \tag{12.8}$$

Recently, Huang (2018) developed a simple and pretty accurate correlation to determine the saturated pressure for a given temperature in degree Celsius (°C), which is as follows:

$$P_{sat} = \frac{\exp\left[34.494 - \dfrac{4924.99}{T+237.1}\right]}{(T+105)^{1.57}} \tag{12.9}$$

Example 12.2 (General Engineering): Compare the values of saturation pressure of water at 35°C obtained from the steam tables, improved Magnus equation, and Huang's equation.

Step 1: Given data
Temperature, $T = 35°C$

To be determined: Saturation pressure, P_{sat}

Assumptions: The temperature remains constant.

Approach: We will simply substitute the value of temperature in the improved Magnus equation and Huang's equation to determine the saturation pressure.

Step 2: Calculations
From the steam table, the value of the saturation pressure at 35°C is

$$P_{sat} = 5629 \text{ Pa}$$

Using the improved Magnus equation,

$$P_{sat} = 610.94 \exp\left(\frac{17.625T}{T+243.04}\right) = 610.94 \times \exp\left(\frac{17.625 \times 35}{35+243.04}\right) = 5618 \text{ Pa}$$

Using the Huang's equation,

$$P_{sat} = \frac{\exp\left[34.494 - \dfrac{4924.99}{T+237.1}\right]}{(T+105)^{1.57}} = \frac{\exp\left[34.494 - \dfrac{4924.99}{35+237.1}\right]}{(35+105)^{1.57}} = 5629 \text{ Pa}$$

It is to be noted that temperature of warmer air can hold more vapor than colder air; it is therefore important to specify the temperature while presenting the RH data.

12.5   Specific Volume (V_s)

Defined as the volume occupied by a unit mass of dry air, specific volume is an important parameter in psychrometrics. Mathematically,

$$V_s = \frac{V_a}{m_a} \tag{12.10}$$

Using the ideal gas law,

$$P_a V_a = \frac{m_a}{MW_a} RT_a \tag{12.11}$$

$$V_s = \frac{V_a}{m_a} = \frac{RT_a}{P_a MW_a} \tag{12.12}$$

$$V_s = 287 \frac{T_a}{(P_{\text{atm}} - P_v)} \tag{12.13}$$

The above equation can be simplified further. From Eq. (12.6), we can write the expression for P_v as

$$P_v = \frac{\omega P_{\text{atm}}}{(\omega + 0.622)} \tag{12.14}$$

Now, substituting this expression in Eq. (12.13) and simplifying will yield

$$V_s = 461.5 \frac{(\omega + 0.622)}{P_{\text{atm}}} T_a \tag{12.15}$$

Example 12.3 (Environmental Engineering): Determine the specific volume of air in a hatchery at 30°C with a humidity ratio of 8 $\frac{\text{g water}}{\text{kg dry air}}$.

Step 1: Given data
Dry bulb temperature, $T = 30°C$; Humidity ratio, $\omega = 8 \frac{\text{g water}}{\text{kg dry air}}$

To be determined: Specific volume, V_s.

Assumptions: The pressure in the barn is 1 atm and the temperature remains constant.

Approach: To determine the specific volume, we will directly use the mathematical relation between the humidity ratio and specific volume.

Step 2: Calculations
The specific volume is given by

$$V_s = 461.5 \frac{(\omega + 0.622)}{P_{\text{atm}}} T_a$$

$$V_s = 461.5 \times \frac{0.008 + 0.622}{101,325} \times (273 + 30)$$

$$V_s \approx 0.869 \frac{\text{m}^3}{\text{kg}} \tag{12.3A}$$

12.6   Enthalpy of Air–Vapor Mixture (*h*)

In addition to the physical properties of air and vapor, the thermodynamic and thermal properties are also important in psychrometric analysis. The properties of air and water vapor are summarized in Table 12.1.

Because moist air is a combination of dry air and water vapor, we can write the enthalpy of the air–vapor mixture as

$$H = H_a + H_v \tag{12.16}$$

In terms of specific enthalpies,

$$H = h_a m_a + h_v m_v \tag{12.17}$$

Property	Value	Description
Specific heat of dry air ($C_{p,a}$)	$1.005 \dfrac{\text{kJ}}{\text{kg} \cdot \text{K}}$	Average value
Specific heat of water vapor ($C_{p,v}$)	$1.86 \dfrac{\text{kJ}}{\text{kg} \cdot \text{K}}$	Average value
Specific enthalpy of dry air (h_a)	$1.005 \, T_a \dfrac{\text{kJ}}{\text{kg}}$	At reference temperature of 0°C and 1 atm
Specific enthalpy of water vapor (h_v)	$2{,}501.4 + (1.86T_v) \dfrac{\text{kJ}}{\text{kg}}$	At reference temperature of 0°C and 1 atm

TABLE 12.1 Thermodynamic Properties of Air and Water Vapor Needed for Psychrometric Calculations

In most of the psychrometric applications, the mass of the dry air m_a remains constant (m_v can change, however) and, therefore, we can normalize the enthalpy in terms of m_a. Therefore, after dividing the equation with m_a, we obtain

$$h = \frac{H}{m_a} = h_a + v_v \frac{m_v}{m_a} = h_a + \omega h_v \tag{12.18}$$

After substituting the values of specific enthalpies of vapor and air from Table 12.1

$$h = 1.005 T_a + \omega(2501 + 1.86T_v) \tag{12.19}$$

Since the temperatures of vapor and air are the same, then the above equation can be written as

$$h = 1.005 T_a + \omega(2{,}501 + 1.86T_a) \tag{12.20}$$

where temperature is in °C.

Example 12.4 (General Engineering): Determine the enthalpy of air–vapor mixture at 34°C that contains 20 g of water per kg dry air.

Step 1: Given data
Dry bulb temperature, $T = 34$°C; Humidity ratio, $\omega = 20 \, \dfrac{\text{g water}}{\text{kg dry air}}$

To be determined: Specific enthalpy of the air–vapor mixture, h.

Assumptions: The pressure in the barn is 1 atm and the temperature remains constant.

Approach: To determine specific enthalpy, we will directly use the mathematical relation between the humidity ratio, dry bulb temperature, and specific enthalpy.

Step 2: Calculations
The specific enthalpy can be calculated as

$$h = 1.005 T_a + \omega(2501 + 1.86T_a)$$

$$h = (1.005 \times 34) + [0.020 \times (2501 + 1.86 \times 34)]$$

$$h = 85.57 \, \frac{\text{kJ}}{\text{kg}}$$

$$\text{(E12.4A)}$$

12.7 Temperature

Based on the kinetic theory of gases, the molecules in the air are in continuous moving in random directions and collide with other molecules in their path. The air temperature is a measure of the average kinetic energy of these randomly moving molecules. Obviously, the molecules in the air at higher temperatures have greater kinetic energy than the molecules in the air at lower temperatures. In the context of psychrometrics, air temperature is expressed in three ways which are described as follows.

12.7.1 Dry Bulb Temperature (T_{db})

It is the actual temperature of the air that is measured by a thermometer or any suitable sensor. This of course excludes the effects of radiation and excessive moisture around the temperature sensor. Because it is very easy to measure, it is mostly used in day-to-day applications. For example, the thermostat in our houses, thermocouples we use at work, and the weather reports we watch every day involve dry bulb temperatures (Fig. 12.2).

12.7.2 Adiabatic Saturation Temperature (T_{AS})

The temperature at which water completely saturates the air in its vicinity by evaporation is called an adiabatic saturation temperature. To derive an expression to determine this temperature, let us consider a large body of water placed in a perfectly insulated environment (a long, insulated half-filled pipe) over which a stream of air is allowed to flow (Fig. 12.3). As the air with a mass flow rate of $\dot{m}_{a,1}$ (with relevant psychrometric properties) comes in contact with the surface of water, heat and mass transfer occurs such that the sensible heat of the air is used to evaporate the water. As a result, the air at the exit is completely saturated (RH = 100%) along with a drop in its temperature such that air temperature becomes the same as the temperature of the water. Besides, to maintain the same water level, we will also replace the water that is lost via evaporation as the makeup water, $\dot{m}_w$.

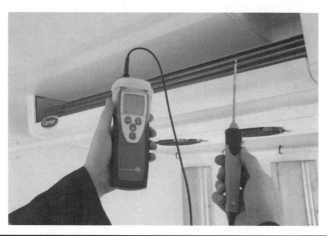

FIGURE 12.2 A popular digital dry bulb thermometer. (Image courtesy of Testo North America. Reprinted with permission.)

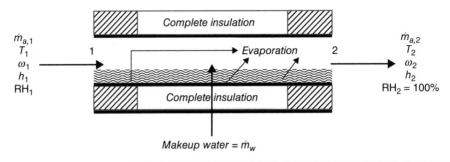

FIGURE 12.3 Schematic describing evaporation of water from an insulated chamber due to air flow.

Now let us perform the mass and energy balances as follows:

Air balance:

$$\dot{m}_{a,1} = \dot{m}_{a,2} = \dot{m}_a \tag{12.21}$$

Vapor balance:

$$\omega_1 \dot{m}_a + \dot{m}_w = \omega_2 \dot{m}_a \tag{12.22}$$

i.e.,

$$\dot{m}_w = (\omega_2 - \omega_1)\, \dot{m}_a \tag{12.23}$$

Energy balance:

$$\dot{m}_a h_1 + h_w \dot{m}_w = \dot{m}_a h_2 \tag{12.24}$$

Writing the enthalpies for air and vapor in terms of specific heat and humidity ratio:

$$h_1 = h_{a,1} + \omega_1 h_{v,1} = C_{P,a} T_1 + \omega_1 h_{v,1} \tag{12.25}$$

$$h_2 = h_{a,2} + \omega_2 h_{v,2} = C_{P,a} T_2 + \omega_2 h_{v,2} \tag{12.26}$$

Substituting the expressions for enthalpies, h_1, and h_2 and $\dot{m}_w$ in the energy balance equation, and cancelling out $\dot{m}_a$, we obtain

$$C_{P,a} T_1 + \omega_1 h_{v,1} + h_w (\omega_2 - \omega_1) = C_{P,a} T_2 + \omega_2 h_{v,2} \tag{12.27}$$

After rearranging

$$\omega_1 (h_{v,1} - h_w) = \omega_2 (h_{v,2} - h_w) + C_{P,a} (T_2 - T_1) \tag{12.28}$$

If we are providing the makeup water at a temperature T_2, the above equation simplifies to

$$\omega_1 (h_{v,1} - h_{w,2}) = \omega_2 \lambda + C_{P,a} (T_2 - T_1) \tag{12.29}$$

where λ is the latent heat of vaporization, and the adiabatic saturation temperature is written as

$$T_2 = T_{AS} = T_1 - \left[\frac{\omega_2 \lambda - \omega_1 (h_{v,1} - h_{w,2})}{C_{P,a}} \right] \tag{12.30}$$

12.7.3 Wet Bulb Temperature (T_{wb})

The temperature of the air is sometimes expressed as wet bulb temperature (T_{wb}). It is the minimum temperature of the air that can be cooled to saturation via evaporation of water. It involves the measurement of the air temperature by a thermometer whose bulb is in contact with a wet sock (Fig. 12.4) and subjected to an air flow (>2 m/s).

The resulting adiabatic evaporation of water from the wet sock will decrease its temperature causing a temperature gradient between the air and the water. Equilibrium conditions will eventually be established when the heat transfer rate from air to water equals the evaporation rate of water. The temperature of the wick at this equilibrium is called the wet bulb temperature. Usually, the wet bulb temperatures are lower than the dry bulb temperatures, unless the air is completely saturated (relative humidity is 100%). For most of the psychrometric applications in biological and agricultural engineering involving air–water mixtures, the constant wet bulb temperature and adiabatic saturation temperature are almost the same. However, this is not the case for other systems and caution must be exercised to separate the differences between adiabatic saturation temperature and wet bulb temperature.

12.7.4 Dew Point Temperature (T_{dp})

It is the temperature at which the air attains saturation (relative humidity is 100%). When a packet of air is cooled, its relative humidity increases and reaches its maximum. Any further cooling of air will result in the water getting condensed out of the air to

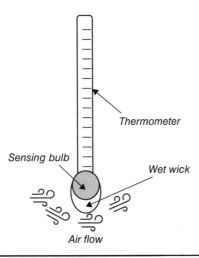

FIGURE 12.4 Schematic describing the concept of wet bulb which involves cooling of the sensing bulb due to the evaporation of the water from the surrounding wick due to air flow.

form what we commonly refer to as dew. In other words, whenever the dry bulb temperature and dew point temperatures are the same, the air is said to have reached 100% relative humidity.

12.8   The Psychrometric Chart

The properties of air–vapor mixtures can be determined using the individual mathematical equations presented as earlier. However, in the early 1900s Willis Carrier invented an ingenious method to determine the psychrometric properties directly from a graph. This plot is called a psychrometric chart and is very useful and appealing to the practicing engineers because of its visual ease of use and simplicity. The general syntax of the plot is depicted in Fig. 12.5.

The x-axis corresponds to the dry bulb temperate (T_{db}) (°C) while the y-axis on the right denotes the humidity ratio (ω) ($\dfrac{\text{kg water vapor}}{\text{kg dry air}}$) (Fig. 12.5). The curved y-axis on the left represents the wet bulb temperature axis, and the slant lines crossing the wet bulb temperature axis are the wet bulb temperatures (T_{db}) (°C). The wet bulb temperature lines when extended beyond the wet bulb axis are called the enthalpy lines (h) ($\dfrac{\text{kJ}}{\text{kg dry air}}$). Similarly, the curved lines parallel to the wet bulb temperature axis are the

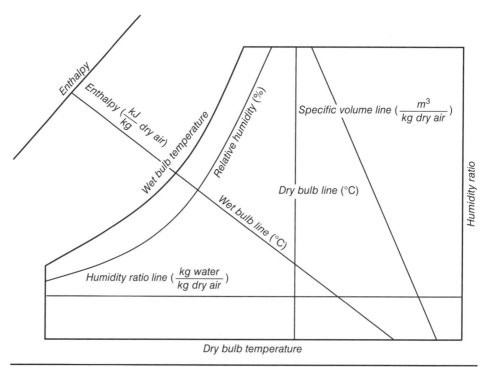

FIGURE 12.5   Illustration of a psychrometric chart depicting the physical and thermodynamic properties of moist air.

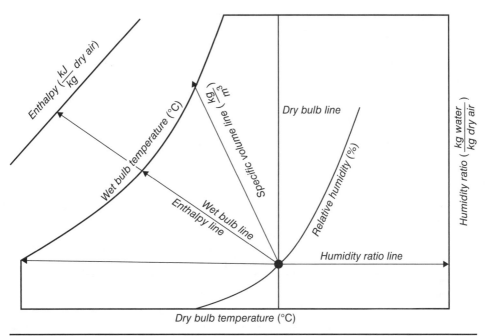

FIGURE 12.6 A psychrometric chart allows to determine all the seven properties of moist air by knowing only two properties.

relative humidity curves (RH) (%). Finally, the second set of slanted lines represent the specific volume (V_s) ($\frac{\text{m}^3}{\text{kg dry air}}$).

Psychrometric charts are available online or from ASHRAE. However, we must note that the psychrometric properties are a function of air pressure. Therefore, depending on the elevation, it is suggested to obtain the correct psychrometric charts to match the air pressures.

As we can see, using only two properties, we can use a psychrometric chart to determine all the remaining properties of the air–vapor mixture. If we know dry bulb temperature and the relative humidity (which are easily available), we can read the remaining properties at the point of intersection of T_{db} and RH, as shown in Fig. 12.6.

The following example will illustrate the step-by-step procedure to determine the psychrometric properties graphically.

Example 12.5 (Environmental Engineering—Indoor animal agriculture): During a recent visit to a broiler facility, an engineer measured the air temperature and relative humidity to be 36°C and 80%, respectively. Determine the state of the air in the facility.

Step 1: Schematic
The given problem and the solution is shown in Fig. 12.7.

Step 2: Given data
Dry bulb temperature, T_{db} = 36°C; Relative humidity, RH = 80%

To be determined: Remaining psychrometric properties.

Assumptions: The pressure in the broiler facility is 1 atm and the temperature remains constant.

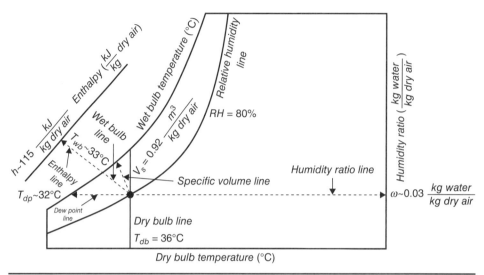

FIGURE 12.7 Graphical illustration for the determination of humidity ratio, dew point temperature, wet bulb temperature, enthalpy, and specific volume using dry bulb temperature and relative humidity.

Approach: On a psychrometric chart at 1 atm, we will locate the point of intersection (reference point) between the curves passing through $T_{db} = 36°C$ and RH = 80% and read the remaining properties by extending the lines along each property. For example, we will draw a horizontal line to the right to intersect the y-axis (to the right side) to determine the humidity ratio.

Step 3: Calculations

3(a). From the reference point (Fig. 12.7), we draw a horizontal line to the right to intersect the y-axis (to the right side) to determine the humidity ratio, $\omega = 0.03 \dfrac{\text{kg water}}{\text{kg dry air}}$.

3(b). To determine the dew point, draw a line (along the constant humidity ratio) to lower the temperature until we reach 100% saturation. This value is approximately 32°C.

3(c). The line from the reference point intersecting the wet bulb axis is the wet bulb temperature, which is $T_{wb} \approx 33°C$.

3(d). If we extend the $T_{wb} = \sim 33°C$ line to the enthalpy axis, we observe that the enthalpy of the air–vapor mixture is $h \approx 115 \dfrac{\text{kJ}}{\text{kg dry air}}$.

3(e). The specific volume is the oblique line that passes through the intersection point. In this case, the line drawn by us is between the lines corresponding to $V_s = 0.9$ and $V_s = 0.95$. After visual interpolation, the value is approximately $0.92 \dfrac{\text{m}^3}{\text{kg}}$.

The state of the air in the facility is summarized in Table 12.2.

12.9 Energy Requirement for Heating of Air

The psychrometric chart can directly be used to determine the sensible heating requirement of a given air–vapor mixture. Imagine a stream of air with a mass flow rate of $\dot{m}_a$ is to be heated from a temperature T_1 to a temperature T_2. If we perform an energy balance around the system, we obtain:

$$\dot{m}_a h_1 + \dot{q} = \dot{m}_a h_2 \tag{12.31}$$

Property, symbol (units)	Value (approximate due to slight error in reading)
Dry bulb temperature, T_{wb} (°C)	36
Relative humidity, RH (%)	80
Humidity ratio, $\omega \left(\dfrac{\text{kg water}}{\text{kg dry air}} \right)$	0.03
Dew point temperature, T_{dp} (°C)	32
Wet bulb temperature, T_{wb} (°C)	33
Enthalpy, $h \left(\dfrac{\text{kJ}}{\text{kg dry air}} \right)$	115
Specific volume, $V_s \left(\dfrac{\text{m}^3}{\text{kg}} \right)$	0.92

TABLE 12.2 Properties of Moist Air in the Broiler Facility as Calculated from the Psychrometric Chart

which can be simplified as

$$\dot{q} = Q \frac{\Delta h}{V_S} \tag{12.32}$$

Because

$$\dot{m}_a = Q\rho = \frac{Q}{V_s} \tag{12.33}$$

Example 12.6 (Agricultural Engineering—Heating requirement in a grain storage): The air flowing in a large grain storage facility at 5°C and 30% RH is heated to 30°C by flowing the air over a heating coil (Fig. 12.8).

Determine the size of the heater for an air flow rate of $3 \dfrac{\text{m}^3}{\text{min}}$ assuming a heater efficiency of 80%.

Step 1: Schematic
The description of the given problem is summarized in Fig. 12.8.

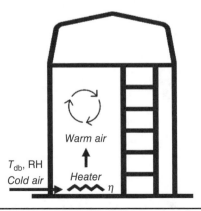

FIGURE 12.8 Schematic of the grain storage with the heater described in Example 12.6.

Step 2: Given data

Dry bulb temperature of the air entering the heating coil, $T_{db,2} = 5°C$; Relative humidity of the inlet air, $RH_1 = 30\%$; Target dry bulb temperature of the air exiting the heating coil, $T_{db,2} = 30°C$; The volumetric flow of air, $Q = 3 \dfrac{m^3}{min}$; Efficiency of the heater, $\eta = 0.8$

To be determined: Heater capacity with an efficiency of 80%.

Assumptions: The pressure is 1 atm and the temperature remains constant.

Approach: This involves simple heating of the air to increase its sensible heat. On a psychrometric chart at 1 atm, we will locate the point of intersection (reference point) between the curves passing through $T_{db,1} = 5°C$ and $RH = 30\%$ and read the remaining properties by extending the lines along each property (Fig. 12.9). Subsequently, to increase the sensible heat of the air, we will move right on the constant humidity ratio line until we reach 30°C and read the enthalpy at 30°C. The energy requirement can be estimated using Eq. (12.32).

Step 3: Calculations

The relevant psychrometric properties at 5°C and 30% RH are

$$\omega_1 = 0.0016 \frac{kg\ water}{kg\ dry\ air}; h_1 = 9.1 \frac{kJ}{kg\ dry\ air}; V_{s,1} = 0.79 \frac{m^3}{kg\ dry\ air}$$

Now on the psychrometric chart, if we move along the constant humidity ratio line (Fig. 12.9), the enthalpy and RH for target bulb temperature of 30°C and humidity ratio of $0.0016 \frac{kg\ water}{kg\ dry\ air}$ will be $35 \frac{kJ}{kg\ dry\ air}$ and about 6.5%.

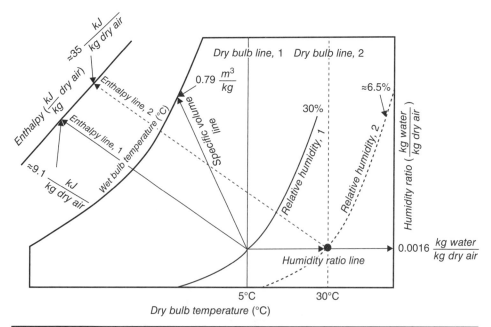

FIGURE 12.9 Graphical illustration for the determination of enthalpies and specific volume during the heating of air at a constant humidity ratio.

Therefore, the energy requirement will be

$$\dot{q} = \frac{\left(Q \dfrac{\Delta h}{V_s} \right)}{\eta} \tag{E12.6A}$$

Substituting the relevant data, we obtain

$$\dot{q} = \frac{\left(\dfrac{3}{60} \times \dfrac{(35 - 9.1)}{0.79} \right)}{0.8} = 2.05 \text{ kW}$$

12.10   Cooling with Dehumidification

For cases where air is cooled along with the removal of moisture, both sensible and latent heats must be considered. Consider a hot stream of air at T_1 with given psychrometric properties is cooled to a temperature of T_2 which is below its dew point resulting in condensation of vapor (Fig. 12.10). By performing a mass and energy balances, we obtain

Air balance:

$$\dot{m}_{a,1} = \dot{m}_{a,2} = \dot{m}_a \tag{12.34}$$

Vapor balance:

$$\omega_1 \dot{m}_a = \dot{m}_w + \omega_2 \dot{m}_a \Rightarrow \dot{m}_w = \dot{m}_a (\omega_1 - \omega_2) \tag{12.35}$$

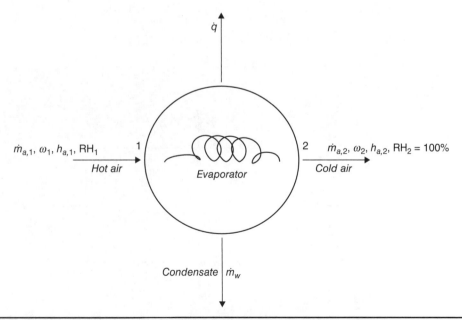

FIGURE 12.10   Schematic of process in which the air is cooled and dehumidified.

Energy balance:

$$h_{a,1} \dot{m}_a = h_{a,2} \dot{m}_a + \dot{q} + h_w \dot{m}_w \qquad (12.36)$$

Substituting Eq. (15.26, Chapter 15) and rearranging will yield

$$\dot{q} = \dot{m}_a (h_{a,1} - h_{a,2}) - h_w \dot{m}_a (\omega_1 - \omega_2) \qquad (12.37)$$

Example 12.7 (Agricultural Engineering—Cooling and dehumidification):   Air flowing at 3 $\dfrac{m^3}{s}$ in a large building is to be cooled from a temperature of 40°C (with a 70% RH) to 25°C using a 210-kW air handler. Determine the mass rate of water removed and the efficiency of the unit (Fig. 12.11).

Step 1: Schematic
The system is described in Fig. 12.11.

Step 2: Given data
Flow rate of the air, $Q_1 = 3 \dfrac{m^3}{s}$; Dry bulb temperature of the inlet air, $T_{db,1} = 40°C$;

Relative humidity of the inlet air, $RH_1 = 70\%$; Dry bulb temperature of the outlet air, $T_{db,2} = 25°C$;
Rating of the air handler, $\dot{q} = 210$ kW

To be determined: Mass of the water removed and the efficiency of the air handler.

Assumptions: The pressure in the building is 1 atm and the temperatures remain constant.

Approach: After we determine the reference point (the intersection point for the lines associated with $T_{db,1} = 40°C$ and $RH_1 = 70\%$), we will move toward left (cooling) until we reach the 100% saturation line (dew point). We will continue to move along the wet bulb axis to drop the temperature further to 25°C simultaneously expelling the water from the air via condensation and causing the humidity ratio to decrease (Fig. 12.12). Once we reach 25°C, we will determine the enthalpy and the humidity ratio for this new state of the air and perform the energy requirement calculations.

Step 3: Calculations
The relevant psychrometric properties at 40°C and 70% RH are

$$\omega_1 = 0.03 \; \frac{kg \; water}{kg \; dry \; air}; \; h_{a,1} = 126 \; \frac{kJ}{kg \; dry \; air}; \; V_s = 0.93 \; \frac{m^3}{kg \; dry \; air}$$

The relevant psychrometric properties at 25°C and 100% RH are

$$\omega_2 = 0.02 \; \frac{kg \; water}{kg \; dry \; air}; \; h_{a,2} = 76 \; \frac{kJ}{kg \; dry \; air}$$

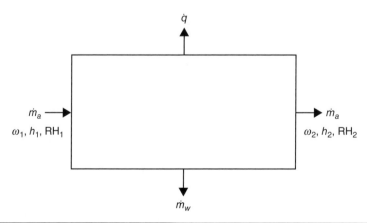

FIGURE 12.11   Schematic of the cooling process of the building described in Example 12.7.

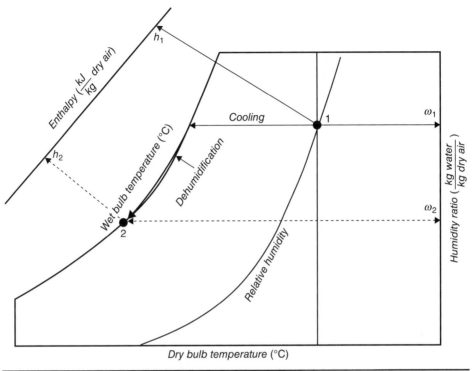

FIGURE 12.12   Line diagram of the psychrometric chart describing the cooling and dehumidification processes described in Example 12.7.

The rate of water removed will be

$$\dot{m}_w = \dot{m}_a(\omega_1 - \omega_2) = \frac{Q_1}{V_{s,1}}(\omega_1 - \omega_2) \tag{E12.7A}$$

$$\dot{m}_w = \frac{3}{0.93}(0.03 - 0.02) = 0.032 \ \frac{\text{kg}}{\text{s}}$$

In terms of volumetric flow and specific volume, the rate of energy removed will be

$$\dot{q} = \frac{Q_1}{V_{s,1}}(h_{a,1} - h_{a,2}) - h_w \frac{Q_1}{V_{s,1}}(\omega_1 - \omega_2) \tag{E12.7B}$$

Additionally, if we assume that the condensed water is saturated at an average temperature (40°C and 25°C) of 32.5°C, $h_w \approx h_f$ at 32.5°C, which is 136 $\frac{\text{kJ}}{\text{kg}}$.
   Substituting all the data we obtain

$$\dot{q} = \frac{3}{0.93} \times (126 - 76) - 136 \times \frac{3}{0.93} \times (0.03 - 0.02)$$

$$\dot{q} = 156.9 \ \frac{\text{kJ}}{\text{s}} = 156.9 \text{ kW}$$

The efficiency of the unit is $\dfrac{156.9}{210} = 74.7 \approx 75\%$.

12.11   Analysis of Air–Vapor Mixtures

Some situations require mixing of two streams of air with different properties to obtain an air stream with certain desirable properties. We can perform mass and energy balances to determine the properties of the mixture. Imagine that stream A with a mass flow rate $\dot{m}_A$, humidity ratio ω_A, and enthalpy h_A is mixed with stream B flowing at a mass flow rate of $\dot{m}_B$, and possessing a humidity ratio ω_B, and enthalpy h_B to obtain the mixture, stream C associated with $\dot{m}_C$, humidity ratio ω_C, and enthalpy h_C (Fig. 12.13).

The overall mass balance around the system yields,

$$\dot{m}_A + \dot{m}_B = \dot{m}_C \tag{12.38}$$

The vapor mass balance yields,

$$\dot{m}_A \omega_A + \dot{m}_B \omega_B = \dot{m}_C \omega_C \tag{12.39}$$

Finally, the energy balance will be

$$\dot{m}_A h_A + \dot{m}_B h_B = \dot{m}_C h_C \tag{12.40}$$

From Eqs. (12.36), (12.37), and (12.38), we can write the enthalpy and the humidity ratio of the equivalent mixture as

$$h_C = \frac{\dot{m}_A h_A + \dot{m}_B h_B}{\dot{m}_A + \dot{m}_B} \tag{12.41}$$

$$\omega_C = \frac{\dot{m}_A \omega_A + \dot{m}_B \omega_B}{\dot{m}_A + \dot{m}_B} \tag{12.42}$$

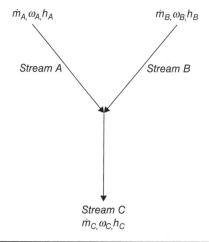

$\dot{m}_A, \omega_A, h_A$ $\dot{m}_B, \omega_B, h_B$

Stream A   Stream B

Stream C
$\dot{m}_C, \omega_C, h_C$

FIGURE 12.13   Mixing of two stream of air to obtain an equivalent stream with desirable properties.

Example 12.8 (Bioprocessing Engineering—Mixing of two sources of air for drying): In a drying operation, an ambient air, flowing at 30,000 LPM at 28°C with a wet bulb temperature of 22°C is mixed with a 10,000 LPM of hot air stream at 60°C at 30% relative humidity. Determine the relative humidity and other properties of the mixture (Fig. 12.14).

Step 1: Schematic
The problem is described in Fig. 12.14.

Step 2: Given data

Stream A: Flow rate, $Q_A = 30,000$ LPM; Dry bulb temperature, $T_{db,A} = 28°C$; Wet bulb temperature, $T_{wb,A} = 22°C$

Stream B: Flow rate, $Q_B = 10,000$ LPM; Dry bulb temperature, $T_{db,B} = 60°C$; Relative humidity, $RH_B = 30\%$

To be determined: RH of the equivalent stream and other properties of the mixture.

Assumptions: The pressure is 1 atm and the temperature remains constant.

Approach: We will determine the enthalpy and humidity ratio of both the streams from the given data using the psychrometric chart. Next, the enthalpy and humidity ratio of the mixture can be determined easily using Eqs. (12.41) and (12.42). Subsequently, the RH and all other properties can also be read from the chart.

Step 3: Calculations
From the given data for stream A, the enthalpy and the humidity ratio can be read as

$$h_A = 64 \ \frac{kJ}{kg \ dry \ air}; \ \omega_A = 0.014 \ \frac{kg \ water}{kg \ dry \ air}, \ V_{s,A} = 0.87 \ \frac{m^3}{kg}$$

Similarly, for stream B, the enthalpy and the humidity ratio can be read as

$$h_B = 162 \ \frac{kJ}{kg \ dry \ air}; \ \omega_A = 0.04 \ \frac{kg \ water}{kg \ dry \ air}; \ V_{s,B} = 1.0 \ \frac{m^3}{kg}$$

The enthalpy of the mixture is

$$h_C = \frac{\dot{m}_A h_A + \dot{m}_B h_B}{\dot{m}_A + \dot{m}_B} \tag{E12.8A}$$

$$h_C = \frac{\left[\left(\dfrac{\dfrac{30,000}{1000 \times 60}}{0.87}\right) \times 64\right] + \left[\left(\dfrac{\dfrac{10,000}{1000 \times 60}}{1.0}\right) \times 162\right]}{\left[\dfrac{\dfrac{30,000}{1000 \times 60}}{0.87} + \dfrac{\dfrac{10,000}{1000 \times 60}}{1.0}\right]} = 86 \ \frac{kJ}{kg \ dry \ air}$$

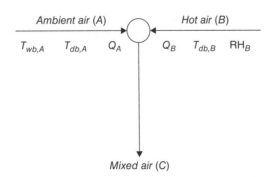

FIGURE 12.14 Schematic to represent mixing of two air streams described in Example 12.8.

The humidity ratio of the mixture is

$$\omega_C = \frac{\dot{m}_A \omega_A + \dot{m}_B \omega_B}{\dot{m}_A + \dot{m}_B} \tag{E12.8B}$$

$$\omega_C = \frac{\left[\dfrac{\left(\dfrac{30,000}{1000 \times 60}\right)}{0.87} \times 0.014\right] + \left[\dfrac{\left(\dfrac{10,000}{1000 \times 60}\right)}{1.0} \times 0.04\right]}{\left[\dfrac{30,000}{1000 \times 60}}{0.87} + \dfrac{\dfrac{10,000}{1000 \times 60}}{1.0}\right]} = 0.019 \, \frac{\text{kg water}}{\text{kg dry air}}$$

Using the values of $\omega_C = 0.019 \, \dfrac{\text{kg water}}{\text{kg dry air}}$ and $h_C = 86 \dfrac{\text{kJ}}{\text{kg dry air}}$, we can obtain the properties of the mixture as

$$\text{RH}_C = 55\%, \, T_{\text{db},C} = 35°C, \, V_{S,C} = 0.9 \, \frac{\text{m}^3}{\text{kg}}, T_{\text{wb},C} = 27°C, \text{ and } T_{\text{dp},C} = 25°C$$

12.12   Drying

Another application of the psychrometric chart is drying. Drying is associated with removal of water by flowing hot air over the given material. During the process, the hot air removes the moisture from the material resulting in a decrease of the dry bulb temperature while simultaneously increasing the humidity ratio of the air. On a psychrometric chart, drying follows an adiabatic cooling along the constant wet bulb temperature (Fig. 12.15). This will be dealt with separately in Chap. 13.

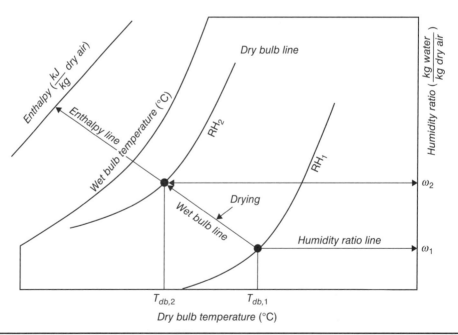

FIGURE 12.15   Drying process on a psychrometric chart can be described on the constant wet bulb temperature line.

Practice Problems for the FE Exam

1. The specific humidity or the humidity ratio is described as the
   a. mass of water vapor present in a unit mass of dry air.
   b. mass of water vapor present in a unit volume of dry air.
   c. inverse of the relative humidity.
   d. None of the above.

2. The humidity ratio for air at 40°C at 80% RH is closest to
   a. 21 g Water/kg dry air.
   b. 17 g Water/kg dry air.
   c. 11 g Water/kg dry air.
   d. 39 g Water/kg dry air.

3. A relative humidity of 100% suggests that the
   a. temperature of the dry bulb exceeded the temperature of the wet bulb.
   b. temperature of the wet bulb exceeded the dry bulb temperature.
   c. dry bulb temperature approached the wet bulb temperature.
   d. None of the above.

4. Which of the following is not true about the saturation pressure of any given liquid?
   a. It depends on the temperature.
   b. It depends on the mass of the liquid present in the container.
   c. It depends on the nature of the liquid.
   d. The evaporation rate of the liquid equals the condensation rate of the gas to liquid.

5. The specific enthalpy of water vapor at 40°C with a humidity ratio of 0.01 kg water/kg dry air is about
   a. 25 kJ/kg.
   b. 41 kJ/kg.
   c. 26 kJ/kg.
   d. None of the above.

6. During a field visit, the dry bulb temperatures were found to be close to the dew point. This suggests that the relative humidity of the air is close to
   a. 50%.
   b. 100%.
   c. 0%.
   d. None of the above.

7. The specific volume of the air in a poultry barn at 35°C with a humidity ratio of 0.029 kg water/kg dry air is

   a. $0.91 \dfrac{m^3}{kg}$.

   b. $0.81 \dfrac{m^3}{kg}$.

   c. $0.85 \dfrac{m^3}{kg}$.

   d. None of the above.

8. The statement to the effect that constant wet bulb temperature is about the same as the adiabatic saturation temperature is

   a. valid only for air–water mixtures under normal conditions.

   b. valid for air–water mixtures for all conditions.

   c. always correct.

   d. None of the above.

9. On a psychrometric chart, drying follows the constant

   a. wet bulb temperature curve.

   b. dry bulb temperature curve.

   c. RH curve.

   d. specific humidity curve.

10. On a psychrometric chart, dehumidification follows the constant

    a. wet bulb temperature curve.

    b. dry bulb temperature curve.

    c. RH curve.

    d. None of the above.

Practice Problems for the PE Exam

1. During a recent visit to a broiler facility, an engineer measured the air temperature and relative humidity to be 31°C and 75%. Determine the state of the air in the facility.

2. A swine house (50 m × 15 m × 5 m) is to be maintained at 25°C and 60% RH. Calculate the amount of moisture present in the house. Additionally, calculate the enthalpy of the air.

3. Air in a sweet potato storage facility at 10°C and 20% RH is heated to 25°C by flowing the air over a heating coil. Determine the size of the heater for an airflow rate of $5 \dfrac{m^3}{min}$ assuming a heater efficiency of 72%.

4. Air flowing at $1 \frac{m^3}{s}$ in a warehouse is to be cooled from a temperature of 35°C (with an 80% RH) to 27°C using a 125-kW air handler. How much water is removed in the process? Also, calculate the efficiency of the unit.

5. In a grain drier, air at 25°C at 15,000 LPM with 50% RH is mixed with a 12,000 LPM of hot air stream at 50°C at a wet bulb temperature of 36°C. Determine the properties of the mixture.

CHAPTER 13

Principles of Drying

Chapter Objectives

The focus of this chapter is to

- Introduce the concept of drying of biological materials.
- Apply the principles of psychrometrics to solve simple drying problems.
- Perform calculations related to dying rates and drying times.

13.1 Motivation

Agricultural and biological engineers routinely employ drying as a tool for enhancing the quality of biological materials. From a process standpoint, drying is a combination of heat, momentum, and mass transfer operations. Therefore, engineers are expected to understand the fundamental engineering processes involved in drying for optimal design of equipment and processes. The principles covered in this chapter will equip students to determine the drying time and temperature and provide engineering recommendations.

13.2 Introduction

Drying involves removal of moisture from the surface and from within the material. It is an important biological and agricultural unit operation especially in forestry, chemical, pharmaceutical, and food processing industries. In fact, drying is one of the oldest forms of food preservation known to humankind. Archaeological evidence suggests that the heat from the sun and the air was used to dry food materials. Depending on the material, the purpose of drying can be either to mitigate microbial growth or to improve the material handling properties, or to decrease the freight costs. This chapter deals with the principles of drying for food and forestry materials. Because drying is all about moisture, we will begin with fundamental definitions of moisture content and its determination followed by the estimation of drying times.

13.3 Moisture Content

It is defined as the amount of water present in a material. If we separate the material into two components, namely water and solids (which may include all nonwater components), we can express the moisture content in two different ways: dry basis and wet basis.

13.3.1 Moisture Content on a Dry Basis (*X*)

It is defined as the ratio of mass of water to the mass of the solids and is expressed as a percentage

$$X = \left(\frac{M_w}{M_s}\right) \times 100\% \qquad (13.1)$$

where M_w and M_s represent the masses of water and solids, respectively. Many biological materials, including lignocellulosic biomasses, food, and manures, have water as the main component. Therefore, the value of X may be greater than 100%.

13.3.2 Moisture Content on a Wet Basis (*x*)

Defined as the ratio of the mass of water to the mass of entire material, moisture content on a wet basis considers entire mass of the material as the basis

$$x = \frac{M_w}{M_w + M_s} = \left(\frac{M_w}{M_{\text{Total}}}\right) \times 100\% \qquad (13.2)$$

Unlike X, the value of x is always $\leq 100\%$.

Determination of moisture content is one of the most fundamental tasks during any drying operation. There are two broad approaches to analyzing the material for its moisture content: direct measurement and indirect measurement. Some of the commonly used direct measurement techniques include oven-drying and Karl Fischer (KF) titration.

The oven-drying method involves overnight heating of a known mass of material in an oven maintained between 103°C and 130°C, depending on the material being analyzed. Subsequently, the material is weighed again and the difference between the original and dried masses is considered to be the mass of the water extracted from the material, which can be expressed in either dry or wet basis. Oven-drying method has been the traditional method and provides reasonably accurate results. Over the years, newer heating protocols that include microwave, infrared, and others have been developed for shorter turnaround times. However, heating at temperatures greater than 100°C may also evaporate some of the volatiles present in the material. Sometimes halogen, infrared, and microwave radiation are also employed to expedite the evaporation of water from the sample.

The Karl Fischer (KF) titration is one of the most accurate methods for the determination of water content in a material. The working principle includes an oxidation reaction of sulfur dioxide with iodine in which one molar concentration of water is consumed. The titration can be automatically performed either via colorimetric or volumetric means. The KF titration is suitable for solids, liquids, and gases as well. However, materials with high and low pH may not be ideal candidates for KF titration.

The indirect methods include measuring the electrical properties (e.g., capacitance, conductivity, and response to radio frequency) of the material and correlating it with its moisture content. Several instruments are available in the industry to measure the moisture content of biological materials on an almost real-time basis.

Example 13.1 (General Engineering–All concentrations): A 500-g sample of chicken manure, after drying in an oven for 24 h at 110°C was weighed to be 100 g (Fig. 13.1). Determine its moisture content on a wet and dry basis.

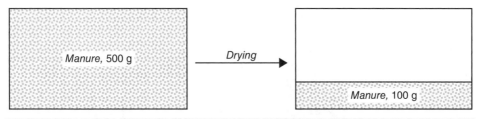

FIGURE 13.1 Schematic representing the loss of water from the manure after drying in Example 13.1.

Step 1: Schematic
The problem is depicted in Fig. 13.1.

Step 2: Given data
Original mass of the manure = 500 g; Mass of the manure after drying = 100 g. Therefore, the mass of the water = 400 g.

To be determined: Moisture content on wet (x) and dry basis (X).

Assumptions: No combustion occurred in the oven and all the water is removed during the heating in the oven.

Approach: We will directly use the equations for moisture contents on a wet and dry basis.

Step 3: calculations
On a wet basis,

$$x = \left(\frac{M_w}{M_w + M_s} \right) = \left(\frac{400}{400 + 100} \right) = 0.8$$

Or, on a wet percentage basis, $x(\%) = \mathbf{80\%}$

On a dry basis,

$$X = \left(\frac{M_w}{M_s} \right) = \left(\frac{400}{100} \right) = 4$$

Or, on a dry percentage basis, $X(\%) = \mathbf{400\%}$

Example 13.2 (General Engineering—All concentrations—Relation between wet and dry basis): Determine the mathematical relation between the moisture contents on the wet and dry basis.

Step 1: Given data
The values of the moisture contents on the wet, x and dry basis, X.

To be determined: The mathematical relation between x and X, i.e., $x = f(X)$.

Step 2: Calculations
The moisture content on a wet basis can be written as

$$\frac{1}{x} = \frac{M_w + M_s}{M_w} \tag{E13.2A}$$

Expanding the above equation

$$\frac{1}{x} = 1 + \frac{M_s}{M_w} = 1 + \frac{1}{X} \tag{E13.2B}$$

Simplifying,

$$x = \frac{X}{1 + X} \tag{E13.2C}$$

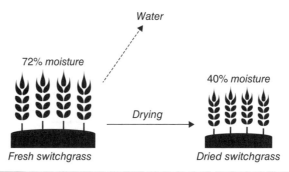

FIGURE 13.2 Schematic for Example 13.3 in which switchgrass is dried from 72% to 40% moisture content (wet basis).

Alternatively,

$$X = \frac{x}{1-x} \tag{E13.2D}$$

Example 13.3 (Bioprocessing Engineering—Drying of lignocellulosic biomass): Two tonnes of freshly harvested switchgrass with 72% moisture content (wet basis) is to be dried for 24 hours to 40% moisture content (wet basis) (Fig. 13.2). Determine how much water is removed in the process.

Step 1: Schematic
The problem is described in Fig. 13.2.

Step 2: Given data
Mass of the switchgrass, M = 2000 kg; Original moisture content = 72%; Moisture content after drying = 40%

To be determined: Mass of the water removed.

Assumptions: Drying removes only moisture and the solid content in switchgrass will remain the same.

Step 3: Calculations
Original moisture content is expressed as

$$0.72 = \frac{M_w}{M_w + M_s} = \frac{(0.72 \times 2000)}{(0.72 \times 2000) + (0.28 \times 2000)} = \frac{1440}{1440 + 560}$$

Let us assume that x kg of water is lost due to drying. Because the mass of the solids is not changed during the drying, the new moisture content is expressed as

$$0.4 = \frac{\text{Mass of new moisture}}{\text{Mass of new moisture} + \text{Mass of the solids}} = \frac{1440 - x}{(1440 - x) + 560} \tag{E13.3A}$$

Solving for x will yield

$$x \approx 1066.67 \text{ kg}$$

The switchgrass will have lost about 1066.67 kg of water after drying for 24 hours.

13.4   Water Activity (a_w)

Water in a biological material is mainly available in two forms: free water (within the pores and cavities) and bound (inside the cell wall) water. The free water is mobile while bound water has less degree of freedom. Although this is an oversimplification

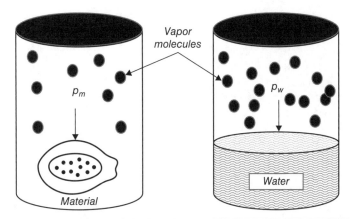

FIGURE 13.3 Schematic illustrating the concept of water activity that describes the extent of partial pressure of water vapor over the biological material relative to the partial pressure over pure water at identical temperature and pressure.

it is perhaps enough to explain the concept. While the moisture content describes all the water available in a material, water activity describes the availability of free water. It is defined as the ratio of partial pressure of the vapor in equilibrium at a given temperature and pressure over the material (p_m) to the partial pressure of vapor over pure water at equilibrium at the same temperature and pressure (p_w) (Fig. 13.3).

Mathematically,

$$a_w = \frac{p_m}{p_w} \tag{13.3}$$

Using the psychrometric definition of relative humidity (RH), we can also write the water activity as

$$a_w = \frac{\text{Equilibrium RH\%}}{100} \tag{13.4}$$

As a measure of free water, the concept of water activity plays an important role in food preservation and lumber processing. Having large amounts of free water will facilitate the growth and proliferation of spoilage microorganisms. Therefore, foods are dried to drive away the moisture including free moisture thereby decreasing the values of water activity. In general, food materials with a water activity less than 0.6 are not susceptible to microbial growth. Similarly, for the processing of lumber, a water activity of less than 0.75–0.8 is considered desirable for minimizing fungal activities on the surfaces.

13.5 Use of Psychrometric Charts Analyzing Drying Processes

Frequently, we are interested in the design parameters such as the drying time, mass rate of the air required for drying, and the rate of water removal. For simple drying operations involving air, psychrometric charts allow us to determine these design parameters graphically as illustrated in the following example problems.

Example 13.4 (Bioprocessing Engineering—Adiabatic drying time of wood): A sample of 500 kg of wood chips at 150% moisture content (dry basis) are being dried. Warm air enters the drier at temperature of 41°C with 10% RH at $2\dfrac{m^3}{s}$ and leaves the drier at 25°C (Fig. 13.4). Determine the drying time to reduce the moisture content to 50% dry basis.

Step 1: Schematic
The problem is illustrated in Fig. 13.4.

Step 2: Given data
Mass of the wood chips $M = 500$ kg; Initial moisture content, $X_1 = 150\%$ (db); Air temperature entering the drier, $T_{db,1} = 41°C$; Relative humidity of air entering the drier, $RH_1 = 10\%$; Temperature of the air leaving the drier, $T_{db,2} = 25°C$; Final moisture content, $X_2 = 50\%$ (db); Air flow rate, $Q = 2\dfrac{m^3}{s}$

To be determined: Drying time to reduce the moisture content of 500 kg of wood chips from 150% to 50% on a dry basis.

Assumptions: The air temperatures, RH, drying rates remain constant and the pressure is 1 atm.

Approach: We will use the psychrometric chart to track the change in humidity ratio to determine the mass of the rate of water removed from the wood chips and subsequently calculate the drying time.

Step 3: Calculations
The relevant properties of the air at 41°C and 10% RH are

$$\omega_1 = 0.005 \,\frac{kg \; water}{kg \; dry \; air} \quad \text{and} \quad V_{s,1} = 0.9 \,\frac{m^3}{kg \; dry \; air}$$

These properties are shown as location 1 on the chart (Fig. 13.5).

Now, we move along the constant wet bulb temperature (bold line) (adiabatic saturation line) and intersect the dry bulb temperature line for the air leaving the drier, that is, at 25°C, we obtain location 2. For location 2, the relevant properties are

$$\omega_2 = 0.0115 \,\frac{kg \; water}{kg \; dry \; air}, RH_2 = 58\%$$

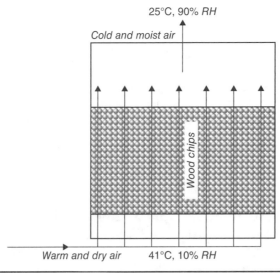

FIGURE 13.4 Schematic of the cross section of the drier processing wood chips in Example 13.4.

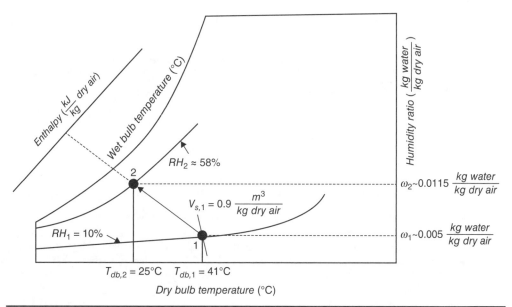

$V_{s,1} = 0.9 \dfrac{m^3}{kg\ dry\ air}$

$RH_2 \approx 58\%$

$\omega_2 \sim 0.0115 \dfrac{kg\ water}{kg\ dry\ air}$

$RH_1 = 10\%$

$\omega_1 \sim 0.005 \dfrac{kg\ water}{kg\ dry\ air}$

$T_{db,2} = 25°C \quad T_{db,1} = 41°C$

Dry bulb temperature (°C)

FIGURE 13.5 Line diagram of the drying process of the wood on a psychrometric chart showing the initial and final states of air.

Therefore, we can determine the rate of loss of water as

$$\dot{m}_{\text{water}} = \dot{m}_a(\omega_2 - \omega_1) = \frac{Q}{V_{s,1}}(\omega_2 - \omega_1) \qquad \text{(E13.4A)}$$

Substituting the relevant values, we obtain

$$\dot{m}_{\text{water}} = \frac{2}{0.9}(0.0115 - 0.005) = 0.014 \frac{\text{kg}}{\text{s}}$$

The initial moisture content,

$$X_1 = \frac{M_{w,1}}{M_s} = 1.5 \qquad \text{(E13.4B)}$$

and

$$M_{w,1} + M_s = 500 \qquad \text{(E13.4C)}$$

where $M_{w,1}$ and M_s are the masses of initial water and solids.
Therefore, solving for $M_{w,1}$ and M_s, we obtain

$$M_s = 200 \text{ kg and } M_{w,1} = 300 \text{ kg}$$

Similarly, the final moisture content,

$$X_2 = 0.5 = \frac{M_{w,2}}{M_s} = \frac{M_{w,2}}{200} \qquad \text{(E13.4D)}$$

Because the mass of the solids is constant during drying.

The mass of water after drying is $M_{w,2} = 100$ kg. Therefore, the mass of water lost by the wood chips is

$$m_{\text{water}} = M_{w,1} - M_{w,2} = 300 - 100 = 200 \text{ kg}$$

Now, the drying time can be calculated as

$$t = \frac{m_{\text{water}}}{\dot{m}_{\text{water}}} = \frac{200 \text{ kg}}{0.014 \dfrac{\text{kg}}{\text{s}}} = 13{,}846.2 \text{ s} \approx 3.85 \text{ h}$$

Example 13.5 (Agricultural Engineering—Water removal rate and heater requirement): One tonne of lumber is being dried in a large drier. Ambient air at 25°C and 40% RH is heated to 70°C and flown in the drier at a flow rate of $0.7\dfrac{\text{m}^3}{\text{s}}$. The air after drying the wood leaves the drier completely saturated (Fig. 13.6). The exit air again passes through another heater to raise the temperature of the air to 90°C and recycles back into the drier. After contact with the wood, the air leaves the drier at 80% RH. Determine the total heater requirement assuming an average efficiency of 80%. Also determine the final mass of the lumber after an hour of drying.

Step 1: Schematic
The problem is described in Fig. 13.6.

Step 2: Given data
Initial mass of the lumber = 1000 kg; Initial air temperature, $T_{db,1} = 25°C$; Initial air's relative humidity, $RH_1 = 40\%$; Temperature of air after heating, $T_{db,2} = 70°C$; Relative humidity after Stage 1 drying, $RH_2 = 100\%$; Temperature of air after Stage 2 heating, $T_{db,3} = 90°C$; Relative humidity after Stage 2 drying, $RH_3 = 80\%$; Flow rate of air, $Q = 0.7 \dfrac{\text{m}^3}{\text{s}}$; Efficiency of the heaters, $\eta = 80\%$

To be determined: Total heating requirement and the mass of the lumber after drying for an hour.

Assumptions: The pressure in the system remains constant at 1 atm.

Approach: We will use the psychrometric chart to track the change in enthalpies and humidity ratios (along the constant wet bulb temperature lines) to determine the heating requirement, the mass of water lost in 1 hour and the final weight of the wood after drying for 1 hour.

Step 3: Calculations
The relevant properties of air at 25°C and 40% RH are

$$\omega_1 = 0.008 \frac{\text{kg water}}{\text{kg dry air}}, \; h_1 = 45 \frac{\text{kJ}}{\text{kg dry air}}, \text{ and } V_{s,1} = 0.855 \frac{\text{m}^3}{\text{kg dry air}}$$

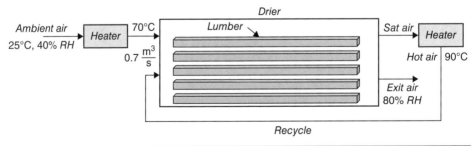

FIGURE 13.6 Schematic for Example 13.5 in which lumber is dried via hot air that is recycled again for a second pass.

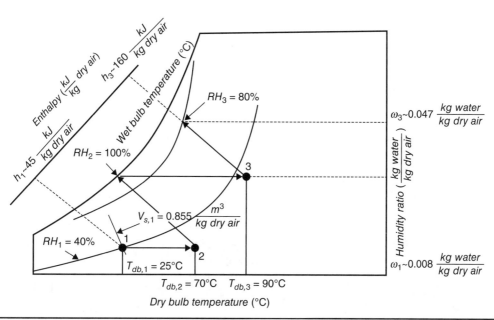

FIGURE 13.7 Line diagram of the drying process of the wood on the psychrometric chart showing the recycling of hot air (Stages 1, 2, and 3).

Heating of the air from 25°C to 70°C along the constant humidity ratio line $\left(\omega_1 = 0.008 \dfrac{\text{kg water}}{\text{kg dry air}} \right)$ will yield location 2 as shown in Fig. 13.7. At this stage (stage 1 drying) the air dries the lumber (along the constant wet bulb temperature line) until the air is completely saturated ($RH_2 = 100\%$) (Fig. 13.7).

Now, the air is again heated to 90°C (along the constant humidity ratio line) to reach location 3. Second state drying occurs (along the constant wet bulb temperature) until the RH ($RH = RH_3$) reaches 80%. The relevant properties at this location are

$$\omega_3 = 0.047 \frac{\text{kg water}}{\text{kg dry air}}, \quad h_3 = 160 \frac{\text{kJ}}{\text{kg dry air}}$$

The total heating requirement can be calculated as

$$\dot{q} = \dot{m}_a (h_3 - h_1) = \frac{Q}{V_{s,1}} (h_3 - h_1) \tag{E13.5A}$$

Substituting the relevant values, we obtain

$$\dot{q} \approx \frac{0.7}{0.855} (160 - 45) = 94.15 \text{ kW}$$

The total heater requirement based on 80% efficiency will be $\dfrac{94.15 \text{ kW}}{0.8} = 117.69 \text{ kW}$.

Similarly, the mass of water removed as a result of drying is

$$\dot{m}_{\text{water}} = \dot{m}_a (\omega_3 - \omega_1) = \frac{Q}{V_{s,1}} (\omega_3 - \omega_1) \tag{E13.5B}$$

Substituting the relevant values, we obtain

$$\dot{m}_{water} \approx \frac{0.7}{0.855}(0.047 - 0.008) = 0.0319\,\frac{kg}{s} = 114.95\,\frac{kg}{h}$$

The mass of the lumber after 1 h = 1000 − 114.95 = 885.05 kg

Note: The values are approximate due to the error in reading the charts. However, several online psychrometric calculators and apps are available for exact values. Also, it should be noted that the air after drying may approach close to saturation will not equal to 100%.

13.6   The Mechanism of Drying

To understand the general process of drying, let us imagine a moisture-rich biological material (food for example) as shown in Fig. 13.8 being dried using a stream of hot air. In the initial phase, the hot air contacts the rich layer of moisture that envelopes the surface of the material. Because water is always present on the surface due to continuous replenishment from within, the drying rate during this phase remains constant. This continues until most of the surface moisture is evaporated.

With the drying process continuing, the moisture within the material will continue to diffuse toward the surface due to the moisture gradient. Because the moisture content within the material is gradually decreasing, the drying rate also decreases along with the decrease in the moisture content and eventually will reach an equilibrium with the surrounding air. Throughout the process, the air provides the heat needed for the evaporation of the moisture making this process a simultaneous heat and mass transfer operation. As a result, the product becomes drier with lowered water activity while the outgoing air will experience a drop in temperature and an increase in the relative humidity.

13.7   Rate of Drying

The rate of drying (r_D) is defined as the mass of water removed from the material per unit time per unit area exposed to drying medium at a given temperature and RH. Mathematically,

$$r_D = -\frac{M\dfrac{dX}{dt}}{A} \tag{13.5}$$

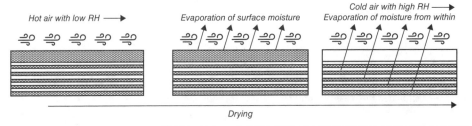

Figure 13.8   Schematic of drying process that includes evaporation of surface water (constant drying) followed by the diffusion and evaporation of water from within the material (variable drying).

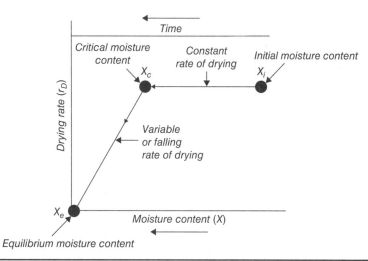

FIGURE 13.9   Graphical representation of drying rate with a reduction in the moisture content during drying of a biological material.

where A is the area of the exposed area of drying (m²), M is the dry mass of the material (kg dry solid), X is the moisture content on the dry basis $\left(\dfrac{\text{kg water}}{\text{kg dry solid}} \right)$, and t is the time (min). In the context of drying, moisture content on a dry basis is frequently used and therefore will be the standard in this chapter.

If we plot the rate of drying against the moisture content, as we previously discussed, during initial drying phase, the rate of drying remains constant until the material reaches a specific moisture content commonly called the critical moisture content (CMC) denoted by X_c, below which the drying rate will become variable (Fig. 13.9). Beyond this stage, the drying rate also called as variable or falling as the drying rate becomes a direct function of the moisture content.

The moisture content will continue to decrease and reaches an equilibrium (with the air) and is aptly called equilibrium moisture content (EMC), denoted by X_e. In other words, for a given air flow rate, temperature, and RH, EMC is the minimum moisture content the material can be lowered to. It is important to note that for a given material, the variable drying rate line may be a combination of two lines with different slopes. However, for the purposes of conceptual understanding, only one single curve is presented.

13.8   Moisture Adsorption–Desorption Isotherm

Consider an isothermal chamber subjected to a constant relative humidity in which a known mass of the food or a biological material is placed. If the moisture gradient between the air and food is positive, the material will continue to adsorb moisture from the surrounding air until an equilibrium is reached. If we continue the experiment at various levels of relative humidity and plot the equilibrium moisture content against the equilibrium relative humidity (also called a_w), we obtain a curve called adsorption isotherm. Similarly, if we repeat the same experiment with decreasing levels of relative

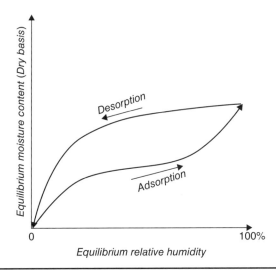

FIGURE **13.10** A representative moisture adsorption isotherm exhibiting a classic hysteresis loop.

humidity, the moisture from the food surface will be desorbed into the surrounding air until equilibrium causing a decrease in the equilibrium moisture content. In this case, the plot between equilibrium moisture content against the relative humidity is now called desorption isotherm (Fig. 13.10). However, it is to be noted that the isotherms for adsorption and desorption are not identical even though the applied equilibrium relative humidity is the same for both processes. Generally, the desorption curve is higher than the adsorption curve. This effect is called hysteresis effect and is theorized to be the result of the changes in the capillary pore structures of the material between the adsorption and desorption processes.

The adsorption isotherm is a helpful tool that helps us to predict the behavior of a given material (food, pharmaceutical agent, or a woody material) under a given RH and temperature. For example, for the given isotherm in Fig. 13.10, we see that when the equilibrium relative humidity exceeds about 70%, the material is prone to adsorb significant amounts of moisture. This information may be used to determine the shelf life, appropriate conditions for storage, and even the right packaging media for the material.

Over the years, several techniques for determining the moisture adsorption–desorption isotherm curves based on hygrometric, manometric, and gravimetric techniques have been developed and described in detail in the literature (Al-Muhtaseb et al., 2002). Once the adsorption isotherm is prepared, it can be mathematically modeled to predict the EMC for a given equilibrium relative humidity. Fortunately, several models have been developed which describe the moisture adsorption with a high degree of accuracy. Some of the commonly used models are summarized in Table 13.1.

13.9 Determination of Drying Rate

The batch rate of drying can be determined via experimentation in which a known mass of the material is exposed to air and continuously monitoring the change in the mass of the material with respect to time. Alternatively, it is also possible to theoretically

Isotherm	Expression	
Langmuir	$EMC = \dfrac{C_L a_w (EMC_m)}{1 + C_L a_w}$	Monolayer adsorption
Brunauer, Emmett, and Teller	$EMC = \dfrac{C_{BET} a_w (EMC_m)}{(1 - a_w)(1 + a_w(C_{BET} - 1))}$	$0.3 \leq a_w \leq 0.4$
Guggenheim, Anderson, and deBoer	$EMC = \dfrac{C_{GAB} K a_w (EMC_m)}{(1 - K a_w)(1 - K a_w + C_{GAB} K a_w)}$	$0.1 \leq a_w \leq 0.9$

EMC = Eqilibrium moisture content $\left(\dfrac{\text{kg water}}{\text{kg solid}}\right)$, EMC_m = Equilibrium moisture content for monolayer $\left(\dfrac{\text{kg water}}{\text{kg solid}}\right)$, a_w = Water activity also called as equilibrium relative humidity, C_L = Langmuir's constant, C_{BET} = BET constant, C_{GAB} = Guggenheim constant, K = correction factor.

TABLE 13.1 Some of the Commonly Used Mathematical Models for Describing Water Adsorption on Food and Biological Materials

predict the drying rate. An excellent approach outlined by Singh and Heldman (2014) involves coupling of convective mass transfer rate with the drying rate (r_D). Assuming the layer of water on a material's surface, the rate of drying can be expressed as

$$r_D = h_{conv} A (C_s - C_\infty) \tag{13.6}$$

where h_{conv}, A, C_s, and C_∞ are the convection mass transfer coefficient $\left(\dfrac{m}{s}\right)$, area (m²), concentration of water vapor on the surface $\left(\dfrac{kg}{m^3}\right)$, and at a location away from the surface $\left(\dfrac{kg}{m^3}\right)$, respectively.

Using ideal gas law, the above equation can be written as

$$r_D = h_{conv} A \frac{MW_v (P_{v,s} - P_{v,\infty})}{RT} \tag{13.7}$$

where MW_v is the molecular weight of water vapor, $P_{v,s}$ is the partial pressure of water vapor at the surface of the moist material, and $P_{v,\infty}$ is the partial pressure of water vapor away from the surface of the moist material.

Now recalling the concept of humidity ratio, ω from Psychrometrics, we can express the partial pressure of the vapor in terms of humidity ratio (after assuming $P_{atm} \gg P_v$) as

$$p_{v,s} = 1.606 \omega_s P_{atm} \text{ and } p_{v,\infty} = 1.606 \omega_\infty P_{atm} \tag{13.8}$$

where the humidity ratios ω_s and ω_∞ are determined at saturation humidity and the actual humidity of the drying air.

Combining the above equations, we obtain the rate of moisture loss in the constant drying phase as

$$r_D = 1.606 h_{conv} A P_{atm} \, MW_{water} \frac{(\omega_s - \omega_\infty)}{RT} \tag{13.9}$$

Example 13.6 (Bioprocessing Engineering—Drying of biomass):   A moist biomass at 20°C is being dried in a tray of dimensions 5 m × 3 m by flowing warm air at 60°C with 20% humidity at 5 m/s over the surface (Fig. 13.11). Determine the drying rate in constant rate phase.

Step 1: Schematic
The problem is illustrated in Fig. 13.11.

Step 2: Given data
Temperature of hot air (far away), $T_\infty = 60°C$; Relative humidity of the hot air (far away), $RH_\infty = 20\%$; Temperature of surface of the biomass, $T_s = 20°C$; Relative humidity of the surface air, $RH_s = 100\%$; Velocity of the air, $v = 5 \frac{m}{s}$; Dimensions of the biomass in the tray 3×5 m and $L_c = 5$ m
From the psychrometric chart, humidity ratio of the air far away at $T_{db,\infty} = 60°C$ and $RH_\infty = 20\%$ is calculated as $\omega_\infty = 0.025 \frac{kg\ water}{kg\ dry\ air}$. Similarly, the humidity ratio of the air at the surface of the moist biomass (assuming an average T_{db} temperature of 40°C and 100% RH) is determined as $\omega_s = 0.049 \frac{kg\ water}{kg\ dry\ air}$.

To be determined: The rate of moisture loss from the biomass.

Assumptions: Ideal gas law is valid, pressure is constant at 1 atm, and the water film of the surface is so thick that constant drying rate is applicable.

Approach: This is similar to convective mass transfer of water via evaporation from a flat plate due to the flow of hot air on the surface. Using the concept we learnt in convection mass transfer we will calculate the mass transfer coefficient from the appropriate correlation and subsequently determine the rate of water loss.

Step 3: Calculations
To calculate the nondimensional heat and flow parameters, we will use an average temperature of 20°C and 60°C to determine the air properties. The relevant properties of air at 40°C are:

Viscosity, $\mu = 0.000019$ Pa·s; Density, $\rho = 1.12 \frac{kg}{m^3}$; Thermal conductivity, $k = 0.027 \frac{W}{m \cdot °C}$; Specific heat, $C_p = 1006 \frac{J}{kg \cdot °C}$; Thermal diffusivity, $\alpha = 2.39 \times 10^{-5} \frac{m^2}{s}$.

In addition, we can also estimate the diffusivity of water in air via the Marrero and Mason's correlation as

$$D_{water,\ air} = 1.87 \times 10^{-10} \times \frac{T^{2.072}}{P} \qquad \text{(E13.6A)}$$

$$D_{water,\ air} = 1.87 \times 10^{-10} \times \frac{(273+40)^{2.072}}{1} = 2.77 \times 10^{-5} \frac{m^2}{s}$$

Now, the Reynolds number is

$$N_{Re} = \frac{\rho v L_c}{\mu} = \frac{1.12 \times 5 \times 5}{0.000019} = 1.473 \times 10^6 \ (>500,000 \text{ and hence turbulent flow})$$

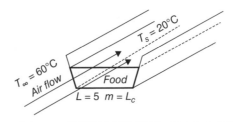

FIGURE 13.11   Schematic of the drying of moist biomass by hot air as described in Example 13.6.

The Schmidt number is

$$N_{Sc} = \frac{v}{D_{water,\,air}} = \frac{\mu}{\rho D_{water,\,air}} = \frac{0.000019}{1.12 \times 2.77 \times 10^{-5}} = 0.612 > 0.6$$

Therefore, we can use the following correlation to determine the Sherwood number, N_{Sh}

$$N_{Sh} = 0.037 N_{Re}^{0.8} N_{Sh}^{0.33} \qquad \text{(E13.6B)}$$

Substituting the relevant data and solving

$$N_{Sh} = 0.037 \times (1.47 \times 10^6)^{0.8} \times 0.612^{0.33} = 2703.8$$

$$N_{Sh} = \frac{h_{conv} L_c}{D_{water,\,air}} \qquad \text{(E13.6C)}$$

$$2703.8 = \frac{h_{conv} \times 5}{2.77 \times 10^{-5}} \qquad \text{(E13.6D)}$$

$$h_{conv} = 0.0149 \frac{m}{s}$$

Now, we can calculate the rate of moisture loss via drying as

$$r_D = 1.606 h_{conv} A P_{atm} MW_{water} \frac{(\omega_s - \omega_\infty)}{RT} \qquad \text{(E13.6E)}$$

$$r_{D,C} = \frac{1.606 \times 0.0149 \times (5 \times 3) \times 101,325 \times 0.018 \times (0.049 - 0.025)}{8.314 \times (273 + 40)}$$

$$r_{D,C} = 0.006 \frac{kg}{s} = 21.6 \frac{kg}{h}$$

13.10   Determination of Drying Time

After determining the drying rates, one can determine the batch drying time as developed by Smith (2011). For constant drying rate (t_c), the rate is independent of the moisture content and, therefore, the equation can be directly integrated to obtain the drying time (min) in the constant drying rate phase (Fig. 13.9, X_i to X_c) as follows:

$$t_c = \frac{W}{r_{D,C} A}(X_1 - X_2) \qquad \text{(13.10)}$$

where X_1 and X_2 are the initial and final moisture contents of the material (%) on a dry basis such that $X_1 > X_c$ and $X_2 \geq X_c$, W is the weight of the material (kg), $r_{D,C}$ is the rate of drying in the constant drying rate phase $\left(\frac{kg}{m^2 \cdot min}\right)$, and A is the area subjected to drying (m^2).

For the variable drying rate (Fig. 13.9, X_c to X_e), however, the drying rate of a function of the moisture content. The drying time in this phase can be determined by

$$t_v = \frac{W}{r_{D,C} A}(X_c - X_e)\left[\ln\left\{ \frac{(X_1 - X_e)}{(X_2 - X_e)} \right\} \right] \qquad \text{(13.11)}$$

where X_c and X_e are the CMC and EMC, respectively, and X_1 and X_2 are the initial and final moisture contents of the material on a dry basis in the variable drying rate phase, respectively, such that $X_1 \leq X_c$ and $X_2 < X_c$.

The total drying time t is the summation of constant and variable drying times and is expressed as

$$t = t_c + t_v \tag{13.12}$$

The following example illustrates the calculation of drying times if the CMC and EMC are known.

Example 13.7 (Bioprocessing Engineering—Determination of drying time of food sample): Partially dried corn with a moisture content of 60% (dry basis) is to be dried to a moisture content of 10% (dry basis). If the constant drying rate was experimentally found to be $0.019 \dfrac{\text{kg}}{\text{m}^2 \cdot \text{min}}$, determine the drying time needed if the dry mass of the corn sample is 45 kg with a specific surface area of $0.22 \dfrac{\text{m}^2}{\text{kg}}$ knowing that the critical and equilibrium moisture contents are 21% and 5%, respectively (Fig. 13.12).

Step 1: Schematic
The problem is illustrated in Fig. 13.12.

Step 2: Given data
Initial moisture content, $X_i = 60\%$; Final (required) moisture content, $X_f = 10\%$; Constant rate period drying rate, $r_{D,C} = 0.019 \dfrac{\text{kg}}{\text{m}^2 \cdot \text{min}}$; Dry mass of the corn sample, $W = 45$ kg; Specific surface area of corn sample, $SSA = 0.22 \dfrac{\text{m}^2}{\text{kg}}$; Critical moisture content, $X_c = 21\%$; Equilibrium moisture content, $X_c = 5\%$

To be determined: Drying time for moisture content to decrease from 60% to 10%.

Assumptions: Temperature, air flow, and RH remain unchanged during the drying process.

Approach: From the given data, the drying will begin at 60% moisture (dry basis) (Location A) in the constant drying rate period and will continue until it reaches the critical moisture content (21% on dry basis) (Location B). Below this, it will undergo a variable drying rate until it reaches its target moisture

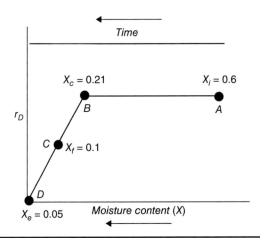

Figure 13.12 Graphical representation of the moisture content during the drying process of the corn in Example 13.7. Total drying time is the summation of constant (AB) and variable (BC) drying times.

content of 10% on dry basis (Location C). Therefore, we will add constant drying rate time (60% to 21%) and variable drying rate time (21% to 10%) (all on dry basis moisture contents).

Step 3: Calculations

The drying time (AB) for decreasing the moisture content from 60% to 21% is given by the constant drying rate (Eq. (13.10)),

$$t_c = \frac{W}{r_{D,C} A}(X_1 - X_2) \qquad \text{(E13.7A)}$$

Noting that $X_1 = X_i$ and $X_2 = X_c$, we can write the above equation as

$$t_c = \frac{W}{r_{D,C} A}(X_i - X_c) \qquad \text{(E13.7B)}$$

Substituting the relevant data, we obtain

$$t_c = \frac{45}{0.019 \times (0.22 \times 45)}(0.6 - 0.21) = 93.3 \text{ min}$$

The drying time (BC) for decreasing the moisture content from 21% to 10% is given by the variable drying rate (Eq. (13.10)),

$$t_v = \frac{W}{r_{D,C} A}(X_c - X_e)\left[\ln\left\{\frac{(X_1 - X_e)}{(X_2 - X_e)}\right\}\right] \qquad \text{(E13.7C)}$$

Noting that $X_1 = X_c$ and $X_2 = X_f$, we can write the above equation as

$$t_v = \frac{W}{r_{D,C} A}(X_c - X_e)\left[\ln\left\{\frac{(X_c - X_e)}{(X_f - X_e)}\right\}\right] \qquad \text{(E13.7D)}$$

After substituting the data

$$t_v = \frac{45}{0.019 \times (0.22 \times 45)}(0.21 - 0.05)\left[\ln\left\{\frac{(0.21 - 0.05)}{(0.10 - 0.05)}\right\}\right]$$

$$t_v = 44.52 \text{ min}$$

Total drying time for the moisture to decrease from 60% to 10% is

$$t = t_c + t_v \qquad \text{(E13.7E)}$$

$$t = 93.3 + 44.52 = 137.84 \text{ min}$$

13.11   Drying Equipment

In its basic form, a drier is simply a closed chamber equipped with heater and an air flow device(s) (usually a fan). A simple batch wood drier is presented in Fig. 13.13.

Depending on the physical properties of the material being processed, several configurations of driers are available in the market. Some of the most common versions are described here.

FIGURE 13.13   A commercial batch drier for dehydrating lumber. (Image courtesy of KILN-DIRECT, USA. Reprinted with permission.)

13.11.1   Tray Drier

As the name suggests, a tray drier is equipped with mobile individual holding trays for the material.

In its simplest design, an internal heater–fan combination moves air resulting in high convective heat and mass transfer coefficients (large Reynolds, Nusselt, and Sherwood numbers) over the material to be dried resulting in shorter drying times (Figs. 13.14 and 13.15). These driers are commonly used in food and pharmaceutical industries. However, being a batch process, tray driers may not be able to handle large quantities of feedstock.

13.11.2   Rotary Dryer

These continuous dryers are suitable for materials of varying particle sizes and therefore highly suitable for dehydrating biomasses and animal feeds. The principle of operation involves movement of a moist material through an inclined rotating cylinder in which hot air interacts with the material continuously (Fig. 13.16a). The cylinder is

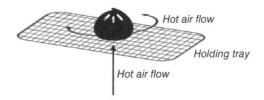

FIGURE 13.14   A schematic of a standard tray drier in which the material placed on a flat tray is dried by the circulation of hot dry air.

FIGURE 13.15   An image of a commercial tray drier manufactured by Commercial Dehydrators America showing the air flow system at the back (a) and the tray after loading with food for drying (b). (Image courtesy of Commercial Dehydrators America. Reprinted with permission.)

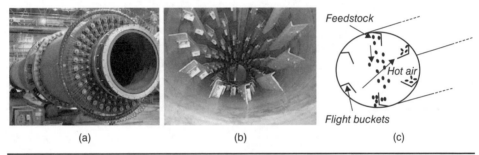

FIGURE 13.16   An example image of a rotary drier (a) and an inside view (b) and the working principle (c). (Image courtesy of Louisville Dryer Company, USA. Reprinted with permission.)

equipped with flight (buckets) along the length on the cylinder (Fig. 13.16b). During rotation, the flights transport the feedstock from bottom to the top of the cylinder and subsequently allow the feedstock to be dropped (like a water flow) from the top, allowing intimate contact between the air flow and the solids (Fig. 13.16c). Although the primary mode of heat transfer is convection, conduction and radiation effects may also be significant.

13.11.3   Fluidized Bed Drier

These are some of the most versatile driers available for granular materials, powders, and materials of small and uniform size, commonly used in the pharmaceutical industry. A fluid, typically, air that also serves as the heat source is flown from the bottom at a velocity high enough to suspend the moist material (Fig. 13.17a). As the hot air flows past the granules, excellent heat and mass transfer occurs between the material and the air that envelops the granules (Fig. 13.17b) resulting in faster and uniform drying of the material.

(a)

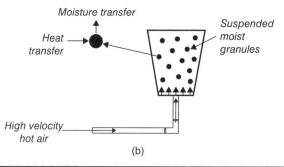

(b)

FIGURE 13.17 An image of a food grade and industrial fluidized bed dryer (a) and a schematic of the working principle (b). (Image courtesy of AVEKA. Reprinted with permission.)

Practice Problems for the FE Exam

1. Which is the following type of water is not removed during drying?

   a. Free water

   b. Capillary water

   c. Chemically bound water

   d. Adsorbed water

2. Water activity is defined as

   a. RH.

   b. $\dfrac{RH}{100}$.

   c. $RH \times 100$.

   d. None of the above.

3. If a material has 75% moisture content on wet basis, the moisture content on dry basis will be

   a. 300%.

   b. 3%.

   c. 30%.

   d. None of the above.

4. The amount of water in kg to be removed from a 75-kg manure to decrease the moisture content from 90% (wb) to 75% (wb) is nearly

   a. 45.

   b. 50.

   c. 40.

   d. None of the above.

5. If the dry bulb temperature of air is 28°C with a relative humidity of 100%. The wet bulb temperature of the air in degrees Celsius is always

   a. less than 28.

   b. equals 28.

   c. greater than 28.

   d. Information given is incomplete.

6. The drying time in a constant rate phase for a food will continue until the moisture content will reach

   a. EMC.

   b. EMC = CMC.

   c. CMC.

   d. 0.

7. The lowest moisture content the food can have at a given temperature and relative humidity is its

   a. EMC.

   b. CMC.

   c. water activity.

   d. moisture content on dry basis.

8. The rate of drying will

   a. increase after the moisture content drops below the CMC.

   b. decrease after the moisture content drops below the CMC.

   c. remain the same as long as the temperature and relative humidity are constant.

   d. None of the above.

9. Which of the following statements is false?

   a. Microbial growth in given food is directly proportional to its water activity.

   b. Constant drying rate curves are almost the same for all materials.

   c. Relative humidity is water activity multiplied by 100.

   d. Adsorption and desorption curves are identical for woody materials.

10. On a psychrometric chart, the process of drying by hot air follows

    a. constant dry bulb temperature.

    b. constant wet bulb temperature.

    c. constant humidity lined.

    d. None of the above.

Practice Problems for the PE Exam

1. One hundred grams of fresh swine manure was dried in an oven at 110°C for 24 hours to yield 7.5 g of dry swine manure. Determine the moisture content on a dry basis.

2. 500 kg of apple slices with 85% moisture content (wet basis) are to be dried to 30% moisture content (wet basis). Calculate the mass of water removed during drying.

3. One thousand kilograms of grain at 150% moisture content (dry basis) is processed in a drier. Three thousand liters per second of air at 45°C (10% RH) enters the drier and leaves the drier at 22. Determine the drying time to reduce the moisture content to 50% dry basis.

4. Biomass (2000 kg) is dried in drying equipment in two stages. Air at 1 m³/s at a temperature of 30 and 30% RH is heated to 60 and flown in the drier. After the first stage of drying, the air leaves the drier completely saturated. The exiting air is again heated to 80 and recycled back into the drier. After the second stage of drying, the air leaves the drier at 65% RH. What will be the total heater requirement assuming an average efficiency of 72%? Also, calculate the final mass of the biomass after an hour of drying.

5. Dewatered manure with a moisture content of 40% (wet basis) is being dried in a dehydrator. Previous experience with the manure suggested that the dehydrator requires 4 h for drying the food from 40% moisture (wet basis) to 25% moisture (wet basis). However, it was desired to further decrease the moisture content to 15% (wet basis). Assuming that the dewatered manure has critical moisture content and an equilibrium moisture content of 30% (wet basis) and 10% (wet basis) respectively, what is the additional drying time assuming the conditions in the dehydrator remain the same?

6. What will be the additional drying time if the equilibrium moisture content was decreased to 5% (wet basis) by changing the environmental conditions in the dehydrator?

CHAPTER 14

Fundamentals of Refrigeration

Chapter Objectives

After learning the content of this chapter, students will be able to

- Calculate the cooling load for a given agricultural system.
- Use pressure–enthalpy plots and determine the heat removed during a cooling operation.
- Design an evaporator for a simple refrigeration system.

14.1 Motivation

Cooling devices and refrigeration systems have become an integral part of lives. Agricultural and biological engineers working in food processing; heating, ventilation, and air conditioning (HVAC); agricultural structures; and bioprocessing industries are expected to design and install refrigeration equipment and predict their performances under various conditions. Therefore, the goal of this chapter is to introduce to the students the basic ideas of refrigeration and illustrate refrigeration cycles via solved numerical examples.

14.2 The Concept of Refrigeration

Refrigeration is a process in which a system (solid, liquid, or gas) is cooled to a temperature lower than its surroundings by transferring the heat from the system to its surroundings.

For example, when a hot pizza at 90°C is exposed to ambient air at 25°C, the pizza will eventually cool down and reach the air temperature. Now, if we somehow extract the heat (we will discuss this later) out of the pizza further and transfer the heat to the air some place far away, the pizza gets much colder than 25°C (Fig. 14.1). This process is called refrigeration. The concept of refrigeration is used in air conditioners used to cool our homes and offices, refrigerators to keep our foods cold, and in food processing industry to store various products in order to extend their shelf lives.

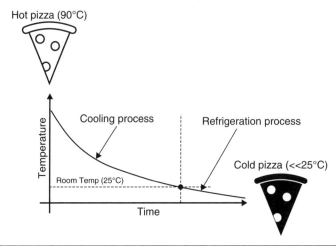

FIGURE 14.1   Schematic to illustrate the differences between cooling and refrigeration.

14.3   Refrigerants

Any fluid that is capable of rapid phase change at relatively lower temperatures is considered as a refrigerant. Pure water boils at 100°C at 1 atm. On the other hand, sulfur dioxide (SO_2) boils at about −10°C at 1 atm and has a latent heat of vaporization of 381.4 kJ/kg. Now, if we flow SO_2 around a body at room temperature, the SO_2 will extract heat from the body and in the process becomes a vapor (Fig. 14.2a). This will result in decreasing the temperature of the body well below the room temperature. When the vapors of SO_2 are condensed to liquid and reused continuously to cool the body further down, it is called refrigeration cycle (Fig. 14.2b).

The history of refrigeration goes back to the early 18th century, with the usage of rubber distillate (caoutchouc) as one of the first refrigerants by John Hauge (circa 1834) (American Society of Heating, Refrigeration, and Air-Conditioning Engineers [ASHRAE]). Over the years, advancements in the fields of chemistry and engineering led to the development of a variety of refrigerants equipped with different thermodynamic and environmental properties. Based on the extent of the hazard, standard 34 published by ASHRAE classifies the refrigerants as flammable, toxic, and toxic-flammable. Flammable refrigerants are subdivided into three num-bered categories namely 1, 2, and 3, where 1 refers to nonflammable and 3 for highly flammable. Similarly, toxic refrigerants are subdivided into two-lettered categories, namely A (low toxicity) and B (high toxicity). For example, a refrigerant rated as A1 is least hazardous relative to B3. It is to be reiterated that refrigerants are toxic and hazardous, and extreme care needs to be exercised during handling, transportation, filling, use, and disposal. Besides, choosing a right refrigerant for a given system is an important part of the engineering design and therefore should be based on several thermodynamic, environmental, and occupational safety metrics as summarized in Fig. 14.3.

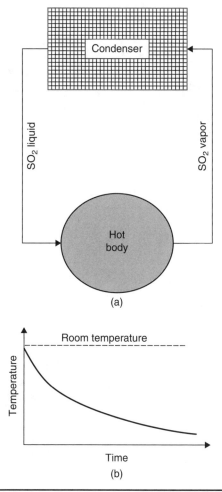

FIGURE 14.2 Illustration of simple sulfur dioxide–based refrigeration cycle in which sulfur dioxide continuously alternates between vapor and liquid phases (a) causing the temperature of body to decrease over time (b).

14.4   Refrigeration Cycle

There are two major types of refrigeration cycles: compression-based and absorption-based. Compression-based systems are widely popular and account for the major share of the HVAC market. We will begin our discussion with compression-based system.

14.4.1   Vapor Compression Refrigeration Cycle (VCRC)

In general, a VCRC consists of four units connected in a loop: evaporator, compressor, condenser, and an expansion valve (Fig. 14.4). Several configurations of each of these components are available in the market and the students are encouraged to look up some of the commonly used models.

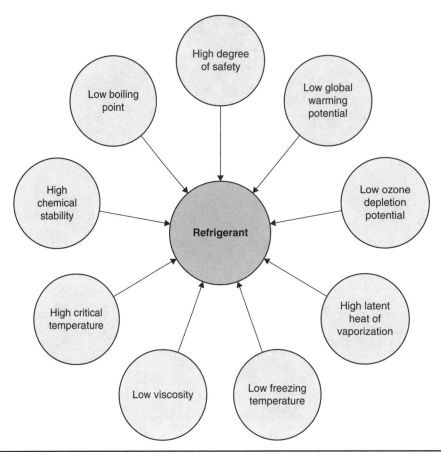

Figure 14.3 Criteria for selecting a refrigerant are based on optimizing thermodynamic, economic, safety, and environmental factors.

Step 1: The refrigerant (gas/liquid mixture), after contacting the evaporator coils, will absorb the heat from the air (that flows across the coil) and vaporize into a saturated gas (sometimes superheated vapor as well) (location 1).

Step 2: Next, to be able to condense this gas, we need to increase its temperature. Therefore, the next task is to flow the refrigerant vapor into a compressor, where its temperature and pressure are increased (location 2).

Step 3: The superheated vapor now is allowed to flow into a condenser (usually located outside the cooling system), in which the vapors contact the high-surface area coil exposed to a stream of flowing air. At this stage, due to the temperature gradient, heat transfer occurs, and the superheated vapors condense into a saturated liquid while simultaneously transferring the heat (heat rejection) to the flowing air (location 3).

Step 4: The saturated liquid refrigerant is passed through an expansion valve (also called metering device) whose task is to decrease the pressure and the temperature of the saturated liquid. This is accomplished by passing the high-pressure liquid

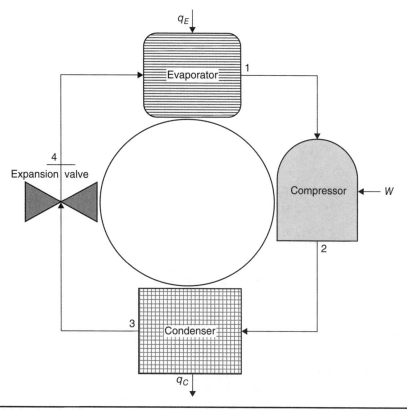

FIGURE 14.4 A typical flow diagram of a vapor compression refrigeration cycle.

through a small opening (recall Bernoulli's equation) and converting the refrigerant into a liquid–vapor mixture, which again is recycled back into the evaporator to complete the loop (location 4).

The entire process can also be depicted on a pressure–enthalpy and temperature–entropy plots as in Fig. 14.5. A typical pressure–enthalpy plot for difluoromethane (R32) is also presented for reference (Fig. 14.6). In addition, the pressure–enthalpy plot for ammonia (R717) is presented in this book's online appendix.

14.4.2 Vapor Absorption Refrigeration Cycle (VARC)

The general working principle of a VARC is similar to that of a VCRC. However, from a design perspective, the major difference is the absence of the compressor in a VARC. Instead, an absorption chamber–pump–generator setup is substituted in lieu of the compressor (Fig. 14.7). The vapors from the evaporator are passed through a heat exchanger-equipped absorber chamber wherein the vapor will interact with the absorber liquid (a portion of the heat is rejected here as well). Subsequently, the solution is pumped at high pressure to a heated chamber commonly called the generator. Within the generator, the refrigerant separates off as a vapor and transferred to the condenser. A secondary loop is provided to transfer the low-strength refrigerant-absorbent solution to the absorber chamber. This concept was originally

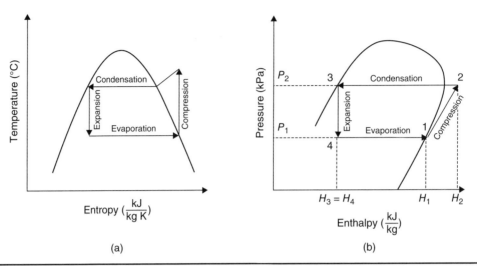

FIGURE 14.5 Graphical representation of thermodynamic processes in a standard vapor compression refrigeration cycle via temperature–entropy (a) and pressure–enthalpy (b) plots.

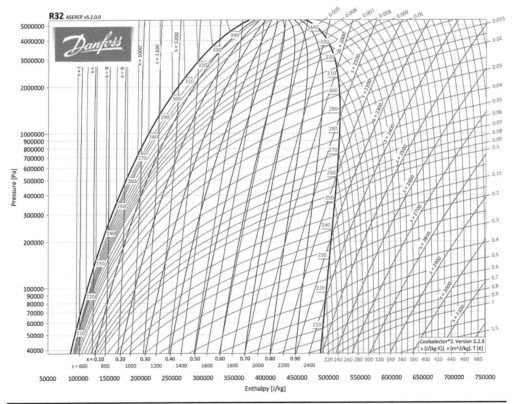

FIGURE 14.6 Pressure–enthalpy plot of difluoromethane (R32). (Image courtesy of Danfoss, USA. Reprinted with permission.)

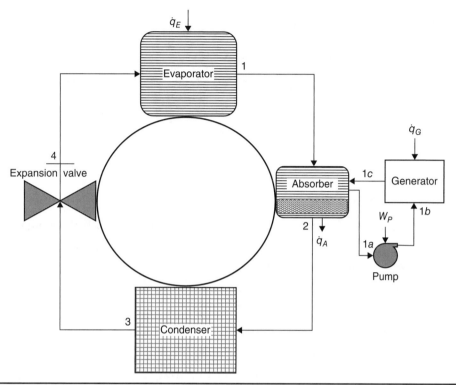

FIGURE 14.7 A typical flow diagram of a vapor absorption refrigeration cycle.

presented by Michael Faraday as adsorption refrigeration which was later developed as absorption refrigeration by Ferdinand Carré in 1858. There are several absorber–liquid combinations available. Some of the most commonly used pairs are ammonia in water and lithium chloride in water.

The VARC systems typically are quiet and experience less wear and tear due to the absence of the compressor. Besides, VARC systems are highly suitable when recycled heat from unit operations and other forms of heat such as geothermal, and solar are available.

It is, however, to be noted in real operating conditions, several losses such as pressure drops and energy losses, coupled with the impurities present in the refrigerant, may result in an equipment being underdesigned. Therefore, a reasonable factor of safety is built into the design and selection of refrigeration systems.

The refrigeration systems we just discussed may also be used as a heating device rather than cooling. When the system absorbs heat from a colder location and expels the heat to a warmer location, it is called a heat pump. Referring to Fig. 14.4, the energy produced in the condenser can now act as a heat source. Heat pumps are widely used in standard residential HVAC units and are very effective when the temperature gradients are lower.

14.5   Quantifying Refrigeration Capacity

14.5.1   Ton of Refrigeration (TOR)

We need a standard to quantify the refrigeration capacity, or the cooling load required for a system. The industry uses TOR, which is defined as the rate of heat transfer required to convert 907.18 kg (or 2000 lbs) of water at 0°C into ice at 0°C (latent heat = $335 \frac{kJ}{kg}$) within 24 hours. Mathematically,

$$1 \text{ TOR} = \frac{907.18 \times 335}{24 \times 60 \times 60} = 3.52 \text{ kW}$$

For a VCRC (Figs. 14.4 and 14.5b), if we assign $\dot{m}_R$ as the mass flow rate of the refrigerant $\left(\frac{kg}{s}\right)$, and $\dot{q}_E$, W, and $\dot{q}_C$ as rate of heat transferred in the evaporator (kW), compressor (work performed) (kW), and condenser (kW), we can express the heat transfer rates as follows:

The rate of heat transfer in the evaporator is given by,

$$\dot{q}_E = \dot{m}_R \Delta H = \dot{m}_R [H_1 - H_3] = \dot{m}_R [H_1 - H_4] \tag{14.1}$$

Because $H_3 = H_4$

The rate of work done in the compressor is given by,

$$W = \dot{m}_R \Delta H = \dot{m}_R [H_2 - H_1] \tag{14.2}$$

The rate of heat transfer in the condenser is given by,

$$\dot{q}_C = \dot{m}_R \Delta H = \dot{m}_R [H_2 - H_3] \tag{14.3}$$

14.5.2   Coefficient of Performance (COP)

In the context of refrigeration, it is a measure of the effectiveness of the system. It is the ratio of the desirable effect to the work performed. For a cooling process, it is the rate of heat removed in the evaporator to the rate of work performed by the compressor. Mathematically

$$\text{COP}_C = \frac{\dot{q}_E}{W} \tag{14.4}$$

Therefore, for a VCRC,

$$\text{COP}_C = \frac{[H_1 - H_4]}{[H_2 - H_1]} = \frac{[H_1 - H_3]}{[H_2 - H_1]} \tag{14.5}$$

Because $H_3 = H_4$

For a heat pump however, the COP is the heat generated by the condenser to the work performed, which is expressed as

$$\text{COP}_H = \frac{\dot{q}_C}{W} \tag{14.6}$$

$$\text{COP}_H = \frac{[H_2 - H_3]}{[H_2 - H_1]} \tag{14.7}$$

We can algebraically show that,

$$COP_H = COP_C + 1 \tag{14.8}$$

For simultaneous cooling and heating, the COP is expressed as

$$COP_T = COP_C + COP_H \tag{14.9}$$

$$COP_T = \frac{(H_1 + H_2) - (2H_3)}{H_2 - H_1} \tag{14.10}$$

However, for a VARC, the mathematical form of the COP will be slightly different because there will be several energy sources and sinks. For a typical layout of VARC (Fig. 14.7), if we assign $\dot{q}_G, W_P$, and $\dot{q}_A$ as rate of heat transferred to the generator, pump, and out of the absorber (kW), respectively, the overall energy balance will result in

$$\dot{q}_E + \dot{q}_G + W_P = \dot{q}_C + \dot{q}_A \tag{14.11}$$

The energy supplied to the pump is relatively negligible and therefore may be ignored.

$$\dot{q}_E + \dot{q}_G = \dot{q}_C + \dot{q}_A \tag{14.12}$$

After assuming a perfectly reversible system, we can write as

$$\frac{\dot{q}_E}{T_E} + \frac{\dot{q}_G}{T_G} = \frac{\dot{q}_C}{T_\infty} + \frac{\dot{q}_A}{T_\infty} \tag{14.13}$$

where T_E, T_G, and T_∞ are the operating temperatures (K) of the evaporator, generator, and absorber (the ambient temperature) and assuming that no thermal gradients exist between ambient air and absorber and ambient air and condenser.

Combining Eqs. (14.12) and (14.13) and upon simplification, we obtain the COP for a VARC as

$$COP = \frac{\text{Heat removed}}{\text{Work performed}} = \frac{\dot{q}_E}{\dot{q}_G} = \frac{T_E}{T_G} \frac{(T_G - T_\infty)}{(T_\infty - T_E)} \tag{14.14}$$

Example 14.1 (General Engineering—All concentrations): A storage room in a shrimp processing facility is cooled by 50 TOR via a difluoromethane (R32)-based VCRC. The condenser operates at an average temperature of 37°C while the evaporator is at an average temperature of −12°C. Determine the required rate of refrigerant flow, COP, and the heat expelled by the condenser, assuming that vapor does not get superheated, and liquid does not get subcooled.

Step 1: Schematic
The schematic for the given system is shown in Fig. 14.8.

Step 2: Given data
Applied load on the evaporator, $\dot{q} = 50$ TOR $= 50 \times 3.5$ kW; Condenser temperature, $T_C = 37$°C; Evaporator temperature, $T_E = -12$°C; Refrigerant $= R32$

To be determined: COP, Flow rate of R32, and the total heat expelled by the condenser.

Assumptions: The system operates in a saturated and steady-state condition with no superheating and subcooling.

Procedure: It is given that the vapor and liquid are saturated and no superheating and subcooling occurs. Therefore, we can use the values of the enthalpies for each condition and determine the COP and mass flow rate of R32.

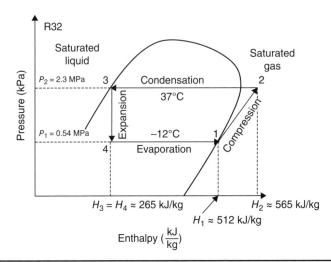

Figure 14.8 Representative pressure–enthalpy diagram for difluoromethane (R32) used for storing frozen shrimp in Example 14.1.

Step 3: Calculations

The refrigeration cycle is constructed on the P–H plot for R32 as shown in Fig. 14.8. From the plot, we can read the enthalpies for saturated vapor, compressed vapor, and condensed liquid as

$$H_1 = 512 \ \frac{kJ}{kg}, \ H_2 \approx 565 \frac{kJ}{kg}, \ H_3 \approx 265 \ \frac{kJ}{kg}$$

Now, from the energy balance around the system the mass flow rate of R32 is

$$\dot{m}_R = \frac{\dot{q}}{H_1 - H_3} = \frac{50 \times 3.5}{(512 - 265)} = 0.7 \ \textbf{kg/s}$$

The COP for the cooling process is given as

$$\textbf{COP}_c = \frac{\text{Heat removed}}{\text{Work performed}} = \frac{H_1 - H_3}{H_2 - H_1} = \frac{(512 - 265)}{(565 - 512)} \approx 4.67$$

Total heat expelled by the condenser is $H_2 - H_3 = 565 - 265 = 300 \ kJ/kg$.

Example 14.2 (Bioprocessing Engineering—Refrigeration with simultaneous superheating and subcooling): A small ice factory is cooled by a 30-kW ammonia-based refrigeration system (R717). The evaporator operates at −10°C and the vapor is superheated to 20°C. The vapor thereafter condenses at 25°C and subcooled further by 5°C. What will be the COP of this system?

Step 1: Schematic

The system is described in Fig. 14.9.

Step 2: Given data

Applied load on the evaporator, $\dot{q} = 10$ kW; Condenser temperature, $T_c = 25°C$ and subcooled by 5°C; Evaporator temperature, $T_E = -10°C$ and superheated by 20°C; Refrigerant = R717

To be determined: COP.

Assumptions: The system operates in a saturated and steady-state condition.

Procedure: It is given that the vapor and liquid are superheated and subcooled. Hence, for superheating process, the temperature = −10°C + 20°C = 10°C. Therefore, on the P–H plot of ammonia (R717), we will extend the evaporator temperature (−10°C) line horizontally until we intersect the 10°C line.

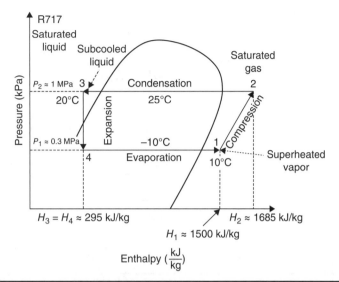

Figure 14.9 Representative pressure–enthalpy diagram for ammonia (R717) used for ice manufacturing in Example 14.2.

Similarly, we will extend the condenser line until we intersect the 20°C constant temperature line (25°C – 5°C = 20°C) as shown in Fig. 14.9.

Step 3: Calculations
The refrigeration cycle is constructed on the P–H plot for R717 as shown in Fig. 14.9. From the plot, we can read the enthalpies for saturated vapor, compressed vapor, and condensed liquid as

$$H_1 \approx 1500 \ \frac{kJ}{kg}, \ H_2 \approx 1685 \ \frac{kJ}{kg}, \ H_3 \approx 295 \ \frac{kJ}{kg}$$

The COP for the cooling process is given as

$$\mathrm{COP}_c = \frac{\text{Heat removed}}{\text{Work performed}} = \frac{H_1 - H_3}{H_2 - H_1} = \frac{(1500 - 295)}{(1685 - 1500)} = 6.51$$

Example 14.3 (Agricultural Engineering—Design of an evaporator for refrigeration): Using the data from Example 14.2, design an evaporator to cool the air from 20°C to –5°C if the overall heat transfer coefficient is given as $2000 \frac{W}{m^2 \cdot °C}$ (Fig. 14.10).

Step 1: Schematic
The schematic for the problem is shown in Fig. 14.10.

Step 2: Given data
Temperature of the evaporator, $T_E = -10°C$; Temperature of the inlet of the air to be cooled, $T_{a,\,in} = 20°C$; Temperature of the outlet of the air to be cooled, $T_{a,\,out} = -5°C$; Overall heat transfer coefficient $U = 2000 \ \frac{W}{m^2 \cdot °C}$

From Example 14.2, we have $H_1 \approx 1500 \ \frac{kJ}{kg}, \ H_2 \approx 1685 \ \frac{kJ}{kg}, \ H_3 \approx 295 \ \frac{kJ}{kg}$

To be determined: The area of the evaporator and the mass flow rate of the refrigerant and air.

Assumptions: The flow is steady state, the temperature in the evaporator will remain constant at –10°C (we will ignore the superheating) and is subjected to a countercurrent flow with air.

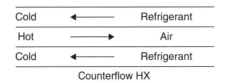

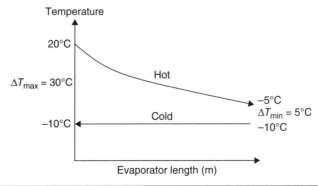

FIGURE 14.10   Schematic describing the change in air temperature in Example 14.3.

Procedure: From the data available from Example 14.2, we will first determine the mass flow rate of the ammonia refrigerant. Subsequently, by assuming constant refrigerant temperature in a counterflow heat exchanger, we will calculate the area and the flow rate of air using the equations that we developed in heat exchanger chapter.

Step 3: Calculations

The mass flow rate of R717 (ammonia) for a 30-kW system is

$$\dot{m}_R = \frac{\dot{q}}{H_1 - H_3} = \frac{30,000}{(1500 - 295) \times 1000} = 0.024 \text{ kg/s}$$

The evaporator is in effect a heat exchanger and using the heat exchanger principles, the heat transfer rate in a heat exchanger is expressed as

$$\dot{q} = UA\Delta T_{\text{LMTD}} \tag{E14.3A}$$

where U is the overall heat transfer coefficient, A is the area, and ΔT_{LMTD} is the log mean temperature difference, which is in turn expressed as

$$\Delta T_{\text{LMTD}} = \frac{\Delta T_{\text{max}} - \Delta T_{\text{min}}}{\ln\left[\dfrac{\Delta T_{\text{max}}}{\Delta T_{\text{min}}}\right]} \tag{E14.3B}$$

From the given data and Fig. 14.11, we can determine the log mean temperature difference as

$$\Delta T_{\text{LMTD}} = \frac{30 - 5}{\ln\left[\dfrac{30}{5}\right]} = 13.95°C$$

The area of the evaporator can be calculated as

$$A = \frac{\dot{q}}{U\Delta T_{\text{LMTD}}} = \frac{30,000}{2000 \times 13.95} = 1.09 \text{ m}^2$$

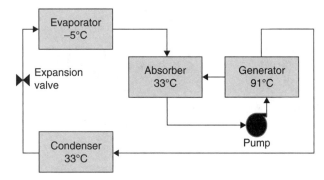

The mass rate of air flow can be calculated as

$$\dot{q} = \dot{m}_h C_{p,h}[T_{h,\text{in}} - T_{h,\text{out}}] \tag{E14.3C}$$

$$\dot{m}_a = \frac{\dot{q}}{C_{p,a}[T_{a,\text{in}} - T_{a,\text{out}}]} = \frac{30{,}000}{1005 \times [20 - (-5)]} = 1.19 \text{ kg/s}$$

Assuming a specific heat of air as 1005 J/kg·K.

Example 14.4 (Environmental Engineering–COP in a vapor absorption refrigeration cycle): An ammonia–water-based vapor absorption refrigerator system consists of an evaporator maintained at –5°C while the condenser chamber and the generator are measured to be at 33°C and 91°C, respectively. Determine the COP of this system.

Step 1: Schematic
The given system is described in Fig. 14.11.

Step 2: Given data
Ammonia–water-based absorption system; Temperature of the evaporator, $T_E = -5°C$; Temperature of the condenser, $T_C = 33°C$; Temperature of the generator $T_G = 91°C$

To be determined: The COP of the system.

Assumptions: The system is ideal and reversible and the temperature in the evaporator, condenser, absorber, and condenser will remain constant. In addition, the temperature of the absorber is the same as the temperature of the condenser which is the temperature of the ambient air (33°C).

Procedure: We will directly use the equation for COD of a VARC.

Step 3: Calculations
The COP for a VARC is

$$\text{COP} = \frac{T_E \ (T_G - T_\infty)}{T_G \ (T_\infty - T_E)} \tag{E14.4A}$$

Substituting the data into the above equation,

$$\text{COP} = \frac{(-5 + 273)}{(91 + 273)} \left\{ \frac{[91 + 273] - (33 + 273)]}{[(33 + 273 - (-5 + 273)]} \right\} = 1.12$$

Note: The general range of the COP is 2 to 5. Therefore, a thorough evaluation of the system is suggested.

Example 14.5 (Agricultural Engineering–Selection of a heat pump): A cold storage is maintained at –25°C in a location whose 24-h average temperature was measured to be 21°C. If the measured heat loss from the cold storage is 15 kW, determine the capacity of the heat pump to expel this heat (Fig. 14.12).

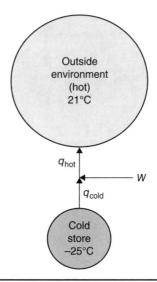

FIGURE 14.12   Schematic of a heat pump described in Example 14.5.

Step 1: Schematic
The schematic for the given system is shown in Fig. 14.12.

Step 2: Given data
Temperature of the cold store, $T_{cold} = -25°C$; Temperature of the surrounding environment, $T_{hot} = 21°C$; Heat loss from the cold store, $\dot{q}_{cold} = 15$ kW

To be determined: The capacity of heat pump.

Assumptions: The system is ideal and reversible and the temperature in the cold store and outside environment will remain constant.

Procedure: We will determine the COP from which the power drawn by the heat pump can be calculated.

Step 3: Calculations
The equation of COP of the refrigerator is given by

$$\text{COP}_C = \frac{\dot{q}_{cold}}{W} \qquad \text{(E14.5A)}$$

Mathematically, the COP can also be expressed as

$$\text{COP}_C = \frac{T_{cold}}{T_{hot} - T_{cold}} \qquad \text{(E14.5B)}$$

Combining the above equations,

$$W = \frac{\dot{q}_{cold}}{\left(\dfrac{T_{cold}}{T_{hot} - T_{cold}}\right)} \qquad \text{(E14.5C)}$$

Substituting the data, we obtain

$$W = \frac{15}{\left(\dfrac{-25 + 273}{(21 + 273) - (-25 + 273)}\right)} = 2.782 \text{ kW}$$

Practice Problems for the FE Exam

1. Which of the following is associated with a refrigeration process?

    a. Cooling of a body to a temperature lower than the surrounding temperature

    b. Cooling of a body to a temperature equal to the surrounding temperature

    c. Condensation around the body

    d. None of the above

2. A ton of refrigeration is equivalent to

    a. 3.5 GW.

    b. 3.5 kW.

    c. 3.5 W.

    d. 3.5 MW.

3. Which is the correct order of operations in a vapor compression refrigeration cycle?

    a. Evaporation → Compression → Condensation → Expansion

    b. Evaporation → Condensation → Compression → Expansion

    c. Condensation → Expansion → Compression → Evaporation

    d. None of the above

4. Which of the operations in a vapor compression cycle is isentropic?

    a. Evaporation

    b. Compression

    c. Condensation

    d. Expansion

5. Which of the operations in a vapor compression cycle is isenthalpic?

    a. Evaporation

    b. Compression

    c. Condensation

    d. Expansion

6. The state of the refrigerant before entering the evaporator is

    a. vapor and liquid.

    b. liquid.

    c. vapor.

    d. superheated vapor.

7. The state of the refrigerant at the exit of the condenser is
   a. vapor and liquid.
   b. liquid.
   c. vapor.
   d. saturated liquid.

8. Which operation is associated with an isobaric process?
   a. Evaporation
   b. Condensation
   c. a and b
   d. None of the above

9. The coefficient of performance for a cooling process is expressed as
   a. $\dfrac{\text{Rate of work by the compressor}}{\text{Heat removal rate of the evaporator}}$.
   b. Heat removal rate of the evaporator × Rate of work by the compressor.
   c. $\dfrac{\text{Heat removal rate of the evaporator}}{\text{Rate of work by the compressor}}$.
   d. None of the above.

10. Which of the following correctly describes the coefficients of performances?
    a. $COP_{hot} = COP_{cold} + 1$
    b. $COP_{hot} = COP_{cold} - 1$
    c. $\dfrac{COP_{hot}}{COP_{cold}} = 1$
    d. None of the above

Practice Problems for the PE Exam

1. A small cold storage is equipped with a 24-kW ammonia-based refrigeration system (R717). The evaporator operates at –5°C and the vapor is superheated to 10°C. The vapor thereafter condenses at 30°C and subcooled further by 5°C. Calculate the COP of this system.

2. For the cold storage described in Problem 1, design an evaporator to cool the air from 23°C to 0°C if the overall heat transfer coefficient is given as $1500\dfrac{W}{m^2 \cdot °C}$.

3. A walk-in freezer at a university laboratory is maintained at –40°C and the ambient average temperature is 26°C. If the measured heat loss from the cold storage is 25 kW, determine the capacity of the heat pump to expel this heat.

CHAPTER 15

Introduction to Adsorption

Chapter Objectives

This chapter introduces the following concepts to students:

- The fundamental processes of adsorption.
- Adsorption isotherms, kinetics, and thermodynamics.
- Batch and continuous adsorption and estimation of maximum adsorption capacity of a packed bed.

15.1   Motivation

Adsorption is a common unit operation in agricultural, environmental, and chemical industries to separate solutes from a mixture. One common example of adsorption that we all can relate to is the activated carbon mask/respirator to filter out contaminants from air. Engineers are expected to design and analyze such equipment using the principles of engineering, chemistry, and physics. This chapter aims to provide suitable background on the principles of adsorption and integrate them with engineering design and analyses of adsorption systems specifically focused from a waste treatment perspective.

15.2   Introduction

Adsorption is a natural phenomenon in which molecules of a fluid adhere to a surface of a solid. To illustrate this idea, let us consider any wall around us. Now, if we measure carefully, the concentration of air molecules on the surface will be slightly higher than the air concentration in the bulk (Fig. 15.1). The air molecules are attaching themselves to the wall's surface because of adsorption. If we repeat the experiment with water and any solid surface, the concentration of water molecules will be higher on the surface of the solid due to adsorption.

In adsorption terminology, the species that adheres is called the *adsorbate or sorbate*. The *surface* on which the adhesion occurs is called *adsorbent or sorbent* (Fig. 15.1). In our example, air is the adsorbate, and the wall is the adsorbent. Although Fig. 15.1 shows adsorption of gas (air) molecules on solid (wall) surface, adsorption can also occur between two gas or liquid molecules. Enzymatic reactions in solutions

419

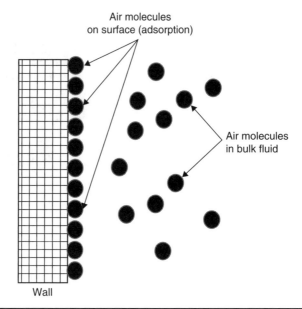

FIGURE 15.1 Conceptual illustration of adsorption phenomena in which air molecules (adsorbate) naturally adhere to a solid surface (adsorbent).

involve adsorption of the substrates on the enzyme surfaces as the first step in the subsequent biochemical reactions.

Now, you might be asking yourself why the molecules adhere to the surfaces. To understand this in a simplistic way, let us again consider any solid surface (Fig. 15.2). The molecules of the adsorbent in the bulk are balanced by other molecules around it. However, the molecules closer to the surface of the adsorbent are not completely balanced. Therefore, they can attract other molecules to the surface.

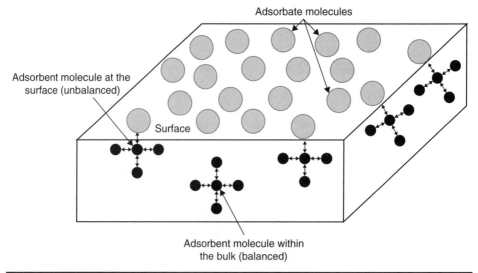

FIGURE 15.2 General mechanism of adsorption on a solid surface.

Because adsorption occurs on the surface (surface phenomenon), the surface area of an adsorbent is extremely important. Therefore, engineers have developed materials equipped with very surface areas. One common example is activated carbon. As the name suggests, when a carbonaceous material (e.g., wood) is treated in the absence of oxygen at high temperature (>400°C) followed by activation with chemicals (e.g., zinc chloride) or gases (e.g., steam) and reheating at high temperatures, will result in a solid called activated carbon. Activated carbon is equipped with a highly porous network and specific surface areas usually greater than $1200 \dfrac{m^2}{g}$ (Fig. 15.3). To put this in perspective, a typical football field has an area of about 5400 m². In other words, 5 g of activated carbon possesses more area than a football field and we can easily imagine the number of molecules 5 g of activated carbon can adsorb if the entire surface area is made available for adsorption.

Now let us understand the mechanism of adsorption. In our previous example, when the air molecules reach the surface, they are attracted to the molecules of the adsorbent (solid) due to van der Waals type of forces resulting in adsorption on the surface. This type of adsorption is called *physical adsorption or physisorption*. This type of adsorption is nonspecific. If we consider air to consist of nitrogen and oxygen, both nitrogen and oxygen adsorb on the surface and the wall generally has no preference for either nitrogen or oxygen. Physisorption is usually weak (enthalpy < 40 kJ/mol) and reversible and most importantly, the molecule is chemically unchanged.

Now imagine the wall has a special affinity only toward oxygen. When the air molecules meet the surface, the wall only attracts oxygen molecules and reacts with them resulting in formation of covalent bonding between oxygen molecules and the molecules of the wall's surface. Such type of adsorption is called *chemical adsorption or chemisorption.* Chemisorption is stronger (enthalpy >> 40 kJ/mol), highly specific, irreversible (under normal conditions), and alters the chemical makeup of the adsorbate molecule due to sharing of electrons. A common example of chemical adsorption is the rusting of iron (chemisorption of oxygen). Besides, the heterogeneous catalytic reactions are based on chemisorption.

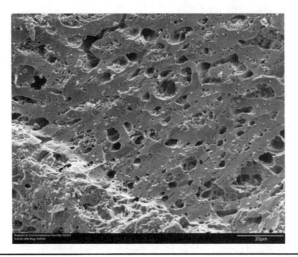

Figure 15.3 A section of activated carbon showing the pores that provide surface area capable of adsorbing molecules.

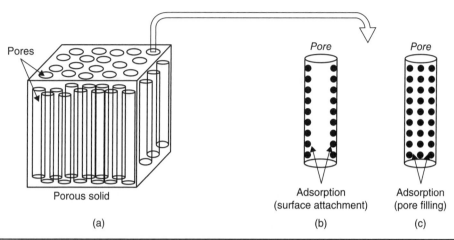

Figure 15.4 Schematic of a porous solid (a) for the illustration of the differences between adsorption (surface phenomenon) (b) and absorption (bulk phenomenon) (c).

We also must note that adsorption is different from absorption. To illustrate this, let us visualize a porous solid as shown in Fig. 15.4a in which the pores are depicted as uniform and parallel to each other. Of course, this is oversimplification, but it serves as a good model to understand the differences between adsorption and absorption. As discussed previously, adsorption is purely a surface phenomenon. Therefore, the adsorbate molecules only attach themselves to the surface of the pore walls (Fig. 15.4b). Absorption, on the other hand, is a bulk phenomenon in which the fluid molecules fill the pores in the bulk of the solid (Fig. 15.4c).

There are several industrial applications of adsorption. One common application we all immediately relate to is the silica adsorbents placed in clothing for removing moisture (Fig. 15.5).

Figure 15.5 A sample image of the silica gel packet, a commonly used adsorbent for moisture adsorption. (Image courtesy of WISESORBENT Technology, NJ, USA. Reprinted with permission.)

Adsorption is used to remove contaminants from air and wastewater, selectively separate off-flavor compounds from juices and beverages. Adsorbents are also used as decoloring agents in sugar processing. Some gas masks are equipped with specially designed activated carbons capable of adsorbing toxic and hazardous gases. Additionally, all the heterogeneous catalytic chemical reactions begin with chemisorption of reactants on the catalyst surface.

15.2   Factors Affecting Adsorption

The adsorbent's ability to adsorb depends on a combination of several factors which are briefly discussed here.

15.2.1   Surface Area and Pore Sizes

Adsorption is a surface phenomenon, and hence the specific surface area (m^2/g) plays an important role in adsorption of any given species. Generally, materials with larger surface areas are equipped with higher number of active site and therefore can adsorption higher amounts of solute. However, the effect of surface area must be considered in light of the pore size of the adsorbent. Adsorbents are usually equipped with micro (<2 nm), meso (2–50 nm), and macro (>50 nm) pores which contribute to the specific surface area of the adsorbent. Therefore, the adsorption capacity will depend on the relative proportion of micropores and the molecular size of the species being adsorbed. For example, if the largest dimension of a given adsorbate species is greater than 2 nm, the presence of micropores in the adsorbent matrix will not contribute to the adsorption despite possessing a large specific surface area. Hence, to be able to remove a mixture of species of varying molecular sizes, a combination of micro (<2 nm), meso (2–50 nm), and macro (>50 nm) pores is preferred. In fact, techniques to synthesize materials with hierarchical porosities are readily available in the literature.

15.2.2   Temperature

Adsorption in general is an exothermic process. Nonetheless, depending on the exact nature of the adsorption, temperature can either inhibit or promote the adsorption process. For purely physical adsorption, higher temperatures inhibit adsorption because of the dominating effects of desorptive forces. However, for surfaces involved in chemisorption, higher temperatures can enhance the adsorption. Additionally, for porous materials, an increase in temperature will enhance the transport of the species into the pore structure thereby increasing the adsorption capacity.

15.2.3   Point of Zero Charge (PZC)

The PZC is one of the most important factors that can directly influence the adsorption capacity of any surface. The PZC is the pH of the solution at which the surface of the adsorbent attains a neutral charge. In other words, at PZC the positive charges are balanced by the negative charges. When the pH of the solution is lower than the PZC of the surface, the surface develops a positive charge and can actively interact with negatively charged species. On the other hand, surfaces with PZC lower than the solution pH will promote adsorption of positively charged species.

15.2.4   Surface Chemistry of the Adsorbent

The chemical makeup of adsorbent surface is another important factor when chemisorption dominates the overall adsorption process. Several studies have shown that the presence of acidic functional groups can negatively impact the adsorption capacity of carbon-based adsorbents for removal of organic compounds in water. During adsorption, water outcompetes and is preferentially adsorbed on acidic surfaces resulting in decrease of adsorption of organic compounds. On the other hand, removal of ammonia and other hydrophilic species is significantly enhanced by surfaces decorated with oxygenated functional groups due to the irreversible chemical reaction between ammonia and carboxylic groups.

15.3   Quantitative Analysis of Adsorption

Now that we understand the basics of adsorption, let us describe the process mathematically. It will be helpful to use a practical example of removing a contaminant from a batch of wastewater by adding an adsorbent (Fig. 15.6a). After adding the adsorbent,

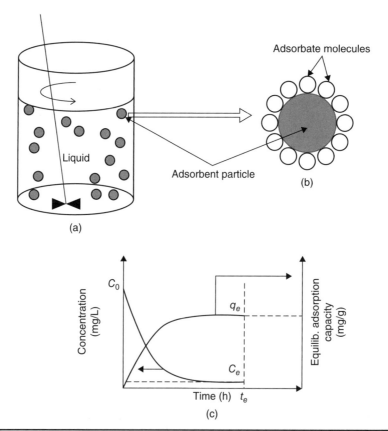

Figure 15.6 Schematic representing the adsorption process that involves contacting the adsorbent with wastewater (a) resulting in the transfer of the contaminant to the surface of the adsorbent (b) and a decrease in its concentration in the liquid phase (c).

the contaminant will diffuse toward the adsorbent and begins to adsorb on the surface (Fig. 15.6b). As a result, the concentration of the contaminant in the wastewater will begin to decrease and will reach an equilibrium concentration once the adsorbent surface is saturated (after an equilibrium time) (Fig. 15.6c). Similarly, the adsorption capacity of the adsorbent will increase from zero to an equilibrium adsorption capacity.

After assigning V for the volume of liquid (L), C_0 and C_e for the initial and equilibrium concentration of the contaminant $\left(\dfrac{\text{mg}}{\text{L}}\right)$, M for the mass of the adsorbent added (g), q_e for equilibrium adsorption capacity $\left(\dfrac{\text{mg}}{\text{g}}\right)$, t_e for equilibrium adsorption time (h), performing a mass balance around the system, we obtain

$$VC_0 - VC_e = Mq_e \tag{15.1}$$

Or

$$q_e = \frac{V}{M}(C_0 - C_e) \tag{15.2}$$

Example 15.1 (General Engineering—All concentrations): Butyric acid is a malodorous compound commonly present in animal wastewater. An experiment was performed to test the efficacy of an engineered activated carbon to remove acetic acid from water at 25°C. One gram of activated carbon was contacted with 100 mL of butyric acid solution whose initial concentration was 100 ppm. Data was collected at various time intervals and the concentration of the butyric acid in the water was measured by a gas chromatograph. The collected data is presented as follows:
[Time (min), concentration (ppm)]: [0, 100], [10, 80], [30, 50], [60, 40], [120, 35], [180, 35], [240, 35].

Determine the adsorption capacity of the engineered activated carbon at equilibrium for butyric acid.

Step 1: Schematic
The problem statement is illustrated in Fig. 15.7a and the data is plotted in Fig. 15.7b.

Step 2: Given data
Initial concentration, $C_0 = 100 \dfrac{\text{mg}}{\text{L}}$; Equilibrium concentration, $C_e = 35 \dfrac{\text{mg}}{\text{L}}$; Mass of the engineered activated carbon (adsorbent), $M = 1$ g; Volume of the liquid, $V = 100$ mL; Temperature, $T = 25$°C; and the data for concentration of butyric acid at various times.

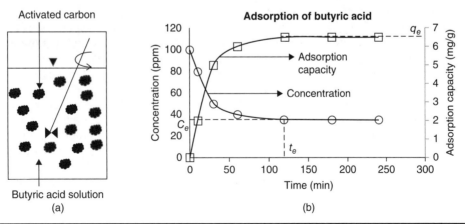

FIGURE 15.7 Schematic of the adsorption experiment (a) for Example 15.1 wherein butyric acid concentration in the liquid decreases with time due to adsorption (b).

To be determined: Adsorption capacity at equilibrium, q_e.

Assumptions: The engineered activated carbon was fresh and is free from butyric acid, and no reaction occurs between butyric acid and water.

Approach: We will first plot the data to visually analyze the trend and subsequently use Eq. (E15.1A) to determine the equilibrium adsorption capacity, q_e.

Step 3: Calculations
The raw data are plotted in Fig. 15.7b. The concentration of butyric acid decreases and remains constant (35 pm) from 120 min onward. Therefore, we will choose 2 hours as the equilibrium time, t_e and 35 ppm as the equilibrium concentration. Now using the mass balance equation, we can determine the adsorption capacity as

$$q_e = \frac{V}{M}(C_0 - C_e) \tag{E15.1A}$$

Substituting the values in the above equation, we obtain

$$q_e = \frac{0.1\ \text{L}}{1\ \text{g}}\left(100\ \frac{\text{mg}}{\text{L}} - 35\ \frac{\text{mg}}{\text{L}}\right) = 6.5\ \frac{\text{mg}}{\text{g}}$$

Based on the data analysis, we can infer that the engineered activated carbon can remove 6.5 mg of butyric acid for every 1 g of activated carbon added to the liquid at equilibrium at 25°C.
   Additionally, we can also plot the adsorption capacity at any time as

$$q_t = \frac{V}{M}(C_0 - C_t) \tag{E15.1B}$$

The adsorption capacity at various times is also plotted on a secondary y-axis in Fig. 15.7b. We can visually confirm that the adsorption capacity reaches an equilibrium at 6.5 ppm at 120 min.

15.4   Adsorption Isotherms

In our previous discussion, we determined the equilibrium adsorption capacity for a given concentration. However, all things being equal, the equilibrium adsorption capacity is a variable that generally increases with an increase in the concentration of a given species. The mathematical relationship between the concentration of the species and the amount of the species adsorbed at equilibrium at a constant temperature is called an adsorption isotherm.

   During the early 1900s, Irving Langmuir, an American engineer and Nobel laureate, developed a theoretical model for an adsorption isotherm by considering the surface to consist of several active sites for adsorption and assuming that (1) all active sites are identical, (2) each site can accommodate only one molecule, and (3) once adsorbed, the molecules do not interact with each other.

   With those assumptions, consider a surface with active sites exposed to a fluid containing a species whose concentration is C (Fig. 15.8). As adsorption proceeds, a portion of active sites will be occupied by the molecules. Let us call the proportion of the active sites occupied by the molecules as θ. Therefore, by difference, the proportion of the vacant active sites will be $1 - \theta$.

   Now, the rate of adsorption is proportional to the proportion of the vacant sites on the surface of the adsorbent and the concentration of the molecules in the bulk fluid around the adsorbent. Mathematically,

$$r_{\text{Ads}} \propto (1 - \theta)C \tag{15.3}$$

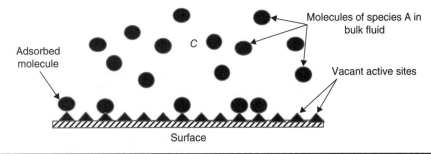

Figure 15.8   Schematic describing Langmuir's adsorption phenomenon (physical).

or

$$r_{Ads} = k_{Ads}(1-\theta)C \tag{15.4}$$

where k_{Ads} is the adsorption rate constant.

Similarly, the rate of desorption is proportional to the proportion of the occupied sites by the molecules, that is, θ

$$r_{Des} \propto \theta \tag{15.5}$$

or

$$r_{Des} = k_{Des}\theta \tag{15.6}$$

where k_{Des} is the desorption rate constant.

At equilibrium, the rate of adsorption is the same as the rate of desorption and $C = C_e$. Therefore,

$$k_{Ads}(1-\theta)C_e = k_{Des}\theta \tag{15.7}$$

If we define $\dfrac{k_{Ads}}{k_{Des}} = b$, the adsorption equilibrium constant, and rearrange the above equation as

$$bC_e = \theta(1+bC_e) \tag{15.8}$$

$$\theta = \frac{be}{1+bC_e} \tag{15.9}$$

The proportion of surface sites occupied can be defined as the ratio of amount of the species adsorbed on the surface at equilibrium (q_e) to the amount of the species that can be adsorbed if all sites were fully occupied (maximum adsorption, q_{max}). Therefore,

$$\theta = \frac{q_e}{q_{max}} \tag{15.10}$$

Combining Eqs. (15.9) and (15.10), we obtain

$$q_e = \frac{bC_e q_{max}}{1+bC_e} \tag{15.11}$$

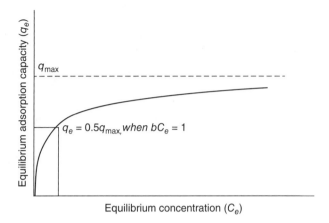

Figure 15.9 Graphical representation of Langmuir's adsorption model (mathematical).

This is the famous Langmuir's adsorption equation or adsorption isotherm equation because the process occurs at a constant temperature. Langmuir's adsorption isotherm can be graphically presented as shown in Fig. 15.9.

Analyzing the equation and the adsorption isotherm, we can derive some important conclusions which are as follows.

Case (i): If the concentration of the species is very low, that is,

$$bC_e \ll 1 \tag{15.12}$$

The adsorption equation becomes

$$q_e \approx bC_e q_{max} \tag{15.13}$$

Since b and q_{max} are constants

$$q_e \propto C_e \tag{15.14}$$

Therefore, at very low concentrations, the equilibrium adsorption capacity is linearly related to the equilibrium concentration, which can also be seen from Fig. 15.9.

Case (ii): If the concentration of the species is very high, that is,

$$bC_e \gg 1 \tag{15.15}$$

The adsorption equation becomes

$$q_e \approx q_{max} \tag{15.16}$$

Which suggests that the equilibrium adsorption capacity is independent of the equilibrium concentration at very high concentrations.

Case (iii): For a special case, when

$$bC_e = 1 \tag{15.17}$$

The adsorption equation becomes

$$q_e = 0.5q_{max} \tag{15.18}$$

Using algebra, the Langmuir's adsorption isotherm can easily be linearized as

$$\frac{1}{q_e} = \left(\frac{1}{bq_{max}}\right)\frac{1}{C_e} + \frac{1}{q_{max}}$$ (15.19)

The linearized form of the equation has been traditionally used to determine the maximum adsorption capacity, q_{max} and adsorption equilibrium constant, b from the slope and the intercept of the experimental data, $\frac{1}{q_e}$ vs $\frac{1}{C_e}$.

   Note on linearization: Mathematically, it may be easier to linearize the isotherm and kinetic equations to solve and obtain the relevant parameters. However, it is important to note that linearization sometimes leads to incorrect values of the parameters and may impact the design of the adsorption system. Fortunately, several software packages including OriginLab, Microsoft Excel, and Systat Sigma Plot can easily handle nonlinear regression. Therefore, it is recommended to test both linear and nonlinear approaches to compare the results. Examples 15.2 and 15.3 illustrate the procedure.

Example 15.2 (Bioprocessing Engineering—Determination of maximum adsorption capacity):   Biochar was tested in laboratory to remove p-cresol from swine wastewater. The experiment was conducted at 25°C in an agitated reactor in which 100 mL solutions of p-cresol of various concentrations was mixed with 1 g of biochar (Fig. 15.10). The liquid samples were periodically collected and analyzed via a mass spectrometer to determine the remaining p-cresol concentration in the solution and the data is summarized next (Table 15.1). Construct an adsorption isotherm and determine the maximum adsorption capacity of the biochar and the Langmuir adsorption equilibrium constant.

Step 1: Schematic
The problem is depicted in Fig. 15.10.

Step 2: Given data
Mass of the adsorbent, $M = 1$ g; Volume of the solution, $V = 100$ mL; Temperature of the liquid, $T = 25°C$; Number of experiments (initial concentrations) = 4

To be determined: Adsorption isotherm, maximum adsorption capacity, and Langmuir adsorption equilibrium constant, all at 25°C.

Assumptions: The agitation is adequate that the concentration of p-cresol is uniform within the reactor, reactor walls are nonreactive, and do not adsorb p-cresol.

Approach: We will first determine the adsorption capacity for all sampling periods ($t = 0$ to $t = 70$ min) and determine the adsorption equilibrium capacity (q_e) and equilibrium concentration (C_e) for each experiment. Subsequently, we will construct the isotherm at 25°C by plotting the q_e versus C_e data. Finally, using the data we will calculate the maximum adsorption capacity (q_{max}), and Langmuir adsorption equilibrium constant (b) using nonlinear and linear curve fitting of the isotherm data.

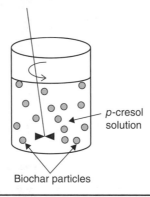

p-cresol
solution

Biochar particles

FIGURE 15.10   Schematic for adsorption of p-cresol on biochar in Example 15.2.

Experiment 1		Experiment 2		Experiment 3		Experiment 4	
Time (min)	Conc (mg/L)	Time (min)	Conc (mg/L)	Time (min)	Conc (mg/L)	Time (min)	Conc (mg/L)
0	55.6	0	105.5	0	199.4	0	494.8
10	52.8	10	100.2	10	188.9	10	458.6
20	48.8	20	93.8	20	181.7	20	463.2
30	46.4	30	90.8	30	178.2	30	456.77
40	45.1	40	89.7	40	172.3	40	448.93
50	43.4	50	87.5	50	170.6	50	443.77
60	44.1	60	87.3	60	173	60	444.57
70	42.89	70	86.5	70	173.1	70	436.64

TABLE 15.1 Experimental Data Collected during an Adsorption Experiment for Removal of p-Cresol from Wastewater

Step 3: Calculations

Construction of adsorption isotherm

The adsorption capacity of biochar after any given time is determined via

$$q_t = \frac{V}{M}(C_0 - C_t) \tag{E15.2A}$$

where q_t is the adsorption capacity after time, t (min), C_0 and C_t are the initial concentration and the concentration $\left(\frac{mg}{L}\right)$ after time, t, M is the mass of the adsorbent added (g), and V is the volume of the fluid (L).

Therefore, adsorption capacity of the biochar for experiment 1 (whose initial concentration is 55.6 ppm) after 10 min can be calculated as

$$q_{10\ min} = \frac{V}{M}(C_0 - C_{10\ min}) = \frac{0.1\ L}{1\ g}\left(55.6 - 52.8\right)\frac{mg}{L} = 0.28\ \frac{mg}{g}$$

Similarly, as described above, the adsorption capacities and equilibrium concentrations for all the experiments (bolded data) after all sampling times are calculated and tabulated as follows:

	Experiment 1		Experiment 2		Experiment 3		Experiment 4	
Time (min)	Conc (ppm)	Adsorption Capacity (mg/L)	Conc (ppm)	Adsorption Capacity (mg/L)	Conc (ppm)	Adsorption Capacity (mg/L)	Conc (ppm)	Adsorption Capacity (mg/L)
0	55.6	0	105.5	0	199.4	0	494.8	0
10	52.8	0.28	100.2	0.53	188.9	1.05	458.6	3.62
20	48.8	0.68	93.8	1.17	181.7	1.77	463.2	3.16
30	46.4	0.92	90.8	1.47	178.2	2.12	456.77	3.803
40	45.1	1.05	89.7	1.58	172.3	2.71	448.93	4.587
50	43.4	1.22	87.5	1.8	170.6	2.88	443.77	5.103
60	44.1	1.15	87.3	1.82	173	2.64	444.57	5.023
70	42.89	1.271	86.5	1.9	173.1	2.63	436.64	5.816

TABLE 15.2 Analysis of the Experimental Data to Determine the Equilibrium Adsorption Capacity for Each Concentration

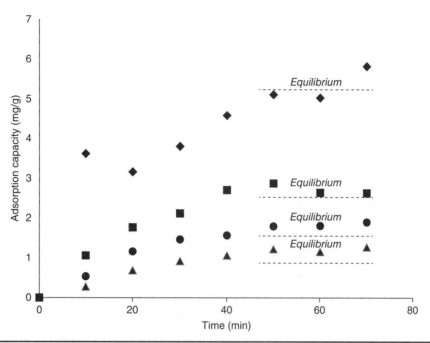

Figure 15.11   Adsorption versus time plot for the raw experimental data in Example 15.2.

In addition, the data is plotted as shown in Fig. 15.11.

From Table 15.2 and Fig. 15.11, we can observe that the equilibrium is attained around 50 min and the equilibrium concentrations (C_e) and the equilibrium adsorption capacities (q_e) are 43.4 mg/L and 1.22 mg/g, 87.5 mg/L and 1.8 mg/g, 170.6 mg/L and 2.88 mg/g, and 443.77 mg/L and 5.103 mg/g, respectively. Now, the isotherm can be constructed by plotting q_e versus C_e as shown in Fig. 15.12.

Determination of maximum adsorption capacity

Linearization method: The linear form of the Langmuir equation is

$$\frac{1}{q_e} = \left(\frac{1}{bq_{max}} \right) \frac{1}{C_e} + \frac{1}{q_{max}}$$ (E15.2B)

Therefore, the plot of $\dfrac{1}{q_e}$ versus $\dfrac{1}{C_e}$ will result in slope and intercept of $\dfrac{1}{bq_{max}}$ and $\dfrac{1}{q_{max}}$, respectively. The data is shown in Table 15.3.

The plot of $\dfrac{1}{q_e}$ versus $\dfrac{1}{C_e}$ is presented in Fig. 15.13.

From the intercept, we can determine the maximum adsorption capacity (q_{max}) as

$$\frac{1}{q_{max}} = 0.1662 \Rightarrow q_{max} = 6.01 \, \frac{mg}{g}$$

Similarly, from the slope of the curve, the Langmuir adsorption equilibrium constant (b) as

$$\frac{1}{bq_{max}} = 29.439 \Rightarrow b = 0.0056 \, \frac{L}{mg}$$

Nonlinear method: The nonlinear form of the Langmuir equation is given by

$$q_e = \frac{bC_e q_{max}}{1 + bC_e}$$ (E15.2C)

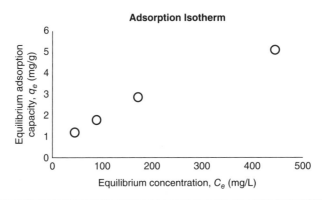

FIGURE 15.12 Langmuir's adsorption isotherm plot for the experimental data in Example 15.2.

Using the C_e and q_e data presented in Table 15.3, we can use Excel Solver to determine the values of q_{max} and b as

$$q_{max} = 8.96 \frac{mg}{g} \quad \text{and} \quad b = 0.003 \frac{L}{mg}$$

Considering that nonlinear fitting is more accurate, we may select the maximum adsorption capacity as $8.96 \frac{mg}{g}$.

$C_e \left(\dfrac{mg}{L} \right)$	$q_e \left(\dfrac{mg}{g} \right)$	$\dfrac{1}{C_e} \left(\dfrac{L}{mg} \right)$	$\dfrac{1}{q_e} \left(\dfrac{g}{mg} \right)$
43.4	1.22	0.023	0.81
87.5	1.8	0.011	0.55
170.6	2.88	0.005	0.34
443.77	5.103	0.002	0.19

TABLE 15.3 Equilibrium Adsorption Data for Linearization Method

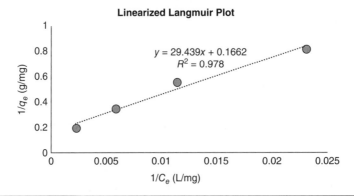

FIGURE 15.13 Linearized Langmuir's adsorption isotherm plot for the experimental data in Example 15.2.

It is also important to recognize that some of the assumptions we made are not always valid. We assumed that all the sites are identical. However, based on the evidence from surface characterization techniques, we now know that the surface can be highly heterogeneous. Similarly, mono-layer adsorption is valid in some cases, but vast majority of surfaces in fact allow for multilayer adsorption. Further, adjacent molecules frequently interact with each other thereby invalidating the assumption. Despite these limitations, Langmuir's model works very well for chemisorption and is a first step in any adsorption analysis. Nonetheless, over the years, numerous adsorption isotherm models have been developed and are available in the literature. Some of the common models used in biological and agricultural engineering are discussed here.

15.5 Freundlich Adsorption Isotherm

Prof. Herbert Freundlich, a German chemist, proposed an empirical model to describe the adsorption which is presented in Eq. (15.20).

$$q_e = K_F C_e^{\frac{1}{n}} \tag{15.20}$$

where q_e and C_e are equilibrium adsorption capacity $\left(\dfrac{mg}{g}\right)$ and equilibrium concentration $\left(\dfrac{mg}{L}\right)$, respectively while K_F and $\dfrac{1}{n}$ represent constants called Freundlich parameters $\left(\dfrac{mg}{g}\right)\left(\dfrac{L}{mg}\right)^{(1/n)}$ and surface heterogeneity (dimensionless), respectively. The interpretation of n is summarized in Table 15.4.

Freundlich's model is often linearized as

$$\ln(q_e) = \ln(K_F) + \frac{1}{n}\ln(C_e) \tag{15.24}$$

By plotting $\ln(q_e)$ versus $\ln(C_e)$, we can easily obtain the Freundlich parameters. Freundlich's model, although empirical, accounts for the heterogeneity of the surface and molecules adsorbing as multilayers on the surface. However, a high concentration, Freundlich's model is found to be inconsistent.

Surface Heterogeneity	Inference	Equation #
$\dfrac{1}{n} < 1$	Favorable	(15.21)
$\dfrac{1}{n} = 1$	Linear	(15.22)
$\dfrac{1}{n} > 1$	Unfavorable	(15.23)

TABLE 15.4 Interpretation of the Values of Freundlich's Surface Heterogeneity (*n*)

Example 15.3 (Bioprocessing Engineering—Calculation of adsorption capacity):   Using the data presented in Example 15.2, determine the Freundlich adsorption parameters.

Step 1: Schematic
The problem depicted in Fig. 15.10 for Example 15.2 is the same for this example.

Step 2: Given data
The processed data from Example 15.2 is tabulated in Table 15.5.

To be determined: Freundlich's adsorption parameters.

Approach: We will determine the Freundlich adsorption parameters using both linearization and nonlinearization methods. For the linearization method, the slope and the intercept of the plot between ln (q_e) and ln (C_e) will yield K_f and $\frac{1}{n}$. For nonlinearization, we will directly use the Excel Solver function.

Step 3: Calculations
Linearization method: The linearized form of Freundlich's adsorption isotherm is given by

$$\ln (q_e) = \ln (K_F) + \frac{1}{n} \ln (C_e) \tag{15.3A}$$

and the plot between ln (q_e) and ln (C_e) is presented in Fig. 15.14.
From the plot, the slope and the intercept are given by 0.622 and –2.1639, respectively. Therefore,

$$\text{Slope} = \frac{1}{n} = 0.622$$

$$\text{Intercept} = \ln (K_F) = -2.1639 \Rightarrow K_F = 0.115 \frac{L}{mg}$$

C_e (mg/L)	q_e (mg/g)
43.4	1.22
87.5	1.8
170.6	2.88
443.77	5.103

TABLE 15.5   The Equilibrium Adsorption Data from Example 15.2

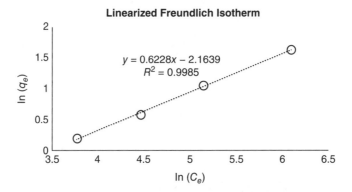

FIGURE 15.14   The linearized Freundlich adsorption plot for the experimental data.

Nonlinearization method:
Using the Excel Solver, the data in Table 15.6 are fit to the Freundlich's equation. The obtained values are

$$\frac{1}{n} = 0.62 \text{ and } K_F = 0.116 \frac{\text{L}}{\text{mg}}$$

Therefore, both linear and nonlinear approaches seem to provide similar results.

15.6 Brunauer–Emmett–Teller (BET) Adsorption Isotherm

In 1938, three American scientists—S. Brunauer, P. Emmet, and E. Teller—revised Langmuir's adsorption isotherm to include multilayer adsorption on surface. The model, predominantly used to describe physisorption of gases on solid surfaces, takes into account the fact that adsorption continues after the saturation of the surface by adsorption of molecules on the top of the adsorbed surface (Fig. 15.15) such that the proportion of sites occupied by all the layers equals 1 ($\sum_1^n \theta_n = 1$, where n is the number of layers).

Popularly called the BET model, it is considered a gold standard to analyze the surface areas of porous materials and the procedure is aptly called the BET method. A version of the BET model for adsorption of liquids on surfaces as developed by Ebadi et al. (2009) is presented as follows:

$$q_e = \frac{q_{max}K_1C_e}{(1 - K_2C_e)(1 - K_2C_e + K_1C_e)} \tag{15.25}$$

where q_e, q_{max}, and C_e are the equilibrium adsorption capacity $\left(\frac{\text{mg}}{\text{g}}\right)$, maximum adsorption capacity $\left(\frac{\text{mg}}{\text{g}}\right)$, and equilibrium concentration $\left(\frac{\text{mg}}{\text{L}}\right)$, respectively, while K_1 and K_2 are the equilibrium adsorption constants $\left(\frac{\text{L}}{\text{mg}}\right)$ for first and top layers.

Of course, while operating at low concentrations, that is, when $K_2C_e \ll 1$, the BET model will be reduced to the familiar form of Langmuir's model.

$$q_e \approx \frac{q_{max}K_1C_e}{(1 + K_1C_e)} \tag{15.26}$$

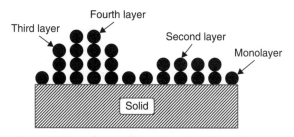

FIGURE 15.15 The simplified schematic representing BET adsorption process.

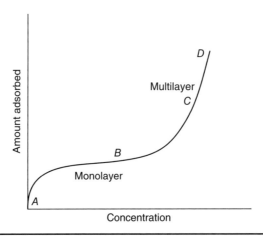

FIGURE 15.16 Sample adsorption isotherm of the data confirming to BET adsorption mechanism depicting mono (AB) and multilayer (CD) adsorption.

At high concentrations, however, the adsorption capacity will reach infinity. The typical shape of the BET model is shown in Fig. 15.16.

The adsorption models we described are equally valid for gas-phase adsorption after replacing the concentrations with gas pressures as follows:

Model	Mathematical Form	Equation #
Langmuir	$\theta = \dfrac{bP}{1 + bP}$	(15.27)
Freundlich	$\theta = K_F P^{\frac{1}{n}}$	(15.28)
BET	$\theta = \dfrac{C\left(\dfrac{p}{p_{sat}}\right)}{\left(1 - \dfrac{p}{p_{sat}}\right)\left(1 + \dfrac{p}{p_{sat}}(C - 1)\right)}$	(15.29)

P = Pressure (Pa)

b = Langmuir's adsorption equilibrium constant $\left(\dfrac{1}{Pa}\right)$

C = BET constant

K_F = Freundlich's parameter

$\dfrac{1}{n}$ = surface heterogeneity

p and p_{sat} are vapor and saturated vapor pressure (Pa) of the gas

$\theta = \dfrac{V_e}{V_{max}} = \dfrac{\text{Volume of gas adsorbed at equilibrium (m}^3)}{\text{Maximum volume of the gas at monolayer (m}^3)}$

TABLE 15.6   Adsorption Isotherm Models for Adsorption of Gases on Surfaces

The shape of the adsorption isotherm is an excellent tool to predict the surface morphology of a given adsorbent material. Brunauer classified adsorption based on their shapes (Fig. 15.17). The salient features of each type are summarized in Table 15.7.

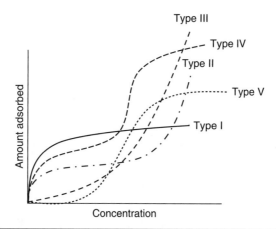

FIGURE 15.17   General types of adsorption isotherms depending on their shapes (Brunauer et al., 1940).

Adsorption Type	Features	Examples
Type 1	• Monolayer coverage • Represents chemisorption • Adsorption increases rapidly initially and plateaus at high concentration	Oxygen on activated carbon
Type 2	• Multilayer adsorption • Physisorption • Begins as a monolayer and turns into multilayer	Nitrogen on activated carbon
Type 3	• Strongly convex pattern • Not commonly observed • Suggests strong interaction between adsorbent molecules relative to adsorbent–adsorbate interaction	Moisture on dry fruits
Type 4	• Multilayer and capillary condensation • Presence of hysteresis due to different condensation and evaporation rates	Water on activated carbon
Type 5	• Similar to type 3 with weak interaction between adsorbent and adsorbate	Water on zeolite

TABLE 15.7   Representative Features of Various Types of Adsorption Isotherm Models (Kurk and Jaroniec, 2001)

15.7   Design of Batch Adsorption Systems Using Adsorption Isotherm Data

The isotherm data can be combined with the mass balance for a batch process to determine the mass of the adsorbent needed for a given application. For example, if Langmuir's isotherm fits the model best, we write the isotherm and the batch equation as

$$q_e = \frac{bC_e q_{max}}{1 + bC_e} \tag{15.30}$$

$$q_e = \frac{V}{M}(C_0 - C_e) \tag{15.31}$$

Combining the above equations, we can express the mass of the adsorbent needed for a given fractional removal as per the design equation:

$$M = \frac{V\,(C_0 - C_e)\,(1 + bC_e)}{bC_e q_{max}} \tag{15.32}$$

Similarly, for situations Freundlich's model is appropriate, the mass can be determined via the following design equation.

$$M = \frac{V\,(C_0 - C_e)}{K_F C_e^{\frac{1}{n}}} \tag{15.33}$$

In either case, the values of b, q_{max}, K_F, and $\frac{1}{n}$ are from the respective isotherm analysis, initial (C_0) and the target (equilibrium) concentration C_e, and volume as usually known.

Example 15.4 (Environmental Engineering–Calculating the mass of adsorbent needed):   Phenols are an undesirable group of compounds that are present in wastewater. Therefore, you as an engineer were asked to decrease the concentration of phenol from 50 to 22 mg/L from a 1,000,000 L wastewater effluent tank using a commercial adsorbent at 25°C (Fig. 15.18). The data sheet from the adsorbent manufacturers indicated that the adsorption model is Langmuir type as described by

$$q_e = \frac{1.4C_e}{1 + 0.02C_e} \tag{E15.4A}$$

How much of the adsorbent would you recommend?

Step 1: Schematic
The problem is described in Fig. 15.18.

Step 2: Given data
Initial concentration of phenol, $C_i = 50\ \frac{mg}{L}$; Target (equilibrium) concentration of phenol, $C_e = 22\ \frac{mg}{L}$; Volume of wastewater to be treated, $V = 1,000,000$ L; Temperature of wastewater, $T = 25°C$; Suggested adsorption model at 25°C is $q_e = \frac{1.4C_e}{1 + 0.02C_e}$

To be determined: Mass of the adsorbent needed, M.

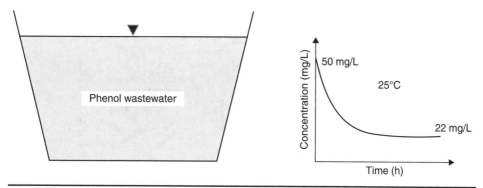

FIGURE 15.18   Schematic for the adsorptive removal of phenol in Example 15.4.

Assumptions: The fresh adsorbent is free of phenol, the temperature remains constant during treatment, the adsorbent only removes phenol, the conditions under which the adsorption isotherm was prepared are the same as that of treatment, and the adsorption occurs in batch mode.

Approach: This is the simple application of the batch adsorption design equation. We will combine the mass balance with the given adsorption isotherm equation to determine the mass of the adsorbent required.

Step 3: Calculations
Equating the mass balance with the given adsorption equation,

$$\frac{V}{M}(C_0 - C_e) = \frac{1.4C_e}{1 + 0.02C_e} \tag{E15.4B}$$

Substituting the relevant data,

$$\frac{1,000,000}{M}(50 - 22) = \frac{1.4 \times 22}{1 + (0.02 \times 22)} \tag{E15.4C}$$

The mass of adsorbent needed will be, $M \approx 1309.1$ kg.

15.8   Kinetic Analysis of the Adsorption Data

One of the important engineering considerations while selecting an adsorbent is its kinetics. The kinetic analysis will inform us about the mechanism of adsorption, the rate-limiting steps. Typically, within a porous adsorbent, the adsorption proceeds in a series of steps including the transport of the species from the bulk to the surface boundary layer followed by the film diffusion through the boundary layer onto the adsorbent surface. Subsequently, the species is transported within the particle via either surface or pore diffusion (D_s or D_p, respectively), to eventually adsorb on the surface of the adsorbent (Ekhard, 2012). For all practical purposes, in well mixed systems, the transport of the species from bulk liquid to the surface of the boundary layer is very fast because of the intimate contact between the adsorbent and the fluid. Similarly, from a thermodynamics point of view, the process of adsorption between species and the surface is also very fast. Therefore, either the film diffusion (within the boundary layer) or pore diffusion will eventually control the rate of adsorption (Fig. 15.19).

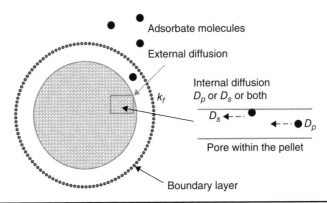

FIGURE 15.19   Illustration of various transport processes within spherical adsorbent pellet.

Based on mass balance around a batch reactor, the film diffusion can be expressed as

$$\ln\left(\frac{C}{C_0}\right) = -\frac{m a k_f}{V} t \tag{15.34}$$

where C_0 (mg L^{-1}) and C (mg L^{-1}) are the concentrations of the species (initial and after time t), respectively; m (g), a (m^{2}g^{-1}), k_f (m min^{-1}), and V (L) are the mass of adsorbent, specific surface area of the adsorbent, film mass transfer coefficient, and the volume of the fluid, respectively.

The diffusional effects within the pore can be modeled as follows:

$$\frac{q}{q_e} = \frac{6}{R}\sqrt{\frac{D_p t}{\pi}} + K \tag{15.35}$$

where q_e (mg g^{-1}) and q (mg g^{-1}) are adsorption at equilibrium and at any time, t (min) respectively; R (m) is the radius of the particle, D_p (m^2 min^{-1}) is the pore diffusivity; and K is the intraparticle diffusion constant, respectively.

The slope of the linear portion of the curve between $\ln(C_t/C_0)$ and time (Eq. (15.34)) provides an estimate of the film mass transfer coefficient (k_f) while the slope of the linear portion of q_t/q_e versus $\sqrt{t}$ (Eq. (15.35)) is the internal (pore) diffusivity (D_p). Once the film mass transfer coefficient and the pore diffusivity are determined, the Biot number can be used to determine the rate-limiting step. For porous adsorbents such as activated carbons, Khraisheh et al. (2002) proposed that when Biot number ($L_c k_f / D_p$) exceeds 100, the system may be considered as pore-diffusion limiting.

15.9   Adsorption Thermodynamics

Another approach to obtain additional insight into the mechanism of adsorption is to use experimental data to evaluate the thermodynamics of adsorption. To accomplish this, the change in the enthalpy and entropy is calculated from the Gibbs free energy equation:

$$\ln(K_d) = \frac{\Delta G^0}{RT} = -\frac{\Delta H^0}{RT} + \frac{\Delta S^0}{R} \tag{15.36}$$

where k_d is the adsorption distribution coefficient and is defined as $K_d = \dfrac{q_e}{C_e}$, ΔG^0 is the Gibbs free energy, ΔH^0 is the change in enthalpy, ΔS^0 is the change in the entropy, R is the universal gas constant, and T is the absolute temperature (K).

By plotting the experimental data, $\ln (K_d)$ versus $\dfrac{1}{T}$ to obtain the slope and intercept as $\dfrac{-\Delta H^0}{R}$ and $\dfrac{-\Delta S^0}{R}$, respectively, we can infer the adsorption as endothermic if $\Delta H^0 > 0$ and spontaneous if $\Delta G^0 < 0$. Similarly, the change in entropy $\Delta S^0 > 0$ suggests an increase in randomness of the adsorption process at the surface.

15.10 Column Adsorption

Batch adsorption systems are useful for liquid systems when the amount of liquid to be treated is small to moderate. However, in industrial wastewater and air treatment operations, continuous packed bed adsorption columns are routinely used. Typically, the contaminant fluid is flown through a cylindrical column loaded with specially engineered adsorbent pellets (Fig. 15.20a). As the fluid passes through the bed, the contaminant is adsorbed on the surface of the adsorbent and the contaminant-free fluid exits the column.

During the initial stages, assuming that the adsorbent is fresh, the fluid exiting the outlet will be free of contaminant because the bed is adsorbing all the contaminant. However, with the passage of time, sections of the bed get saturated, the mass transfer zone (the section of the bed where adsorption occurs) starts to move toward the outlet of the column. The time at which the mass transfer zone approaches the outlet of the column is called the breakthrough time, t_b. From a process standpoint, the concentration at the breakthrough time may be viewed as the maximum allowable concentration of the contaminant the outlet can carry. Typically, once the breakthrough is reached the column is considered spent and replaced with a new column with fresh batch of adsorbent.

After the breakthrough time, the concentration of the contaminant at outlet will begin to increase, and eventually reaches the original inlet concentration C_0 after an equilibrium time t_e (at which the entire bed is used up) (Fig. 15.20b). For all practical purposes, a column is said to have broken through when the outlet concentration reaches 5% to 10% of the inlet concentrations.

From a design standpoint, we are interested to know how much of the contaminant is adsorbed on the bed and what is the total adsorption capacity of the bed. If C_0 is the inlet concentration of the contaminant $\left(\dfrac{mg}{L}\right)$, Q is the flow rate (LPM), and t is the time (min), the total mass of the contaminant that entered the column in a given time t is QC_0t which can also be written as

$$M_{in} = \int_0^t QC_0\,dt \tag{15.37}$$

The mass of contaminant that exited the outlet is given by

$$M_{out} = \int_0^t QC\,dt \tag{15.38}$$

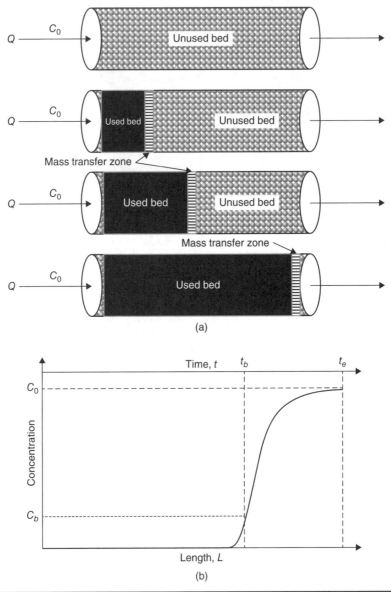

FIGURE 15.20 Adsorption of a species through a packed bed. As the bed begins to get exhausted (a) the concentration in the outlet begins to increase (b).

The mass of the contaminant accumulated in the bed is $M_{accu} = M_{in} - M_{out}$, which after simplification becomes

$$M_{accu} = QC_0 \int_0^t \left(1 - \frac{C}{C_0}\right) dt \qquad (15.39)$$

Therefore, the mass of contaminant accumulated on the column up to breakthrough, M_b is

$$M_b = QC_0 \int_0^{t_b} \left(1 - \frac{C}{C_0}\right) dt \qquad (15.40)$$

where t_b is the breakthrough time.
The total equilibrium adsorption capacity of the column is

$$M_e = QC_0 \int_0^{t_e} \left(1 - \frac{C}{C_0}\right) dt \qquad (15.41)$$

where t_e is the equilibrium time.
And the portion of column used up at breakthrough is given by

$$L_{\text{used}} = \left(\frac{M_b}{M_e}\right) L \qquad (15.42)$$

Example 15.5 (Environmental Engineering—Adsorption column analysis): It is proposed to remove acid orange 7, a textile dye from a wastewater stream. Laboratory experiments were conducted using a 25-cm long and 2.5-cm diameter adsorption column loaded with 10 g of sodium hydroxide-activated biochar. A 100 mg/L of dye solution was pumped through the adsorbent bed at 12 mL/min and the concentration of the dye at the outlet was measured via spectrophotometer every 30 min as shown next (Fig. 15.21).

[Time (min), Concentration (mg/L)]: [0, 0], [30, 0], [60, 0], [90 ,0], [120, 0], [150, 0], [180, 5], [210, 10], [240, 14], [270, 20], [300, 28], [330, 40], [360, 60], [390, 80], [420, 85], [450, 88], [480, 90], [510, 97], [540, 99], [570, 99], [600, 99].

Determine the mass of the dye adsorbed at breakthrough point if breakthrough occurs when the outlet concentration reaches 10 mg/L. Also determine the length of the unused column and the total adsorption capacity of the column.

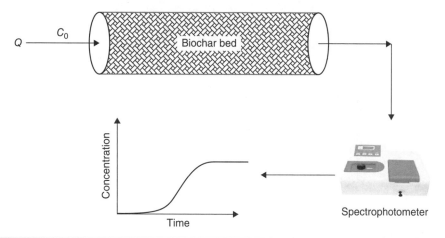

Figure 15.21 Schematic for the adsorptive removal of acid orange 7 through a packed bed column in Example 15.5.

Step 1: Schematic
The problem is described in Fig. 15.21.

Step 2: Given data
Mass of the adsorbent, $M = 10$ g; Length of the column, $L = 25$ cm; Diameter of the column, $D = 2.5$ cm; Initial concentration of the dye, $C_0 = 100 \frac{mg}{L}$; The flow rate of the dye, $Q = 12 \frac{mL}{min}$; Outlet concentration versus time data is also provided in the problem statement.

To be determined: Mass of the dye removed at the breakthrough point, length of the unused column, and total adsorption capacity of the column.

Assumptions: The flow is steady state, the acid orange 7 dye does not degrade with time, and adsorption is the only removal mechanism.

Approach: We will plot the change in the outlet concentration with time and determine the breakthrough time when the concentration in the outlet reaches 10 mg/L. Subsequently, we will integrate the area under the curve to obtain the mass of the dye adsorbed on the column up to the breakthrough point. The total adsorption capacity is the total area above the curve (when outlet concentration approaches the inlet concentration). The unused portion of the column is simply the difference between the total column length and the length of the column at breakthrough point.

Step 3: Calculations
From the concentration data plotted in Fig. 15.22 and the given data, we can locate that the time to achieve the breakthrough (outlet concentration = 10 mg/L) is 210 min.

Therefore, the mass of contaminant accumulated on the column up to breakthrough, M_b is given by

$$M_b = QC_0 \int_0^{210 \text{ min}} \left(1 - \frac{C}{C_0} \right) dt \tag{E15.5A}$$

which is the area above the breakthrough curve, which is represented by the area OABC, multiplied by QC_0. We can obtain the value within the integral by numerically integrating the equation between 0 and 210 min in Excel. As shown in the calculations in Table 15.8, the area of OABC is 207 min. Therefore, the mass adsorbed on the column at the breakthrough point is

$$M_b = QC_0 \int_0^{210 \text{ min}} \left(1 - \frac{C}{C_0} \right) dt = 0.012 \times 100 \times 207 = 248.4 \text{ mg}$$

Similarly, the total adsorption capacity of the column at equilibrium is

$$M_e = QC_0 \int_0^{t_e} \left(1 - \frac{C}{C_0} \right) dt \tag{E15.5B}$$

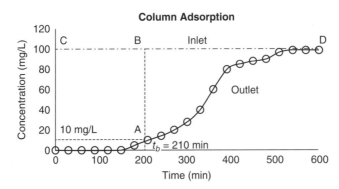

FIGURE 15.22   The breakthrough plot between outlet concentration and time for the given data in Example 15.5.

t (min) Time	C₀ (mg/L) Inlet	C (mg/L) Outlet	C/C₀	1 − (C/C₀)	Average of (1 − (C/C₀))	Average of (1 − (C/C₀)) *delta t	
0	100	0	0	1			
30	100	0	0	1	1	30	
60	100	0	0	1	1	30	
90	100	0	0	1	1	30	
120	100	0	0	1	1	30	
150	100	0	0	1	1	30	
180	100	5	0.05	0.95	0.975	29.25	
210	100	10	0.1	0.9	0.925	27.75	**207**
240	100	14	0.14	0.86	0.88	26.4	
270	100	20	0.2	0.8	0.83	24.9	
300	100	28	0.28	0.72	0.76	22.8	.
330	100	40	0.4	0.6	0.66	19.8	
360	100	60	0.6	0.4	0.5	15	
390	100	80	0.8	0.2	0.3	9	
420	100	85	0.85	0.15	0.175	5.25	
450	100	88	0.88	0.12	0.135	4.05	
480	100	90	0.9	0.1	0.11	3.3	
510	100	97	0.97	0.03	0.065	1.95	
540	100	99	0.99	0.01	0.02	0.6	
570	100	99	0.99	0.01	0.01	0.3	
600	100	99	0.99	0.01	0.01	0.3	
					Total area	**340.65**	

TABLE 15.8 Spreadsheet Outlining the Numerical Integration Procedure to Determine Mass of Acid Orange 7 Adsorbed on Sodium Hydroxide-Activated Biochar

The area within the integral is given by the area of ODC multiplied by QC_0. In Excel, numerical integration between 0 and 600 min yields 340.65 min. Therefore, the total mass adsorbed on the column at equilibrium is

$$M_e = 0.012 \times 100 \times 340.5 = 408.78 \text{ mg}$$

The length of the column used until breakthrough is

$$L_{\text{used}} = \left(\frac{M_b}{M_e}\right) L = \left(\frac{248.4}{408.78}\right) \times 0.25 = 0.152 \text{ m}$$

$$L_{\text{unused}} = 0.25 - 0.152 = 0.098 \text{ m}$$

15.11   Kinetic Modeling of Column Adsorption

Some of the widely used models in kinetic analysis are summarized in the following sections.

15.11.1   Bohart–Adams Model

One of the most popular models to describe the adsorption kinetics was proposed by Bohart and Adams in 1920 to describe the adsorption of chlorine on charcoal and assumed no axial dispersion and an ideal rectangular isotherm. The general form of the model as presented by Bohart–Adams is

$$\frac{\partial q}{\partial t} = k_{BA} C [q_e - q] \tag{15.43}$$

where k_{BA} is the Bohart–Adams constant $\left(\dfrac{L}{mg \cdot min}\right)$, C is the concentration $\left(\dfrac{mg}{L}\right)$ at any time (min), q_e and q are the maximum adsorption capacity at equilibrium $\left(\dfrac{mg}{g}\right)$, and adsorption capacity $\left(\dfrac{mg}{g}\right)$ at any time, t. In its simplified form, the model is expressed as

$$\ln\left(\frac{C_0}{C} - 1\right) = k_{BA}\left(\frac{N_0 L}{v} - C_0 t\right) \tag{15.44}$$

where C_0 is the inlet concentration $\left(\dfrac{mg}{L}\right)$, C is the concentration $\left(\dfrac{mg}{L}\right)$ at any time t (min), k_{BA} is the Bohart–Adams constant $\left(\dfrac{L}{mg \cdot min}\right)$, L is the length of the bed (m), N_0 is the adsorption per unit volume of the bed $\left(\dfrac{mg}{L}\right)$, v is the fluid's superficial velocity $\left(\dfrac{m}{min}\right)$.

A plot between $\ln\left(\dfrac{C_0}{C} - 1\right)$ and t will yield $k_{BA} C_0$ as the slope and $\dfrac{k_{BA} N_0 L}{v}$ as the intercept, from which the k_{BA} and N_0 can be determined.

15.11.2   Yoon–Nelson Model

In 1984, Yoon and Nelson proposed a simple model with fewer parameters to analyze the gas adsorption kinetics and predict the half-life of the bed saturation (or 50% breakthrough) as follows:

$$\ln\left(\frac{C}{C_0 - C}\right) = k_{YN}(t - \tau) \tag{15.45}$$

where C_0 is the inlet concentration $\left(\dfrac{mg}{L}\right)$, C is the concentration $\left(\dfrac{mg}{L}\right)$ at any time t (min), k_{YN} is the Yoon–Nelson constant $\left(\dfrac{1}{min}\right)$, and τ is the time for 50% breakthrough (min).

15.11.3 Thomas Model

Proposed by Henry Thomas in 1944, this model also assumes that adsorption does not involve axial dispersion. In addition, it also assumes that the adsorption on the surface occurs via Langmuir's first-order kinetic model. The most common version is expressed as

$$C = C_0 \frac{1}{1 + \exp\left(\dfrac{m q_e k_T}{Q} - k_T C_0 t\right)} \tag{15.46}$$

and in its linear form as

$$\ln\left(\frac{C_0}{C} - 1\right) = \frac{m q_e k_T}{Q} - k_T C_0 t \tag{15.47}$$

where C_0 is the inlet concentration $\left(\dfrac{mg}{L}\right)$, C is the concentration $\left(\dfrac{mg}{L}\right)$ at any time t (min), m is the mass of the adsorbent (g), q_e is the maximum adsorption capacity at equilibrium $\left(\dfrac{mg}{g}\right)$, k_T is the Thomas rate constant $\left(\dfrac{L}{mg \cdot min}\right)$, and Q is the flow rate (LPM).

> **Example 15.6 (Environmental Engineering—Determination of half life):** Model the adsorption data in Example 15.5 via Yoon–Nelson model and predict the adsorption time for 50% fractional removal of the acid orange 7.
>
> **Step 1: Schematic**
> See schematic in Example 15.5.
>
> **Step 2: Given data**
> The concentration–time data is provided in Example 15.5.
>
> **To be determined:** Removal at 50% breakthrough.
>
> **Assumptions:** The flow is steady state, the acid orange 7 dye does not degrade with time, and adsorption is the only removal mechanism.
>
> **Step 3: Calculations**
> The linearized version of the Yoon–Nelson model is given by
>
> $$\ln\left(\frac{C}{C_0 - C}\right) = k_{Y-N}(t - \tau) \tag{E15.6A}$$
>
> $$\ln\left(\frac{C}{C_0 - C}\right) = k_{Y-N} t - k_{Y-N} \tau \tag{E15.6B}$$
>
> Using the data, a plot was generated between $\ln\left(\dfrac{C}{C_0 - C}\right)$ and t between 180 and 600 min as shown in Fig. 15.23.
> The slope of the line is
>
> $$k_{YN} = 0.0191 \frac{1}{min}$$

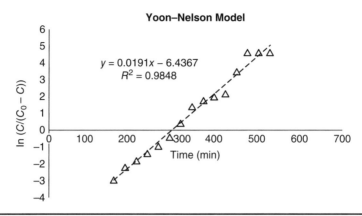

FIGURE 15.23   The linearized Yoon–Nelson plot to determine the 50% breakthrough time for the given data in Example 15.6.

The intercept of the line is

$$-6.4367 = -k_{YN}\tau \qquad\qquad (E15.6C)$$

which leads to

$$\tau = \frac{6.4367}{0.0191} = 337 \text{ min.}$$

Practice Problems for the FE Exam

1. Which of the following statements is not true?
   a. Adsorption is a surface phenomenon.
   b. Adsorption increases with an increase in surface area.
   c. Adsorption decreases with a decrease in particle size.
   d. All of the above are true.

2. Which of the following is an adsorption-based process?
   a. Enzymatic reaction
   b. Uptake of nutrients by cells
   c. Reaction on a solid catalyst
   d. Odor removal by activated carbon

3. Chemisorption generally
   a. increases with temperature.
   b. decreases with temperature.
   c. first decreases and then increases when temperature reaches 100°C.
   d. is independent of temperature.

4. Adsorption is least dependent on

   a. density of the adsorbent.

   b. temperature of the adsorbent.

   c. point of zero charge of the adsorbent.

   d. surface area of the adsorbent.

5. Which of the following is not true about chemisorption?

   a. High specificity

   b. Low enthalpy

   c. Irreversible

   d. Involves chemical reactions

6. In an adsorption experiment, 10 g biochar was mixed with 1 L solution of 50 ppm isovaleric acid. After 30 min, the concentration of the isovaleric acid was measured as 11 ppm. The adsorption capacity of biochar after 30 min is closest to

   a. 4 mg/g.

   b. 10 mg/g.

   c. 7 mg/g.

   d. None of the above.

7. The equilibrium adsorption capacity is linearly related to the equilibrium concentration when the

   a. concentration of the adsorbing species is very high.

   b. concentration of the adsorbing species is very low.

   c. temperature of the adsorbing species is very high.

   d. temperature of the adsorbing species is very low.

8. The equilibrium adsorption capacity is independent of the equilibrium concentration when the

   a. concentration of the adsorbing species is very high.

   b. concentration of the adsorbing species is very low.

   c. temperature of the adsorbing species is very high.

   d. temperature of the adsorbing species is very low.

9. While using the Freundlich adsorption isotherm, the adsorption is considered unfavorable when the surface heterogeneity factor, $1/n$

   a. $1/n > 1$.

   b. $1/n < 1$.

   c. $1/n = 1$.

   d. None of the above.

10. In an adsorption process, if the change in enthalpy, $\Delta H° > 0$, the adsorption reaction is considered

   a. exothermic.

   b. endothermic.

   c. spontaneous.

   d. nonspontaneous.

Practice Problems for the PE Exam

1. A new adsorbent, LTB 20, is prepared from waste biomass via a low-temperature thermochemical process (400°C), processed for 20 min. The adsorbent was tested for the removal of acetic acid in batch reactor at various initial concentrations at 27°C using 1 g of adsorbent and 100 mL of acetic acid solutions. Using the following data, determine the maximum adsorption capacity of LTB 20.
   [Equilibrium adsorption capacity (mg·g^{-1}), Equilibrium concentration (mg·L^{-1})]:
   [(2.18, 1.13); (4.78, 1.51); (9.34, 5.69); (18.83, 14.59); (29.62, 199.93)].

2. If LTB 20 is to be used to treat 5000 L of wastewater containing 100 mg/L of acetic acid at 27°C, using your results from Problem 1, determine the mass of LTB 20 needed for 30% fractional removal.

3. Determine the Freundlich adsorption parameters for LTB 20 using the data from Problem 1.

4. An activated carbon bed is employed to treat propanal-laden air. Laboratory-scale experiments were performed wherein 200 ppmv of propanal was flown through an 18-cm long adsorption column of 2.5 cm diameter packed with 5 of activated carbon at 0.5 LPM. The outlet air sample was collected at regular intervals and analyzed for the concentration of propanal. Using the data presented next, determine adsorption capacity of the column.

Time (min)	0	10	20	30	40	50	60	70	80	90	100
Outlet (ppmv)	0	0	0	0	0	0	0	0	0	0	0

Time (min)	110	120	130	140	150	160	170	180	190	200	210	220
Outlet (ppmv)	10	20	40	60	80	100	160	190	197	199	200	200

5. For the data presented in Problem 4, determine the time needed to attain a 50% breakthrough.

6. Biochar derived from swine manure was used to remove *p*-cresol acid from wastewater. Experiments were conducted in which 1 g of biochar was equilibrated

with 100 mL of *p*-cresol solution of various concentrations at 45°C. The following data were collected:

Time (min)	Experiment 1 Concentration (mg/L)	Experiment 2 Concentration (mg/L)	Experiment 3 Concentration (mg/L)	Experiment 4 Concentration (mg/L)
0	56.5	107.6	201.9	491.6
10	49.1	102.4	186.4	460.9
20	46.8	96.2	178.9	461.8
30	46.7	93.2	176.7	453.2
40	45.4	94.7	174.5	446.4
50	45.3	91.7	171.4	436.0
60	42.3	88.6	168.0	434.5
70	40.6	86.6	166.2	433.7
80	43.2	86.6	165.5	432.8
90	42.0	87.0	162.8	423.2
100	41.0	85.2	162.4	420.2
110	41.4	86.3	161.6	417.9
120	41.5	85.3	162.0	419.4

Determine the equilibrium capacities of biochar for each concentration tested.

7. Using the given data in Problem 6, determine the theoretical maximum adsorption capacity using the linear and nonlinear Langmuir adsorption isotherm.

8. For the data in Problem 6, fit the nonlinear Freundlich model and determine the model parameters.

9. For the data in Problem 6, fit the nonlinear BET model and determine the model parameters.

10. Let us suppose that you are tasked with analyzing the efficacy of the biochar for the removal of butyric acid as described in Problem 6. Specifically, your manager wants you to determine if the adsorption of butyric acid was physical or chemical. How do you distinguish between physical and chemical adsorption? What would be your approach?

References

1. Cengel, Y.A., and A.J. Ghajar (2016). *Heat and Mass Transfer: Fundamentals and Applications*, Fifth Edition, McGraw Hill.
2. Incropoera, F.P., D.P. Dewitt, T.L. Bergman, and A.S. Lavine (2017). *Principles of Heat and Mass Transfer*, Eighth Edition, John Wiley and Sons, Singapore Pte Ltd.
3. Henderson, S.M., R.L. Perry, and J.H. Young (1997). *Principles of Process Engineering*, The American Society of Agricultural and Biological Engineers, USA.
4. Cengel, Y.A., and J.M. Cimbala (2006). *Fluid Mechanics: Fundamental and Applications*, First Edition. McGraw Hill.
5. Smith, P.G (2003). *Introduction to Food Process Engineering*, Kluwer Academic Publishers, USA.
6. Heldman, D.R., and R.P. Singh (2009). *Introduction to Food Engineering*, Fourth Edition, Academic Press, Elsevier, USA.
7. Kumar, D.S. (2015). *Fluid Mechanics and Power Engineering*, Ninth Edition, S.K. Kataria and Sons, India.
8. Kays, W.M., and A.L. London (1984). *Compact Heat Exchangers*, McGraw Hill.
9. Churchill, S.W., and H.H.S Chu (1975). Correlating equations for laminar and turbulent free convection from a horizontal cylinder, *International Journal of Heat and Mass Transfer*, 18(9):1049–1053.
10. Churchill, S.W., and H.H.S Chu (1975). Correlating equations for laminar and turbulent free convection from a vertical plate. *International Journal of Heat and Mass Transfer*, 18(11):1323–1329.
11. Churchill S.W. (1983). Free Convection around immersed bodies. *Heat Exchanger Design Handbook*, Editor: E.U. Schlünder, Hemispheres Publishing, USA.
12. Whitaker, S. (1972). Forced convection heat transfer correlations for flow in pipes, past flat plates, single cylinders, single spheres, and for flow in packed beds and tube bundles, *AIChE Journal*, 18(2):361–371.
13. Churchill, S.W., and M. Bernstein (1977). A correlating equation for forced convection from gases and liquids to a circular cylinder in crossflow. *Transactions of the ASME*, 99:300–306.
14. Winterton, R.H.S (1998). Where did the Dittus and Boelter equation come from? International Journal of Heat and Mass Transfer. 41(45):809–810.
15. The American Society of Heating, Refrigerating and Air-Conditioning Engineers Fundamentals (1997). I.P Editon.
16. Ebadi, A., J.S.S. Mohammadzadeh, and A. Khudiev (2009). What is the correct form of BET isotherm for modeling liquid phase adsorption? *Adsorption*, 15:65–73.
17. Ekhard, W. (2012). Adsorption technology in water treatment: *Fundamentals, Processes, and Modeling*, Berlin: Walter de Gruyter GmbH & Co.

18. Khraisheh, M.A.M., Y.A. Al-Degs, S.J. M.N. Allen, & Ahmad (2002). Elucidation of controlling steps of reactive dye adsorption on activated carbon. *Industrial & Engineering Chemistry Research*, 41, 1651–1657.

19. Kruk, M., and M. Jaroniec (2001). Gas adsorption characterization of ordered organic—inorganic nanocomposite materials. *Chemistry of Materials*, 13:3169–3183.

20. Zhe Xu, Z., J-G. CAI, B-C. Pan (2013). Mathematically modeling fixed-bed adsorption in aqueous systems. *Journal of Zhejiang University-SCIENCE A Applied Physics & Engineering*, 14(3):155–176.

21. Al-Muhtaseb, A.H., W.A.M. McMinn, and T.R.A. Magee (2002). Moisture Sorption Isotherm Characteristics of Food Products: A Review. *Food and Bioproducts Processing*, 80(2):118–128.

22. Marrero, T.R., and E.A. Mason (1972). Gaseous diffusion coefficients, *Journal of Physical and Chemical Reference Data*, 1:3–118.

23. Huang, J (2018). A simple accurate formula for calculating saturation vapor pressure of water and ice. *Journal of Applied Meteorology and Climatology*, 57(6):1265–1272.

24. Alduchov and Eskridge (1995). Improved magnus form approximation of saturation vapor pressure. *Journal of Applied Meteorology and Climatology*, 35(4):601–609.

25. Brunauer, S., L.S. Deming, W.E. Deming, and E.Teller (1940). On a theory of the van der Waals adsorption of gases. *Journal of the American Chemical Society*, 62:1723–1732.

Online Resources

https://www.engineeringtoolbox.com/

https://science.nasa.gov/ems/01_intro

Jukka, K. 2011. Darcy Friction Factor Formulae in Turbulent Pipe Flow. Lunowa∗Fluid Mechanics Paper 110727.

Index

Note: Page numbers followed by *f* denote figures; by *t*, tables.

455

E